T0228924

ENVIRONMENTAL GEOCHEMISTRY

Site Characterization, Data Analysis and Case Histories

ENVIRONMENTAL GEOCHEMISTRY

Site Characterization, Data Analysis and Case Histories

Edited by

BENEDETTO DE VIVO
Università di Napoli Federico II
Dipartimento di Scienze della Terra, Napoli, Italy

HARVEY E. BELKIN
United States Geological Survey, Reston, USA

ANNAMARIA LIMA
Università di Napoli Federico II
Dipartimento di Scienze della Terra, Napoli, Italy

AMSTERDAM • BOSTON • HEIDELBERG • LONDON
NEW YORK • OXFORD • PARIS • SAN DIEGO
SAN FRANCISCO • SINGAPORE • SYDNEY • TOKYO

ELSEVIER

Elsevier
Radarweg 29, PO Box 211, 1000 AE Amsterdam, The Netherlands
Linacre House, Jordan Hill, Oxford OX2 8DP, UK

First edition 2008

Copyright © 2008 Elsevier B.V. All rights reserved

No part of this publication may be reproduced, stored in a retrieval system or transmitted in any form
or by any means electronic, mechanical, photocopying, recording or otherwise without the prior
written permission of the Publisher

Permissions may be sought directly from Elsevier's Science & Technology Rights
Department in Oxford, UK: phone (+44) (0) 1865 843830; fax (+44) (0) 1865 853333; email:
permissions@elsevier.com. Alternatively you can submit your request online by visiting the Elsevier
web site at http://elsevier.com/locate/permissions, and selecting *Obtaining permission to use Elsevier
material*

Notice
No responsibility is assumed by the publisher for any injury and/or damage to persons or property as a
matter of products liability, negligence or otherwise, or from any use or operation of any methods,
products, instructions or ideas contained in the material herein. Because of rapid advances in
the medical sciences, in particular, independent verification of diagnoses and drug dosages
should be made

Library of Congress Cataloging-in-Publication Data
A catalog record for this book is available from the Library of Congress

British Library Cataloguing in Publication Data
A catalogue record for this book is available from the British Library

ISBN: 978-0-444-53159-9

For information on all Elsevier publications
visit our website at books.elsevier.com

Printed and bound by CPI Group (UK) Ltd, Croydon, CR0 4YY

Transferred to Digital Print 2011

Working together to grow
libraries in developing countries

www.elsevier.com | www.bookaid.org | www.sabre.org

ELSEVIER   BOOK AID
             International   Sabre Foundation

CONTENTS

Contributors

Paola Adamo
Dipartimento di Scienze del Suolo, della Pianta, dell'Ambiente e delle Produzioni Animali, Università degli Studi di Napoli Federico II, Via Università 100, 80055 Portici, Italy.

Stefano Albanese
Dipartimento di Scienze della Terra, Univerità degli Studi di Napoli 'Federico II', Via Mezzocannone 8, 80134 Napoli, Italy.

Louise E. Ander
British Geological Survey, Keyworth, Nottingham, NG12 5GG, UK.

Robert A. Ayuso
U.S. Geological Survey, MS 954 National Center, Reston, Virginia 20192, USA.

Harvey E. Belkin
956 National Center, U.S. Geological Survey, Reston, Virginia 20192, USA.

Neil Breward
British Geological Survey, Kingsley Dunham Centre, Keyworth, Nottingham, NG12 5GG, UK.

Domenico Cicchella
Dipartimento di Studi Geologici e Ambientali, Università del Sannio, Via Port'Arsa 11, 82100 Benevento, Italy.

Jean Michel Crépin
Executive Consultant, Business Development, Bodycote Testing Group, Edmonton, Alberta, T6B 3J4, Canada.

Neil Crout
School of Biosciences, University of Nottingham, University Park, Nottingham, NG7 2RD, UK.

Maria Luisa De Luca
Dipartimento di Scienze della Terra, Università di Napoli "Federico II", 80134 Napoli, Italy.

Benedetto De Vivo
Dipartimento di Scienze della Terra, Univerità degli Studi di Napoli 'Federico II', Via Mezzocannone 8, 80134 Napoli, Italy.

Marcello Di Bonito
Environment Agency, Trentside Offices, West Bridgford, Nottingham, NG2 5FA, UK.

Robert B. Finkelman
Department of Geosciences, University of Texas at Dallas, Richardson, Texas 75083, USA.

Deirdre M. A. Flight
British Geological Survey, Keyworth, Nottingham, NG12 5GG, UK.

Nora K. Foley
U.S. Geological Survey, MS 954 National Center, Reston, Virginia 20192, USA.

Fiona M. Fordyce
British Geological Survey, Murchison House, Edinburgh, EH9 3LA, UK.

Giuseppe Grezzi
Dipartimento di Scienze della Terra, Università di Napoli "Federico II", 80134 Napoli, Italy.

David Hope
CEO, Pacific Rim Laboratories, Surrey, Canada.

Christopher C. Johnson
British Geological Survey, Keyworth, Nottingham, NG12 5GG, UK.

Robert Lessard
QA Manager, Bodycote Testing Group, Edmonton, Canada.

Annamaria Lima
Dipartimento di Scienze della Terra, Università di Napoli "Federico II", Via Mezzocannone, 8, 80134 Napoli, Italy.

Gail Lipfert
Department of Earth Sciences, University of Maine, Orono, Maine 04469, USA.

Robert T. Lister
British Geological Survey, Keyworth, Nottingham, NG12 5GG, UK.

Sarah E. Nice
British Geological Survey, Keyworth, Nottingham, NG12 5GG, UK.

Francesco Pepe
Dipartimento di Ingegneria, Università del Sannio, Piazza Roma 21, 82100 Benevento, Italy.

Shaun Reeder
British Geological Survey, Keyworth, Nottingham, UK.

Reijo Salminen
Geological Survey of Finland, 02151 Espoo, Finland.

Cynde Sears
Sears Consulting LLC, 13117 New Parkland Dr., Oak Hill, VA, 20171, USA.

Barry Smith
British Geological Survey, Kingsley Dunham Centre, Keyworth, Nottingham, NG12 5GG, UK.

Marianne Stuart
British Geological Survey, Keyworth, Nottingham, UK.

Christopher Swyngedouw
Consulting Scientist, Bodycote Testing Group, #5, 2712–37th Avenue N.E., Calgary, Alberta, T1Y 5L3, Canada.

Julian K. Trick
British Geological Survey, Keyworth, Nottingham, UK.

Scott Young
School of Biosciences, University of Nottingham, University Park, Nottingham, NG7 2RD, UK.

Mariavittoria Zampella
Dipartimento di Scienze del Suolo, della Pianta, dell'Ambiente e delle Produzioni Animali, Università degli Studi di Napoli Federico II, Via Università 100, 80055 Portici, Italy.

Baoshan Zheng
State Key Laboratory of Environmental Geochemistry, Institute of Geochemistry, 550002 Guiyang, Guizhou, PR China.

Daixing Zhou
Sanitation and Anti-Epidemic Station of Qianxinan Autonomous Prefecture, 562400 Xingyi, Guizhou, PR China.

Preface

The volume "Environmental Geochemistry: Site Characterization, Data Analysis, Case Histories" contains selected papers presented at the "Workshop: Environmental Geochemistry—Site Characterization, Waste Disposal, Data Analysis, Case Histories" held in Napoli (Italy) on May 4–5, 2006. Participants from private and public institutions of Canada, Finland, Greece, Italy, the UK, and USA, took part.

The theme of the Workshop was multidisciplinary methods of characterizing contaminated sites using modern geochemistry with examples from different countries in Europe, North America, and Asia. Special themes included soil, surface, and ground waters contamination, environment pollution, and human health, and data interpretation and management.

At the more local scale, site characterization and site remediation technologies in soil were considered, as well as, sewage sludge disposal. Case histories of brownfield sites in Italy, UK, and USA were also presented.

It is especially appropriate that this volume be completed in 2007 as it is the hundredth anniversary of the birth of Rachel Carson. Rachel Carson published Silent Spring in 1962, which brought environmental concerns to an unprecedented portion of the American public and the world in general. Silent Spring spurred a dramatic reversal in USA national pesticide policy—leading to a nationwide ban on DDT and other pesticides—and the grassroots environmental movement it inspired led to the creation of the United States Environmental Protection Agency and other similar agencies. Now, we are beginning a transition among governments where the methods and techniques of remediation are developed with a greater understanding of the baselines and economics of the applied cleanup processes. Clean up of polluted sites must be accomplished within the budgets of the communities together with the rigor of the science.

A selection of papers on the general theme on soil, surface, and groundwater contamination, environment pollution and human health, and data interpretation and management are published in a special issue of Geochemistry: Exploration, Environment, Analysis Special Issue "Environmental Geochemistry", edited by B. De Vivo, J. A. Plant, and A. Lima.

A selection of papers more professionally and educationally oriented are included in this Elsevier volume. Fifteen papers arising from the conference are included in this special volume. Their content is briefly summarized below.

Salminen R. summarizes experiences from a number of recently completed regional-scale geochemical surveys. It briefly shows the most essential issues to be taken into account in planning and carrying out geochemical surveys in the field.

Swingedouw C. and Crepin J. M. provide an overview of sampling methods and tools suitable to address most site characterizations. The basic sampling types discussed are the systematic, random, and judgmental sampling approaches. In addition to sampling procedures, sampling bias and sampling errors are introduced, leading to some guidance on sample handling, shipping, and chain-of-custody procedures. The presentation focuses on sampling methods for soil only.

Trick J. K., Stuart M., and Reeder S. describe the tools available to the field sampler for the collection of groundwater samples, methods of on-site water quality analysis, and the appropriate preservation and handling of samples. The authors discuss the merits of different purge methodologies and show how on-site measurements such as pH, specific electrical conductance (SEC), oxidation–reduction potential (ORP), dissolved oxygen (DO), temperature, and alkalinity can be used to provide a check on subsequent laboratory analyses. Techniques for the preservation and analysis of samples and quality assurance and quality control are also presented.

Johnson C. C., Flight D. M. A., Ander E. L., Lister T. R., Breward N., Fordyce F. M., and Nice S. F. discuss the collection of drainage samples from active stream channels for geochemical mapping studies. The authors describe details on the sampling methods used by the British Geological Survey in order to establish a geochemical baseline for the land area of Great Britain, involving the collection of stream sediments, waters, and panned heavy mineral concentrates for inorganic chemical analysis. The authors give detailed sampling protocols and discuss sampling strategy, equipment, and quality control.

Johnson C. C., Ander E. L., Lister T. R., and Flight D. M. A. discuss data conditioning procedures involving the verification, quality control, and data-levelling processes that are necessary to make data fit for the purpose for which it is to be used. The authors describe the methods currently used by the British Geological Survey's regional geochemical mapping project that has been generating geochemical data for various sample media for nearly 40 years.

Swingedouw C., Hope D., and Lessard R. describe analytical organic chemistry employing common gas chromatographic techniques which involve dissolving the analyte in organic solvent, removing the interfering co-extractives by solid-phase extraction and then injecting the purified extract into a gas chromatograph coupled to a detector. The paper provides procedures to extract, isolate, concentrate, separate, identify, and quantify organic compounds. It also includes some information on the collection, preparation, and storage of samples, as well as specific quality control and reporting criteria.

Lima A. describes statistical methods to evaluate background values, namely, statistical frequency analysis and spatial analysis. The author illustrates the application of GeoDAS^{TM} software to perform multifractal inverse distance weighted (MIDW) interpolation and a fractal filtering technique, named spatial and spectral analysis (S–A) method, to evaluate geochemical background at regional and local scale.

Albanese S., Cicchella D., Lima A., and De Vivo B. present a synthesis of the main considerations necessary to undertake urban mapping activities in terms of planning, sampling, chemical analyses, and data presentation. In this context, modern Geographical Information Systems (GIS) represent an indispensable tool for better understanding the distribution, dispersion, and interaction processes of some toxic and potentially toxic elements.

Adamo P. and Zampella M. provide a review of the single and sequential chemical extraction procedures that have been more widely applied to determine the plant and the human bioavailability of potentially toxic metals (PTMs) from contaminated soil and their presumed geochemical forms. Examples of complementary use of chemical and instrumental techniques and applications of PTMs speciation for risk and remediation assessment are illustrated.

Di Bonito M., Breward N., Smith B., Crout N., and Young S. describe some of the current methodologies used to extract soil pore water. In particular, four laboratory-based methods, (i) high-speed centrifugation–filtration, (ii) low- (negative-) pressure Rhizon^{TM} samplers, (iii) high-pressure soil squeezing, and (iv) equilibration of dilute soil suspensions, are described and discussed in detail. Some consideration is then taken to assess advantages and disadvantages of the methods, including costs and materials availability.

Di Bonito M. reviews the improved standards achieved with sewage sludge, touching on, in particular, the British experience in the field of regulating the disposal and reuse of these materials.

Ayuso R. A., Foley N. K., and Lipfert G. present part of an extensive study of the coastal environment in the State of Maine (USA) where the occurrence of elevated levels of arsenic in drinking water (>0.010 mg/L) has prompted multifaceted research to understand the cause of this situation. A detailed Pb isotopic study of pesticides is used to understand the source, distribution, and fate of As and Pb in pesticides, soil, bedrock, and waters.

Pepe F. discusses incineration and the most relevant problems as a very efficient technique for municipal solid waste (MSW) management. The author also discusses the different approaches proposed to mitigate the impact of fly ash disposal.

Sears C. examines the efforts of the town of Greenwich, Connecticut (USA), to clean up and redevelop a large, environmentally contaminated former coal-fired power plant. The author shows how the town made several decisions that ensured that the site would be turned into a community asset by considering multiple options for land use, taking advantage of newly available Federal and state funding for environmental assessment and cleanup, and adopting alternative environmental assessment and strategies.

De Vivo B. and Lima A. document the case history of the Bagnoli brownfield site government remediation project, which is still in progress. The site was the second largest integrated steelworks in Italy and is located in the outskirts of Naples, in an area which is part of the quiescent Campi Flegrei volcanic caldera. Hundreds of surficial and deep boreholes have been drilled, with the collection of about 3000 samples of soils, scums, slags, and landfill materials, and water samples from underground waters. In general, heavy metal enrichments in the cores and water suggest mixing between natural (geogenic) and anthropogenic components. The actual pollution to be remediated is the occurrence of polycyclic aromatic hydrocarbons (PAH), distributed in different spots across the brownfield site.

Albanese S., De Luca M. L., De Vivo B., Lima A., and Grezzi G. report geochemical and epidemiological data as maps that represent the detailed patterns of toxic metal concentrations and some, potentially, related pathologies in the Campania region of Italy.

Belkin H. E., Zheng B., Zhou D., and Finkelman R. B. describe a unique case study of chronic arsenic poisoning caused by the domestic combustion of coal in rural southwestern Guizhou Province, P. R. China. The coal, used by several villages, is enriched in arsenic (>100 ppm) and when burnt in nonvented stoves is absorbed by vegetables hung above and then ingested. Characteristic symptoms of arsenosis, such as hyperpigmentation and keratosis, have been used to define affected populations. Effective collaboration between earth scientists and the local public heath community has mitigated the incidence of this endemic arsenic poisoning.

CHAPTER ONE

FIELD METHODS IN REGIONAL GEOCHEMICAL SURVEYS

Reijo Salminen*

Contents

Abstract

This chapter summarizes experiences from a number of recently completed regional-scale geochemical surveys. The aim is to briefly show the most essential issues to be taken into account in planning and carrying out geochemical surveys in the field. Whether the aim of a geochemical survey is prospecting, environmental assessment, or something else, the main principles in the fieldwork are always the same.

1. INTRODUCTION

Geochemical studies vary enormously in an area. At one extreme, they cover continent-wide areas (Gustavsson *et al.*, 2001; Salminen *et al.*, 2005), based on information from not more than a thousand sites, while at the other, detailed maps, based on several thousands of samples, are produced from a small prospecting target (e.g., Kauranne, 1976; McClenaghan *et al.*, 2001).

* Geological Survey of Finland, 02151 Espoo, Finland

Environmental Geochemistry
DOI: 10.1016/B978-0-444-53159-9.00001-2

© 2008 Elsevier B.V.
All rights reserved.

Studies at different scales differ considerably in the way they are carried out. Not only sampling density, but sampling material, sampling depth, analytical methods, and data processing also essentially depend on the aim of the study, the size of the area to be studied, the objects to be recognized, and the contrast between the anomaly and the surrounding area. The sources of the anomalies detected by different sampling densities are also totally different in nature.

2. SAMPLING MEDIA

Minerogenic stream sediments are the traditional medium in small-scale, regional geochemical mapping, particularly if the aim is ore prospecting. In areas of residual overburden, minerogenic stream sediments have proven to be very useful, providing data from a wide drainage area where the stream has been in contact with the bedrock (Hale and Plant, 1994). The most suitable conditions prevail in areas of temperate climate where the rivers are draining *in situ* weathered bedrock, and mountainous areas where the bedrock is widely exposed.

In glaciated areas, the stream is usually disconnected from the bedrock by till and the stream sediments may thus only reflect the variation of element contents in till. The interpretation of results for prospecting purposes becomes complicated. However, in till-covered mountainous areas such as Scotland (Plant *et al.*, 1984) and Norway (Wennervirta *et al.*, 1971), useful results were obtained by stream sediment geochemistry.

Till has conventionally been exploited as a sampling material only on local-scale prospecting studies. However, results from Scandinavia and adjacent areas (Bølviken *et al.*, 1986; Koljonen, 1992; Reimann *et al.*, 1998; Salminen *et al.*, 1995) have shown beyond doubt that highly informative and easily interpreted results can be obtained from till geochemistry practiced on a regional or reconnaissance scale.

In the 1990s and earlier, environmental applications became important in geochemical mapping. New sampling media such as surface water and terrestrial mosses were tested and became more commonly used in geochemical surveys (Lahermo *et al.*, 1990, 1996; Reimann *et al.*, 1998; Rühling 1994; Salminen *et al.*, 2005; Salminen, 2004; Steinnes *et al.*, 1992). This development also brought some new variation not only in the sample media but also in sampling, analysis, and data management methodologies. In principle, most geochemical mapping data can also be used in environmental geochemical studies.

In exploration geochemistry, the concept of a geochemical background value is used to differentiate anomalies caused by mineralized occurrences from the geogenic anomalies caused by normal nonmineralized bedrock. In environmental geochemistry, a new concept of the geochemical baseline was needed to differentiate contamination derived from a point source from that derived from the general background, which includes both natural geogenic element concentration and diffuse anthropogenic pollution (Salminen and Gregorauskiene, 2000). Environmental geochemical studies have concentrated increasingly on defining baselines rather than on detecting high anomaly points; methods to separate local and regional components (anomaly and baseline) have been developed (e.g., De Vivo *et al.*, 2006).

In an attempt to establish a global, common understanding of continuously varying methodologies in regional geochemical surveys, the sampling media were discussed very thoroughly in the 1980s and 1990s as part of the IGCP 259 (International Geoscience Programme) (International Geochemical Mapping) and IGCP 360 (Global Geochemical Baselines) projects. Furthermore, this discussion has continued in the framework of the IUGS/IAGC (International Union of Geological Sciences/International Association of Geochemistry) Working Group on Global Geochemical Baselines. Darnley *et al.* (1995) concluded the earlier discussions with recommendations that were globally accepted. These recommended media, described below, are considered to be the most representative of the Earth's surface environment, and are the most commonly used in past and current environmental geochemical investigations.

- Stream water (filtered and unfiltered)
- Stream sediment: mineral sediment (<0.15 mm)
- Residual soil: upper 0–25 cm horizon/topsoil without the top organic layer (<2 mm)
- Residual soil: lower (C) horizon/subsoil; a 25-cm layer within a depth range of 50–200 cm (<2 mm)
- Organic soil layer (humus)
- Overbank sediment: upper 0–25 cm horizon (<0.15 mm)
- Overbank sediment: bottom layer (<0.15 mm)
- Floodplain sediment: upper 0–25 cm horizon (<2 mm) and
- Floodplain sediment: bottom layer (<2 mm)

Stream, overbank, and *floodplain sediment* samples generally reflect the average geogenic composition of a catchment basin for most elements, although they are somewhat sensitive to pollution. Stream sediment is the most widely used sample material in regional geochemical surveys throughout the world.

Stream waters reflect the interplay between geosphere/hydrosphere and pollution. At the same time, they have a huge economic value, often being a major source of drinking water in some countries.

Soil samples reflect variations in the geogenic composition of the uppermost layers of the Earth's crust. As a result, it is important in regional surveys to avoid soil sampling at locations that have visible or known contamination. Priority for site selection of soil samples should be given to

- forested and unused lands;
- greenland and pastures; and
- noncultivated parts of agricultural land (in very special cases, where residual soil cannot be found).

Comparison of topsoil and subsoil data gives information about enrichment or depletion processes between the layers. One such process is anthropogenic contamination of the top layer. The <2-mm fraction is used most often in environmental and soil scientific studies, whereas the <0.18-mm and finer fractions have been widely used in mineral exploration programs. The FOREGS (Forum for European Geological Surveys) data was planned to be used to create a link between environmental and mineral exploration databases.

Organic soil layer (humus) samples can be used to determine the atmospheric (anthropogenic) input of elements to the ecosystem over a period of tens of years. To reach this aim, samples are collected in forested areas. To reflect the atmospheric input, the uppermost few centimetres of the organic layer are collected immediately under the green vegetation and litter (max. 3 cm).

3. SAMPLING DENSITY

In geochemical studies, the sampling density varies very much according to the aim of the study. Geochemical studies have been classified according to the sampling density in many ways. Ginzburg (1960) divided it into three classes according to the scale of the study: (i) reconnaissance; (ii) prospecting; and (iii) detailed mapping. Thus, rather than specifying sampling densities, Ginzburg described the phases logically included in geochemical studies: the reconnaissance phase reveals regions where geochemical studies can reasonably be carried out; the prospecting phase delineates mineralized areas or formations; and the detailed phase reveals ore out-crops. Bradshaw *et al.* (1972) subsequently proposed a division into three classes strictly based on sampling density: (i) one sample per 40–80 sq. miles, (ii) one sample per 5–20 sq. miles, and (iii) one sample per 1–2 sq. miles. In their view, each class is specific for a certain type of geologic formation, thus reflecting different geochemi-cal features. Although the selected densities were different, a similar three-class, step-by-step proceeding geochemical study system was later used in Finland (Salminen and Hartikainen, 1986). Salminen (1992) combined the basic ideas of Ginzburg (1960) and Bradshaw *et al.* (1972) as shown in Fig. 1.1.

Modification of the classification scheme is needed because of the demand for geochemical baseline data over larger, even continent-wide, study areas. The sampling density used in such large geochemical survey projects as the Geochemical Atlas of Europe (Salminen *et al.*, 2005) or geochemical maps of the United States (Gustavsson *et al.*, 2001) was around 0.0002 samples per km^2. A new classification of geochemical studies that takes into account the very small-scale studies recently carried out is presented in Fig. 1.2. This kind of classification is valid both in environmental and prospecting geochemistry and is based on the fact that if one

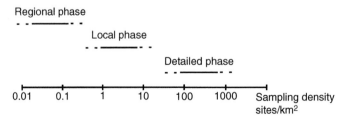

Figure 1.1 Sampling density (sampling sites per km^2) for different scales of geochemical studies according to Salminen (1992).

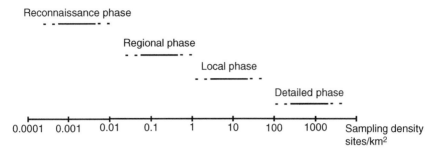

Figure 1.2 Sampling density (sampling sites per km²) for different scales of geochemical studies according to the new classification scheme after the continent-wide geochemical baseline surveys were started.

wants to essentially increase the amount and detail of geochemical information, sampling density must be increased exponentially.

4. SAMPLING NETWORKS

The selection of the sampling grid depends to a large extent on the expected character of the anomaly and the scale of the study.

If one of the sample media is drainage sediment or stream water, the sampling network is based on catchment areas. The size of a catchment can vary, but within a single study, it should be of the same order of size for each sampling site. The samples for the Geochemical Atlas of Europe were collected from two different-sized catchment areas: (i) 100 km², from which stream water and stream sediments were collected (together with soils), and (ii) 1000 km², from which floodplain sediment samples were collected. Thus, these two sites represented different orders of a bigger drainage basin and gave different geochemical information.

In detailed studies, a regular grid is often used when the sampling material, such as soil, is available everywhere. This grid can be regular to every direction or can be a grid in which the distance between sampling sites along lines is much shorter than the distance between lines. The most useful sample network is, however, a regularly irregular grid. The study area is first divided into cells of equal size and shape according to the selected sampling density. The actual sampling site is then selected inside the cell to represent the whole cell area as best as possible. In this case, the preliminary planning is done at the office, but the actual sampling site is defined in the field.

In the FOREGS program (Salminen *et al.*, 2005), the Global Terrestrial Network (GTN) of grid cells (Darnley *et al.*, 1995) was used in planning the sampling grid (Fig. 1.3). Sampling locations were first selected randomly by the aid of computer software, and the final catchment for sampling was defined in the field to best represent the geological landscape of the randomly selected location. The sampling site for each sample media was then selected from the catchment according to the instructions of the field manual (Fig. 1.4). Thus, sample locations were not designed to show the lowest natural background concentrations in the European

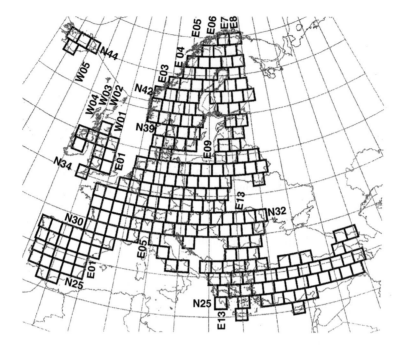

Figure 1.3   Global Terrestrial Network (GTN) cells in Europe (from Salminen *et al.*, 1998).

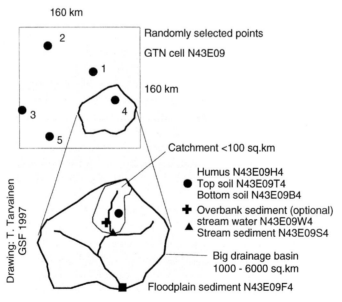

Figure 1.4   Selection of sampling sites, a schematic outline of drainage sampling pattern (from Salminen *et al.*, 1998).

environment but to demonstrate the geochemistry of the surface environment at the end of the twentieth century.

5. QUALITY ASSURANCE IN SAMPLING

Sampling and sample preparation are the most critical processes contributing to the overall uncertainty budget. Uncertainty in chemical analysis, although often significantly less than that associated with sampling and sample preparation, is nevertheless essential to understand and control.

For quality control purposes, it is recommended that 5–10% of samples are collected in duplicate. The whole sampling procedure should be duplicated; for example, separate pits should be dug at a distance of several meters apart when sampling duplicate soil horizons. Duplicate samples should be identified by a special code. During water sampling, blank samples should be prepared from distilled and deionized water. The blank samples should be acidified and handled in the same way as the normal samples and identified by a special code. A blank water sample is normally prepared at the same site as a duplicate sample is collected.

Standard (reference) materials of organic and minerogenic soil and sediment materials should be included in the sample set at the same frequency as the duplicates. These samples can be thoroughly homogenized, normal sample material from the study area divided into subsamples, although commercially available certified reference materials may also be used.

In order to assess the precision of the chemical analysis relative to the variation due to sample preparation and sampling, duplicate samples and repeat measurements should be carried out. The duplicate samples, prepared independently of each other and analyzed randomly along with all other samples, with each duplicate sample also analyzed in duplicate, allows estimation of sampling uncertainty by the analysis of variance (ANOVA) statistical interpretation method.

6. SAMPLING PROCEDURES

The sampling procedures described below are based on experiences from the recent large geochemical survey projects. The FOREGS Geochemical Atlas of Europe (Salminen *et al.*, 2005) was the European contribution to the Global Geochemical Baselines project. At the same time, it was an exercise on how to implement the recommendations given by Darnley *et al.* (1995) for practical mapping work. A field manual (Salminen *et al.*, 1998) was prepared before sampling commenced in order to establish common protocols for all sampling teams. At the same time, another large geochemical mapping project, the Barents Ecogeochemistry project (Salminen *et al.*, 2004), was carried out. A detailed field manual was also prepared for this project, based on the experiences from a large pilot phase of fieldwork (Gregorauskiene *et al.*, 2000). These and other respective field manuals describe in detail how the sampling in the field should be carried out. The most commonly used sample materials are included in this description, but many others are and can be used according to the goals of the exercise. Climatic conditions not covered in the standard texts for temperate climates, such as tropical terrains or deserts, demand that local conditions must be taken into account and sampling methods modified accordingly.

6.1. General aspects of the fieldwork

To avoid any kind of metal contamination, no hand jewellery or medical dressings were worn during sampling. If medical dressings were required, heavy-duty rubber gloves were recommended to be worn at all times to avoid contamination of samples. Metal-free polyethylene or unpainted wooden spades or scoops, metal-free nylon sieve-mesh housed in inert wooden or plastic frames, and metal-free funnels and sample collection containers were used at all stages of sample collection.

6.2. Stream waters

Running stream water was collected from the small, second-order, drainage basins (<100 km^2) at the same site as the active stream sediment. In dry terrains, such as southern Europe, streams have no running water for most of the year. Hence, the sampling, whenever possible, was carried out during the winter and early spring months. Four subsamples of stream water were separately collected from each site.

- unfiltered water for major anion analysis by ion chromatography
- filtered water for inductively coupled mass spectrometry (ICP-MS) and inductively coupled plasma atomic emission spectrometry (ICP-AES) analysis
- unfiltered water for mercury analysis and
- filtered water for dissolved organic carbon (DOC) analysis

High-density Nalgene^{TM} trace element–free bottles were used for sampling for mercury analysis. For the other samples, polyethylene bottles were used. Filtering was carried out using 0.45-μm disposable filters (Schleicher and Schuell^{TM}, pyrogen free) mounted on disposable syringes (Becton and Dickinson^{TM}). Bottles, decanters, syringes, and other equipment were rinsed twice with stream water at the site before sampling. Electrical conductivity (EC) and pH were measured at the sampling site, and alkalinity was also determined by titration.

During sampling, disposable plastic gloves were worn at all times. Further, to avoid any kind of metal contamination, no hand jewellery was allowed, and smoking or having a vehicle running during water sampling was strictly prohibited.

One of the filtered water samples was acidified on the same day as collection by the addition of 1.0 ml of concentrated HNO$_3$ acid using a droplet bottle. The sample for mercury analysis was acidified and preserved by adding 5 ml of a prepared solution of concentrated HNO$_3$ acid and potassium dichromate solution. Bottles were then stored in a refrigerator and sent to the laboratory soon after sampling. A blank water was collected, filtered, and preserved in the same manner as the actual samples after every 20th sample.

Sampling during rainy periods and flood events was avoided. The water sample was always taken before the stream sediment sample, because during the collection of the stream sediment, fine-grained material is agitated and transported into suspension. The water sample was collected from the first, lowermost stream sediment sampling point (the stream sediment sample is a composite of 5–10 subsamples taken over a distance of up to 500 m from the lowermost point).

6.3. Sediments

The active stream sediment sample was collected together with the stream water sample. Wherever possible, wet sieving of stream sediments was recommended. Dry sieving was used as an alternative if wet sieving was not feasible, for example, when sampling seasonal streams in Mediterranean countries. Each stream sediment sample comprised material taken from 5 to 10 points along a 250–500 m length of the stream. Sites were located at least 100 m upstream of roads and settlements. Sampling began from the stream water sampling point, and the other subsamples were collected upstream. A composite sample was made from subsamples taken from beds similar in nature according to standard ISO practices (ISO, 1995). The collected stream sediment sample should include at minimum 0.5 kg <0.15 mm grain size fraction.

According to the catchment basin size distinction made by Darnley et al. (1995), both floodplain and overbank sediments are fine-grained (silty-clay, clayey-silt) alluvial soils from large and small floodplains, respectively. Floodplain and overbank sediments are deposited during flood events in low-energy environments (Alexander and Marriott, 1999; Ottesen et al., 1989); thus, they should be completely devoid of pebbles, which indicate medium-energy environments. The surficial floodplain and overbank sediments are normally affected by recent anthropogenic activities, and may be contaminated. Deeper samples normally show the natural background variation. Floodplain sediment, representing alluvium from the whole drainage basin, was collected from the alluvial plain at the lowermost point (near to the mouth) of a large catchment basin (1000–6000 km^2).

From each sampling site, 2 kg of top floodplain sediment (sampling depth 0–25 cm) was collected. Collecting bottom floodplain sediments (the lowermost 25 cm, actual depth noted on the field sheet) from the exposed section, just above the water level of the river, was optional and the sediments were collected only in Austria, Belgium, France, Germany, Greece, Norway, Portugal, and Spain.

6.4. Soils

Residual or sedentary soil samples were collected from the small, second-order, drainage basin (<100 km^2) at a suitable site above its alluvial plain and slope base where alluvium and colluvium, respectively, are deposited. Residual soil developed either directly on bedrock or on till was accepted. Residual soil from areas with agricultural activities was avoided, since the topsoil is usually affected by human activities. Colluvium or alluvium was not accepted as representing parent material.

Each residual soil sample was a composite of three to five subsamples collected from pits located at a distance of 10–20 m from each other. Two different depth-related samples were taken at each site: a topsoil sample from 0 to 25 cm (excluding material from the organic layer where present) and a subsoil sample from a 25-cm thick section within a depth range of 50–200 cm (the C soil horizon). Living surface vegetation, fresh litter, large roots, and rock fragments (stones) were removed by hand. The subsoil sample was always collected first followed by the topsoil sample, thus avoiding contamination of the subsoil surface from fallen topsoil material.

7. DOCUMENTATION OF FIELD DATA

The upper part of the overburden is continuously subject to a variety of natural processes, which include seasonal changes to the element contents and ratios of elements in sampled material. It is necessary to take these changes into account when interpreting the results. Therefore, certain observations must be recorded during the fieldwork. A special field form is normally designed for collecting the relevant data. Portable computers are recommended to avoid a potential source of error when transferring the data from the forms to computer files at the office.

Observations required depend on the survey strategy, but factors which, in general, may affect the study results may include morphology, topography, vegetation, moisture, position of groundwater level in relation to the sampling depth, soil type and its properties (texture, structure), possible sources of contamination, and type of geological formations.

There is one obligatory observation that must be recorded at the sampling site—the location (coordinates) of the site. This is best achieved using a GPS device from which the data can be downloaded to computer files, thus avoiding errors caused by manual measurement and transcription.

8. PHOTOGRAPHY

Photography is a useful way to document the sampling site. Digital photo archives are easily accessible via the Internet (e.g., http://www.gtk.fi/publ/foregsatlas/ForegsPhotos.htm) or via a local network. At each stream sediment/water sample site, it is recommended that two photographs be taken: the first showing general upstream topography from the lowermost subsite and the second showing the nature of the stream bed at the best subsite. At sites where soils samples are collected, photos of landscape, ground vegetation, and sampling pit should be documented.

9. SAMPLE ARCHIVE

In many cases, sample archives have proven to be very useful. Analytical techniques have been developed, making it possible to determine concentrations of new elements and other parameters and there are often new needs for information about elements that were not determined as part of original studies. Thus, well-organized sample storage and archive is valuable, although sometimes laborious to manage in a proper way. It may also appear expensive, but compared to collecting new samples, it is an economic alternative that is strongly recommended.

REFERENCES

Alexander, J., and Marriott, S. B. (1999). Introduction. *In* "Floodplains: Interdisciplinary Approaches" (S. B. Marriott and J. Alexander, eds.), pp. 1–13. The Geological Society, London, Special Publ. No. 163.

Bølviken, B., Bergström, J., Björklund, A., Kontio, M., Ottesen, R. T., Steenfelt, A., and Volden, T. (1986). "Geochemical Atlas of Northern Fennoscandia. Scale 1:4 000 000." Geological surveys of Finland, Greenland, Norway and Sweden, 19 p. 155 map sheets.

Bradshaw, P. M. D., Clews, D. R., and Walker, J. L. (1972). Exploration geochemistry: A series of seven articles. Reprinted from "Mining in Canada and Canadian Mining Journal." Barringer Research Ltd., Toronto, Ontario, 49 p.

Darnley, A. G., Björklund, A., Bölviken, B., Gustavsson, N., Koval, P. V., Plant, J. A., Steenfelt, A., Tauchid, M., and Xie, X. (1995). A global geochemical database for environmental and resource management. Final report of IGCP Project 259. *In* "Earth Sciences." Vol. 19, 122 pp. UNESCO Publishing, Paris.

De Vivo, B., Lima, A., Albanese, S., and Cicchella, D. (2006). "Atlante Geochimico—Ambientale Della Regione Campania. Geochemical Environmental Atlas of Campania Region." 216 p. Aracne Editrice, Roma. ISBN 88-548-0819-9.

Ginzburg, I. I. (1960). "Principles of Geochemical Prospecting (Translation from Russian)." 311 p. Pergamon Press, Oxford.

Gregorauskiene, V., Salminen, R., Reimann, C., and Chekushin, V. (2000). Field manual for Barents Ecogeochemistry project. *Geol. Surv. Finland*, Report S/44/0000/2/2000, 52 p.

Gustavsson, N., Bølviken, B., Smith, D. B., and Severson, R. C. (2001). Geochemical landscapes of the conterminous United States—New map presenttion for 22 elements. U.S. Geological Survey Professional Paper 1648, 38 p.

Hale, M., Plant, J. A., and Govett, G. J. S. (eds.) (1994). Drainage geochemistry. *In* "Handbook of Exploration Geochemistry", Vol. 6. Elsevier Science Publishers B.V., Amsterdam.

ISO (1995). Water quality-sampling—Part 12: Guidance on sampling of bottom sediments. ISO 5667-12: 1995.

Kauranne, L. K. (Ed.) (1976). Conceptual models in exploration geochemistry—Norden 1975. *J. Geochem. Explor.* **5,** 173–420.

Koljonen, T. (Ed.) (1992). "The Geochemical Atlas of Finland, Part 2: Till." 218 pp. Geological Survey of Finland, Espoo.

Lahermo, P., Ilmasti, M., Juntunen, R., and Taka, M. (1990). "The Geochemical Atlas of Finland, Part 1. The Hydrogeochemical Mapping of Finnish Ground Water." [Suomen geokemian atlas, osa 1. Tiivistelmä: Suomen pohjavesien hydrogeokemiallinen kartoitus], 66 pp. Geological Survey of Finland.

Lahermo, P., Väänänen, P., Tarvainen, T., and Salminen, R. (1996). Suomen geokemian atlas, osa 3: Ympäristögeokemia—purovedet ja sedimentit. *In* "Geochemical Atlas of Finland, Part 3: Environmental Geochemistry—Stream Waters and Sediments." 149 pp. Geological Survey of Finland, Espoo.

McClenaghan, M. B., Bobrowsky, P. T., Hall, G. E. M., and Cook, S. J. (eds.) (2001), 350 p. "Drift Exploration in Glaciated Terrain." The Geological Society, London.

Ottesen, R. T., Bogen, J., Bølviken, B., and Volden, T. (1989). Overbank sediment: A representative sample medium for regional geochemical mapping. *J. Geochem. Explor.* **32,** 257–277.

Plant, J. A., Smith, R. T., Stevensson, A. G., Forrest, M. D., and Hodgson, J. F. (1984). Regional geochemical mapping for mineral exploration in northern Scotland. *In* "Prospecting in Areas of Glaciated Terrain." pp. 103–120. Institution of Mining and Metallurgy, London.

Reimann, C., Äyräs, M., Chekushin, V., Bogatyrev, I., Boyd, R., de Caritat, P., Dutter, R., Finne, T. E., Halleraker, J. H., Jäger, O., Kashulina, G., Lehto, O., *et al.* (1998). "Environmental Geochemical Atlas of the Barents Region." 745 p. NGU-GTK-CKE Spec. Publ., Grytting AS.

Rühling, Å. (Ed.) (1994). "Atmospheric Heavy Metal Deposition in Europe—Estimation Based on Moss Analysis." 53 p. Nordic Council of Ministers, Nord.

Salminen, R. (1992). Scale of geochemical surveys. *In* "Regolith Exploration Geochemistry in Arctic and Temperate Terrains. Handbook of Exploration Geochemistry" (L. K. Kauranne, R. Salminen, K. Eriksson, and G. J. S. Govett, eds.) Vol. 5, pp. 143–164. Elsevier Science Publishers B.V., Amsterdam.

Salminen, R. (Ed.) (1995). "Alueellinen Geokemiallinen Kartoitus Suomessa 1982–1994—Summary: Regional Geochemical Mapping in Finland in 1982–1994." Geologian tutkimuskeskus—Geological Survey of Finland, Tutkimusraportti—Report of Investigation 130. 47 s. 24 appendixes.

Salminen, R., and Gregorauskiene, V. (2000). Considerations regarding the definition of a geochemical baseline of heavy metals in the overburden in areas differing in basic geology. *Appl. Geochem.* **15,** 647–653.

Salminen, R., and Hartikainen, A. (1986). Tracing of gold molybdenum and tungsten mineralization by use of step by step geochemical survey in Ilomantsi, eastern Finland. *In* "Prospecting in Areas of Glaciated Terrain." pp. 201–209. Institute of Mining and Metallurgy, London.

Salminen, R., Tarvainen, T., Demetriades, A., Duris, M., Fordyce, F. M., Gregorauskiene, V., Kahelin, H., Kivisilla, J., Klaver, G., Klein, P., Larson, J. O., Lis, J., *et al.* (1998). "FOREGS Geochemical Mapping Field Manual." 36 pp. Geological Survey of Finland, Guide 47.

Salminen, R., Chekushin, V., Tenhola, M., Bogatyrev, I., Glavatskikh, S. P., Fedotova, E., Gregorauskiene, V., Kashulina, G., Niskavaara, H., Polischuok, A., Rissanen, K., Selenok, L., *et al.* (2004). "Geochemical Atlas of the Eastern Barents Region." 548 p. Elsevier BV, Amsterdam (Reprinted from J. Geochem. Explor., 83).

Salminen, R. (chief-editor), Batista, M. J., Bidovec, M., Demetriades, A., De Vivo, B., De Vos, W., Duris, M., Gilucis, A., Gregorauskiene, V., Halamic, J., Heitzmann, P., Lima, A., *et al.* (2005). "Geochemical Atlas of Europe, Part 1, Background Information, Methodology and maps." 526 p. Geological Survey of Finland, Espoo (also available: http://www.gtk.fi/publ/foregsatlas/).

Steinnes, E., Rambæk, J. P., and Hansson, J. E. (1992). Large scale multi-element survey of atmospheric deposition using naturally growing moss as biomonitor. *Chemosphere* **25,** 735–752.

Wennervirta, H., Bølviken, B., and Nilsson, C. A. (1971). Summary of research and development in geochemical exploration in Scandinavian countries. *Can. Inst. Min. Metall.* **11,** 11–14.

CHAPTER TWO

SAMPLING METHODS FOR SITE CHARACTERIZATION

Christopher Swyngedouw* *and* Jean Michel Crépin[†]

Contents

Abstract

A site characterization is conducted where a hazardous substance has been released, and there is potential for the contamination to reach people or adversely affect the natural ecology. In some cases, site characterization is carried out to provide background

---

* Consulting Scientist, Bodycote Testing Group, #5, 2712–37th Avenue N.E., Calgary, Alberta, T1Y 5L3, Canada
† Executive Consultant, Business Development, Bodycote Testing Group, Edmonton, Alberta, T6B 3J4, Canada

Environmental Geochemistry
DOI: 10.1016/B978-0-444-53159-9.00002-4

© 2008 Elsevier B.V.
All rights reserved.

information, to provide information on site sensitivity, or to monitor incremental impact from anthropogenic activities.

When site characterizations are complete, they provide accurate information about the presence and distribution of target contaminants in relation to the background environment, thereby facilitating cost-effective and efficient remediation.

The product of site characterization is a verified conceptual model. The conceptual model is developed from the analysis of site-operating history, regional geology and hydrology, validated results from prior investigations, and field investigation results. It identifies the contaminant sources, the receptors, and the pathways that contaminants can take to reach the receptors. Information about regional and site-specific geology/ hydrogeology as well as knowledge of the contaminant fate and transport is necessary for making and revising sampling and analytical decisions.

This presentation provides an overview of sampling methods and tools suitable to address most site characterizations. The basic sampling types discussed are the systematic, random, and judgmental sampling approaches.

In addition to sampling procedures, sampling bias and sampling errors are introduced, leading to some guidance on sample handling, shipping, and chain-of-custody procedures.

This presentation will focus on sampling methods for soil only.

1. INTRODUCTION

One of the first steps of planning a sampling program is to define the sampling objectives—why the samples need to be taken. Some reasons for sampling are:

- Show that a requirement applies. For example, a waste may require special treatment if it contains high levels of a certain contaminant (e.g., Pb).
- Define the extend of contamination: when there has been a spill or a release, one may have to define the contamination plume in the groundwater, the downstream transport in a river, or the amount of soil affected by a leak.
- Obtain background (benchmark) information: what were the conditions before a spill?
- Audit: many jurisdictions require a facility to sample themselves to show compliance. The jurisdiction may conduct their on sampling to ensure the facility's analysis and reports are adequate.
- Prove a violation: this is a common reason to sample. Here, the sample and the analytical results may become evidence in a court case.

Types of sampling programs that could be used to meet the above objectives are exploratory sampling, monitoring, and presence/absence sampling.

1.1. Exploratory sampling

Exploratory sampling is a sampling program organized to evaluate phase I assessment, the nature of the disturbance, the possible range of the concentration of the parameters, and the types of disturbance. Control samples are necessary with an exploratory sampling program. Exploratory sampling is used at single industry waste sites for quantitative information, at small waste sites for qualitative information, where there is a known chemical spill, when there is a physical disturbance of previously undisturbed material, and for hot spot sampling.

Exploratory sampling may be used for qualitative assessment of soils where an impact or damage is visible or anticipated (Crépin and Johnson, 1993).

Exploratory sampling makes good use of field and laboratory resources, and the preliminary information is used to design more efficient sampling plans. The exploratory sample can also deal with known impact sites where quantitative data on soil properties are needed ("hot spots").

1.2. Monitoring

Monitoring is intended to provide information on a specific analyte or group of analyte concentrations over a period or within a specific geographic area. Monitoring is usually preceded by exploratory sampling or by the identification of historical data on the analytes of interest at that site. Monitoring is commonly used in long-term groundwater and soil monitoring programs that determine the presence or absence of contamination and any temporal changes; and in monitoring the effects of site remediation.

Long-term monitoring is often required for regulatory purposes. It measures the concentration of an analyte over time, or within a specific geographic area.

1.3. Presence/absence

Presence/absence sampling is used for confirmatory sampling, for example, for a flare pit evacuation. It is also used for exploratory sampling. Presence/absence sampling uses discrete grab samples.

2. Site Characterization

Site characterization is recommended to determine the nature and extent of contamination; or determine the environmental impact, monitor the migration of the contaminants, clean up the contamination, and monitor the effectiveness of the cleanup. What follows are some pertinent points that need to be considered for a site characterization.

Steps in planning and conducting an environmental site characterization:

- Identify the site (strategic mapping)
- Define the objectives and site boundaries
- Collect available existing data and information about that site (desk top review)
- Develop one or more conceptual site models of the site from that existing data
- Perform a reconnaissance site investigation using exploratory sampling. This may include nondestructive geophysical methods and field screening tests to refine the conceptual model

- Develop a detailed site investigation, sampling plan, and data quality objectives
- Collect the field samples and perform the measurements in accordance with the site investigation sampling plan
- Analyze the field and laboratory data to further refine the conceptual model
- Get the conceptual model accepted
- Report
- Remediate (if required)

A conceptual site model is useful in helping to determine the type of environmental samples that is required. A conceptual model emphasizes the type and extent of the contamination, defines the pathways for contaminant migration, and identifies potential receptors (e.g., well users, surface water bodies, and food and feed material) (US EPA, 2002).

2.1. Initial site assessment

An initial site assessment focuses on initial data requirements from a specific perspective, for example, risk. It identifies current and potential future land use, receptors, and points of exposure. It identifies source areas and potential exposure (migration) pathways.

Further, it identifies risk driver chemicals of concern (e.g., benzene). Figure 2.1 provides an example of a site that needs to be characterized.

Sources of contaminants could be point source (punctured barrel or tank or pipeline), a linear source (e.g., diesel in a utilities trench or a contaminated recharging stream), a disturbed source (e.g., pesticides and herbicides), a surface source, and an underground source (leaking gasoline tank). Sources of contamination (point source vs. nonpoint source) are scale dependent, thus influencing the sampling plan that will be used.

Suspected land could be past and present industrial sites, dry cleaning operations, waste processing and storage sites, landfill areas, and any land located near gas stations or storage tanks, in addition to residential areas, agricultural lands, parks, and natural areas.

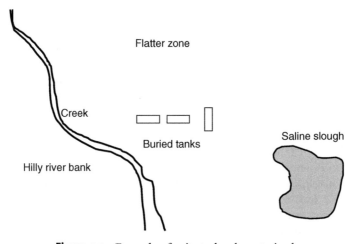

Figure 2.1   Example of a site to be characterized.

Transport mechanisms include mass movement of soil (soil erosion and landslides), wind, rain, surface waters (rivers and lakes), groundwater, and human intervention. The latter may consist of wastewater pipes, drainage ditches, roads, trains, ships, smoke stacks, etc.

Pathways are coming from the surface soils as ingestion, dermal adsorption, or inhalation; from the deep soil through inhalation (ambient air), leaching, and indoor vapor inhalation. When soil and groundwater are considered together, pathways from groundwater are evaluated also. These are water ingestion, inhalation, or ecological contact.

Site boundaries are determined by the land ownership, current and past land use, and by natural site characteristics (topography, soils, geology, hydrology, and biota). Site boundaries are also set by the extent of the contamination, as determined by exploratory sampling.

Existing site information very likely comes from the initial site assessment or phase I Environmental Site Assessment (ESA) done for that site. A phase I ESA normally includes a site inspection, a review of the relevant data, interviews, and a written report. When a phase II ESA is performed, more existing site information is obtained. A phase II contains (limited) soil sampling, done to determine the soil contaminants, or to determine the soil quality on- and off-site. Phase II ESAs have written reports including recommendations and have the extent of contamination delineated.

3. Basic Sampling Types

Before sampling can start however, a sampling strategy needs to be developed. This strategy is based on the conceptual model (hypothesis). Further, the number of samples needed and the required sampling locations are also determined. A further determination is made whether the use of composite or grab samples is appropriate. One needs to know how the data will be used and evaluated. All these are part of the data quality objectives on which a sampling plan is based.

There are many different sampling types. Some of them are random sampling, stratified sampling, and systematic sampling at sites judgmentally chosen. They are, however, all combinations or variations of three basic types of sampling: random sampling (as used in exploratory sampling), systematic sampling, and judgmental sampling (as used in e.g., audit sampling).

3.1. Random sampling

Random sampling makes no assumptions about the distribution of the analytes. It usually costs more because it requires more samples and relies less on specific knowledge. Its usefulness is that it remains a blind study, because no assumptions are made. See Fig. 2.2 for an example of random sampling at a site.

Characteristics of random sampling are that it is only limited by sample size and that the sampling location is determined before going to the field. It is best suited for

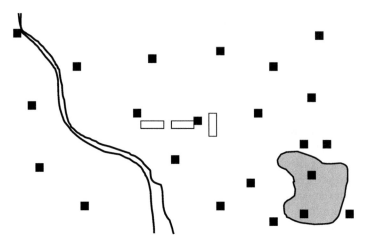

Figure 2.2 Simple random sampling. The sampling units are selected on a random basis. A drawback is that large parts of the sampling site may be left out completely: the buried tanks could be missed altogether.

relatively homogeneous material. Some disadvantages are that it does not provide information on variability and patterns. Depending on the number of sampling locations, random sampling can be deficient for detecting hot spots and for giving an overall picture of the spatial distribution of site contamination. Simple random sampling has quite a number of possible uses. Some of them are postreclamation sampling (e.g., mine and gravel pits), sampling at land-farming operations, compost piles, storage piles, farm land, water bodies, and ambient air.

3.2. Stratified random sampling

In stratified random sampling, the total site is broken down into a number of strata or subpopulations and random samples are taken from each stratum. Figure 2.3 provides an example.

Stratified random sampling requires reasons for stratification. The strata can be defined according to what was found from exploratory sampling and field testing, from the map units (topography, soil classification, land use, etc.), exposure, vegetation types or response, and cleanup procedures. Again, the sampling locations are determined before going to the field, and this for each stratum. The advantages of stratified random sampling are the increased precision, a simpler interpretation, and a reduction in cleanup or treatment costs. It also allows statistics on measurement variability.

Possible uses for stratified random sampling are for spill sites, waste sites, where site-specific management is needed, for atmospheric deposition, in soil sensitivity studies, for effluent and discharge dilution or extinction analysis, and lastly, for modeling.

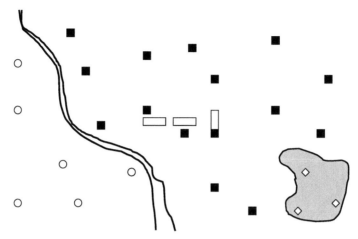

Figure 2.3 Stratified random sampling. This approach divides the site into smaller zones. Samples are collected randomly within each zone, and could even be composited.

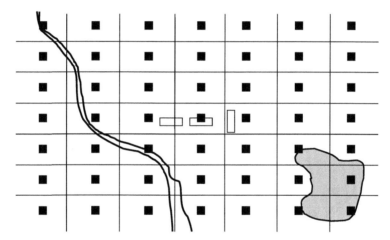

Figure 2.4 Systematic sampling. Systematic sampling divides the site into grid cells. Soil samples are collected from within each of these units, their locations geo-referenced. A drawback may be the missing of "hot spots."

3.3. Systematic sampling (grid)

Systematic sampling makes no assumptions about the distribution of the analytes. The sampling sites are established at a regular and predetermined distance from each other (Fig. 2.4). Systematic sampling is best for relatively homogeneous material that is not stratified. One caution is to avoid periodic trends in spacing that may correspond to the grid spacing. Grid sampling is costly. Major uses of systematic sampling are for contaminated land-based sites, such as landfills. Sometimes grid sampling is used to drill wells for the collection of groundwater.

3.4. Judgmental sampling

Judgmental sampling implements assumptions about the movement and distribution of pollutants with time distance (fate and transport). Figure 2.5 provides an example of judgmental sampling. Although judgmental sampling is inherently biased and limits the usefulness of the data for statistical interpretation, it is routinely used when sufficient knowledge of site history and activities is available.

Frequently, judgment combined with systematic or random sampling is used to take the advantages from each (Fig. 2.6). Table 2.1 summarizes the advantages and

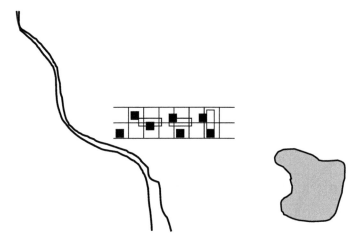

Figure 2.5 Judgmental sampling. The placement of the tanks may have been recorded in documentary sources (e.g., from the initial site assessment). The zone sampled is only that around the known area.

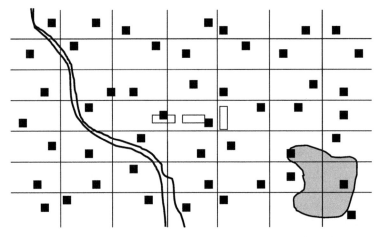

Figure 2.6 Stratified systematic sampling or systematic nonaligned sampling. This approach combines random and systematic sampling. The site is divided into regular spaced regions, and then a sample unit is chosen randomly within each of these regions.

Table 2.1 Summary of basic sampling approaches

Approach	Relative number of samples	Relative bias of the three approaches	Basis of selecting sampling sites
Judgmental	Smallest	Largest	Prior knowledge
Systematic	Large	Small	Consistent grid or pattern
Random	Largest	Smallest	Simple random selection

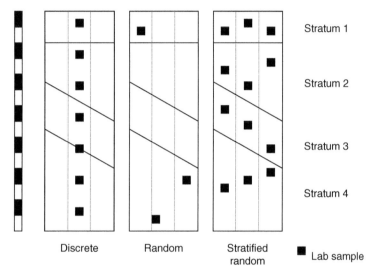

Figure 2.7 Depth sampling. Samples are taken from each of the soil horizons (strata) by any of the three approaches shown here.

some drawbacks of the sampling types relative to the number of samples needed and the amount of bias introduced (US EPA, 2002).

4. SOME FURTHER SAMPLING CONSIDERATIONS

4.1. Depth sampling

When the soil is undisturbed, samples are taken from each of the soil horizons. One could again use random depth sampling or stratified random depth sampling. This one is used when layers are visible. Sampling by depth intervals (discrete sampling) is also used for nutrient management or when the movement of contaminants is investigated. Figure 2.7 depicts these three approaches.

4.2. Types of samples

All sampling efforts take representative samples. A representative sample is a sample of a whole (e.g., of a waste pile) which can be expected to exhibit the expected value of the sampled population.

Two types of samples exist, grab samples and composite samples. Grab samples are collected at a selected position in time or space. Results from grab samples can be used to assess the variability in constituent concentrations. Composite samples are formed by the physical mixing of individual grab samples and thus represent a physical average. Composite samples incorporate natural random variation into the sampling error, making statistical post-processing simpler. Often, composite samples are also chosen for cost-saving reasons. Be aware that samples analyzed for volatiles should not be composited.

4.3. Number of samples

The number of samples taken depends on (CCME, 1993):

- How many distinct areas there are
- How many different analytical methods are needed
- How many samples are needed for each analytical method
- How many control site samples are needed
- What type of quality control (QC) samples are needed
- How many of each type of QC samples are needed
- How many supplementary samples are required (split samples)
- The available funds

4.4. Composite sampling

Characteristics of composite sampling are that the samples (or measurements) are taken following a random approach and then bulked. The best use is for relatively homogeneous material, unless stratified composites are created. A phase I ESA investigation often uses composite samples.

A few cautions are needed (US EPA, 1995). Only a mean is obtained from composite sampling. There is thus no information on variability. Extremes may also influence the obtained mean. Thirdly, the level of confidence is unknown. If one needs more information, then one should use discrete sampling (grab samples).

Composite sampling is often used to reduce the cost of analyzing a large number of samples. Experimental costs are substantially reduced when the frequency of individual samples containing the analytes of interest is low. However, composite sampling also has some limitations that must be considered.

- When considering multiple analytes in a composite, information regarding analyte relationships in individual samples will be lost. Hence, half of the individual samples should be saved before compositing.

- When the objective is an exploratory evaluation, compositing may dilute the analyte to a level below the detection level, thus producing a false negative value. It is thus necessary to adjust the investigative level through division by the number of discrete subsamples within the composite sample (BC-MWLAP, 2001).
- If sampling costs are greater than analytical costs, analyzing each sample individually will be more cost effective.
- If compositing reduces the number of samples collected below the statistical need of the data quality objectives, then those objectives will be compromised.

4.5. Statistical considerations

The number of subsamples required to make up a truly representative composite sample depends on the variability in results of the individual samples. The following expression quantifies the intuitive relationship:

$$n = \left(\frac{t \times s}{D}\right)^2$$

where

- t is the Student's t (2-tailed) statistics for a given probability level and with $(n-1)$ degrees of freedom;
- n is the number of samples required for the composite mean to fall, with its given probability, within the range $\pm D$ around the true mean; and
- s is the standard deviation in the results.

The number of samples required depends therefore on the stringency of the test requirements. As an example, criteria suggested as adequate and realistic for a site characterization soil testing purposes are that, in 8 of 10 sampling occasions (i.e., at an 80% probability level), results for composite samples should have a range no greater than 10% around the mean.

The percent coefficient of variation (%VC) in results is by definition

$$\%CV = \frac{100 \times s}{\text{mean}}$$

And, at an 80% probability level, t lies between 1.3 and 1.5 for $n > 5$. Thus,

$$n \sim 0.02 \times (\%CV)^2$$

If the %CV is high, either a large number n of samples are needed or an improved procedure to reduce the %CV is warranted.

4.6. Sampling bias

There are many sources that contribute to the uncertainty (variability) in environmental data:

- Uneven distribution of the chemical pollutants in the environmental matrix being sampled
- Error from the sampling design, that is, the error inherent in using a portion to represent the whole
- Error from sampling collection and preservation
- Error from the analytical measurements, and
- Error from the data handling

Sampling data contain some degree of uncertainty, and this uncertainty must be considered whenever the data is used. The total variance of measurement data can be expressed in terms of sampling variances due to the measurement and sampling variances due to the sample. Both measurement and sampling uncertainties have to be considered (Taylor, 1997).

$$S_{\text{total}}^2 = S_{\text{measurement}}^2 + S_{\text{sample}}^2$$

This thus requires a sampling plan that reflects the data quality objectives and analytical measurement subjected to the laboratory quality system (Swyngedouw and Lessard, 2007). The measurement uncertainty can be controlled and evaluated (Eurachem, 2000). The sampling variance may contain systematic and random components of error from population representation and sampling protocol. Note that the errors are separate and additive. This means that the laboratory cannot compensate for sampling errors.

Not all factors that influence the reliability and representativeness of data are measurable. Those that are measurable will usually be found if proper data-handling processes are followed. However, there are many immeasurable factors that can severely bias data and that are not readily identified, even by good data-handling and data management procedures. Some of these nonmeasurable errors are:

- Biased sampling (e.g., inappropriate judgmental sampling)
- Contaminated sampling tools
- Sampling the wrong area
- Sampling the wrong matrix
- Switching samples before labeling
- Mislabeling sample containers
- Incorrectly preserving the sample
- Failure to calibrate instruments
- Lack of control samples
- Forgetting equipment or supplies
- Wrong sampling procedures
- Misreading instruments
- Poor documentation

Each of these sources can be prevented or reduced by good sampling efforts. Of course, there are sources of error that cannot be controlled.

4.7. Control (background) samples

Control samples are a necessity in most soil sampling investigations. They should be sampled and processed the same way as other samples. Issues that may affect the validity of control samples are to ensure that the sampling tools are clean and not to treat sampling tools with oil or any other antirust agent.

The background site must be of similar characteristics, that is, same topographical location, similar soil physical characteristics, similar water regimen, and similar parent material and soil forming process.

The number of control samples will depend on the amount of variability in the natural landscape (site) under investigation. The minimum number of control samples is 2. In some cases, it will be necessary to establish control site variability and take a number of samples per stratum.

4.8. Sampling tools

A few of the sampling tools used are blades, tubes, and augers (Byrnes, 1994).

Blades are trowels, shovels, spades, backhoes, knives, etc. They are used when sampling topsoil and shallow subsoil without volatiles. They are easy to clean before taking the next sample. A slice is collected within an open face, permitting observations of soil and roots at the time of sampling.

Tubes are hand pushed near the surface. Most of the tubes are hydraulic. Some have an open or split side. When sampling for volatiles sealed tubes are used. Sampling in tubes also reduces cross-contamination. One needs to watch out for compaction when determining the sampling depth. Tube sampling still allows some examination of the soil.

Augers consist of Dutch augers, screw augers, and hollow stem augers. They can be used on stony soils. They are not suitable, however, when sampling for volatiles. One needs to be watchful for the wall or downhole cross-contamination. Using augers is not recommended when examining soil physical characteristics and stratification.

4.9. Field measurement

Some reasons for screening samples in the field are to improve the selection of samples for laboratory analysis, to avoid excessive site disturbance, to map the site, and to perform the required on-site analysis ("immediates", like pH, EC, and dissolved oxygen).

Field measurements include visual measurements, using portable equipment, or having a mobile laboratory at the site. Vapor probes and immunoassay kits are often being used. Electromagnetic surveys, remote sensing, visual and sensory information, or GPS are also sources of field measurements.

With the exception of required on-site parameters, field data must always be supplemented by data from a regular, accredited laboratory.

4.10. Sample handling

Sample integrity is maintained by sealing the sample container immediately and storing it in a cooler at 4 °C. Label the samples immediately. It is recommended that soil samples not be frozen and that the sample handling in the field is limited. Sometimes, it may be more efficient to let the laboratory composite the samples. Samples are not to be dried, unless the volatility of the organics and inorganics is not an issue. Try not to store contaminated and control samples together (even if trip blanks are used). Pack the samples well to avoid breakage and observe transportation of dangerous goods regulations when shipping the samples. Be aware of same contaminants that may come from the sample containers. Boron and silicon may leach out of glass; stainless steel may contaminate the sample with Cr, Fe, Ni, and Mo. There are phthalate esters in plastic tubing, none in Teflon tubing or polypropylene.

4.11. Documenting sampling

Minimum sampling documentation includes the sampling date and time, the sample identification number, the sampler's name, the sampling site, sampling conditions, the sample type, the preservation used, and the time of preservation. Sometimes, other relevant sampling site observation could be included. The Forum of European Geological Surveys (FOREGS) geochemical project (Salminen *et al.*, 1998) and the IUGS project on Global Geochemical Baselines (Darnley *et al.*, 1995) offer more guidance on the sampling documentation.

Samples are often sent to the laboratory requiring chain-of-custody protocol. Chain-of-custody ensures traceability and custody of the samples, as well as ensuring the integrity of the sample identity. Here sample-related activities are recorded, consisting of sample receipt, storage, preparation, analysis, and disposal. Chain-of-custody also ensures that all laboratory records are assembled and delivered.

5. SUMMARY

This topic, sampling design for site characterization, presents a very general overview of sampling considerations for beginning samplers. More detailed information on statistics could be found in US EPA (2002) and Keith (1988). The IUGS "Blue Book" (Darnley *et al.*, 1995) and FOREGS Geochemical Mapping Field Manual (Salminen *et al.*, 1998) provide practical guidelines.

REFERENCES

BC-MWLAP Government of British Columbia, Ministry of Water, Land and Air Protection (2001). Composite samples: A guide for regulators and project managers on the use of composite samples. Contaminated sites statistical applications guidance document No. 12–10 .

Byrnes, M. E. (1994). "Field Sampling Methods for Remedial Investigations." Lewis Publishers, CRC Press, Boca Raton, Florida, USA.

Canadian Council of the Ministers of the Environment (CCME) (1993). "Guidance Manual on Sampling, Analysis, and Data Management for Contaminated Sites. Volume I: Main Report." CCME, Winnipeg, Manitoba, Canada.

Crépin, J., and Johnson, R. L. (1993). Soil sampling for environmental assessment, in Soil sampling and methods of analysis. *In* "Canadian Society of Soil Science" (M. R. Carter, ed.). Lewis Publishers, CRC Press, Boca Raton, Florida, USA.

Darnley, A. G., Björklund, A., Bolviken, B., Gustavsson, N., Koval, P. V., Plant, J. A., Steenfelt, A., Tauchid, M., and Xuejing, X. (1995). "A Global Geochemical Database for Environmental and Resource Management. Recommendations for International Geochemical Mapping, Final Report of IGCP Project 259. Earth Sciences Series." UNESCO Publishing, Paris.

Eurachem (2000). "Eurachem Guide: Quantifying Uncertainty in Analytical Measurement," 2nd edn. [Online] available: http://www.eurachem.ul.pt/guides/QUAM2000-1.pdf [28 February 2007].

Keith, L. H. (1988). Principles of environmental sampling. *In* "ACS Professional Reference Book" (L. H. Keith, ed.). ACS, Washington, DC, USA.

Salminen, R., Tarvainen, T., Demetriades, A., Duris, M., Fordyce, F. M., Gregorauskiene, V., Kahelin, H., Kivisilla, J., Klaver, G., Klein, H., Larson, J. O., Lis, J., *et al.* (1998). FOREGS geochemical mapping. Field manual. Geological Survey of Finland, Guide 47.

Swyngedouw, C., and Lessard, R. (2007). Quality control in soil chemical analysis. *In* "Soil Sampling and Methods of Analysis" (M. Carter and E. Gregorich, eds.), 2nd edn., CRC Press, Boca Raton, Florida, USA.

Taylor, J. K. (1997). "An Introduction to Error Analysis: The Study of Uncertainties in Physical Measurement." 2nd edn., University Science Books, Sausalito, California, USA.

US EPA United States Environment Protection Agency (1995). EPA Observational Economy Series, Vol. 1: Composite Sampling. EPA-230-R-95–005.

US EPA (United States Environment Protection Agency) (2002). Guidance on choosing a sampling design for environmental data collection for use in developing a quality assurance project plan EPA QA/G-5S.

CHAPTER THREE

CONTAMINATED GROUNDWATER SAMPLING AND QUALITY CONTROL OF WATER ANALYSES

Julian K. Trick,* Marianne Stuart,* *and* Shaun Reeder*

Contents

* British Geological Survey, Keyworth, Nottingham, UK

Environmental Geochemistry
DOI: 10.1016/B978-0-444-53159-9.00003-6

© 2008 Elsevier B.V.
All rights reserved.

Abstract

The objective of groundwater sampling for site characterisation is the collection of samples that represent the underlying conditions at a site and ensuring that sample integrity is maintained from field to laboratory. The authors describe the tools available to the field sampler for the collection of groundwater samples, methods of on-site water-quality analysis and the appropriate preservation and handling of samples. There are a variety of portable sampling devices available for the collection of groundwater; however, each application has different requirements and is dependant on the contaminant(s) of interest and, most importantly, the specification of the borehole to be sampled. A number of different sampling devices and their applicability are presented. Traditionally, to ensure sample representativity, the removal of stagnant water from a monitoring well was accomplished by purging a fixed number of well volumes, generally between three to five volumes, before sample collection. In recent years, research has shown that low-flow purging (pumping at a rate that does not disturb the stagnant water in a well) produces samples that are representative of the formation water. In addition, 'no purge' sampling is becoming an increasingly accepted method of collecting representative groundwater samples for some determinands, in particular VOCs and some metals using diffusion methods. The merits of different purge methodologies are discussed. On-site water-quality measurements are carried out predominantly to monitor effective purging of water at the sampling point before sample collection and to measure unstable parameters that cannot be subsequently reliably determined in the laboratory. On-site measurements such as pH, specific electrical conductance (SEC), oxidation reduction potential (ORP), dissolved oxygen (DO), temperature and alkalinity can be used to provide a check on a subsequent laboratory analysis. Techniques for the preservation and analysis of samples and quality assurance and quality control are also presented.

1. INTRODUCTION

Contaminated sites may pose risks to both the environment and human health. The impacts of contaminated sites in the UK and internationally are managed using a conceptual risk–based assessment model:

$$Source \Rightarrow Pathway \Rightarrow Receptor$$

The *source* is defined by the amount and nature of a potentially hazardous contaminant. The degree to which a source poses a risk depends on the presence of a means of transport (the *pathway*) for the contaminants to the *receptor* (the plants, animals and/or humans and even buildings that may be adversely affected by the contamination). Contaminants can move from the source to the receptor via food, soil, air and water. For humans, the main ways that contaminants can enter our bodies are by ingestion, inhalation and direct contact, for example, by absorption through the skin.

It is important to note that groundwater and surface water may act both as *pathways* (e.g., through percolation through the unsaturated zone, saturated groundwater flow and surface water flow) and as *receptors* (e.g., vulnerable water abstractions, resources or ecological systems). Evaluation of surface water and groundwater as part of contaminated site investigation studies is, therefore, a major concern.

This chapter aims to provide a step-by-step guide for practitioners involved in the collection of contaminated samples by reviewing current groundwater sampling techniques and procedures and highlighting the major sources of uncertainty associated with sample collection. On-site water-quality measurements, quality assurance procedures and sample handling techniques designed to maintain the representativeness of the sample from field to laboratory are also discussed.

2. GROUNDWATER SAMPLING OBJECTIVES

The critical objective of groundwater sampling for site characterisation is to collect representative samples and to ensure that their integrity is maintained from field to laboratory. Sampling and analysis can be expensive, so it is important that a thorough understanding of the site conditions is determined before mobilising a field sampling team. For example, groundwater quality can be variable over quite short distances; therefore, an understanding of the hydrogeology and flow dynamics of a system is important before any water quality sampling is undertaken. Preparing a robust conceptual model of the site in advance will help to guide the type of sample, analysis, and sampling protocol required.

2.1. Planning and preparation

A successful groundwater sampling campaign needs to be planned meticulously before mobilisation of the sampling team. It is important to ensure that all paperwork and relevant information are available to the team. This will include maps of the site detailing borehole locations; borehole details, including purge volumes and completion details; analytical requirements, including bottle types, preservation techniques and on-site measurement requirements; sample sheets and bottle labels; contact details for site supervisors, laboratories, couriers, etc.; data from previous sampling rounds for comparison; and health and safety documentation read and signed by all-field operatives.

3. CHOOSING THE RIGHT PORTABLE SAMPLING DEVICES

There are a variety of portable sampling devices available for the collection of representative groundwater samples; however, each application has different requirements and is dependant on the contaminant(s) of interest and, most importantly, the specification of the borehole to be sampled. The first factors to consider are the depth to the water table and the borehole diameter; other factors including the borehole completion (e.g., open hole, length of screened interval, casing diameter and purge volume required) also need to be considered. In addition, the sampling device should satisfy the following requirements (Schuller *et al.*, 1981):

- the device must not alter the physical or chemical structure of the sample;
- the materials used in the construction of the device must not leach or absorb contaminants to or from the sample;
- the device must be portable and easy to mend in the field;
- the device must be easily cleaned to avoid cross-contamination;
- chemical parameters such as pH, ORP (oxidation–reduction potential) dissolved oxygen (DO) and temperature must not be altered by the pumping mechanism;
- the device should be inexpensive, durable and simple to use.

Comparisons of the different sampling mechanisms available are given by Barcelona *et al.* (1984), Stuart (1984), Nielsen and Yeates (1985) and Pohlmann and Hess (1988). A comprehensive literature review of the effects of sampling devices on water quality is given by Parker (1994).

Portable sampling devices were categorised by Nielsen and Yeates (1985) as grab samplers, suction lift devices and positive displacement mechanisms. A brief review of each category is given below.

3.1. Grab samplers

The most common grab sampler is the bailer, of which there are several types available. Essentially, the bailer is a rigid tube made of PVC, stainless steel or Teflon, with a ball valve at the bottom and an open top. As the bailer is lowered to the required depth, groundwater flows up through the ball valve and out through the open top. Once the bailer is no longer in motion, the pressure of the water column closes the ball valve and seals in the sample. Double valve bailers have valve's top and bottom that are both closed upon reaching the required depth; this stops inflow and mixing of water during the bailer's ascent to the surface. Bailers are generally inexpensive (stainless steel bailers are the exception) and fulfil most of Nielsen and

Yeats's criteria for the 'ideal' sampling tool. They can be dedicated to individual monitoring wells to avoid cross-contamination between boreholes, they are portable, simple to use and relatively easy to clean. They are, however, not suitable for purging large volumes of water, and it can be difficult even with double-ended bailers to determine accurately where the sample was collected. Compared with other sampling devices, the operator is also more at risk of coming into contact with contaminated sample, especially when emptying the bailer (Fig. 3.1).

3.2. Passive diffusion bag sampler

Passive diffusion bag (PDB) sampling takes advantage of the Fick's law of diffusion, which states that compounds will migrate from an area of high concentration to an area of low concentration until equilibrium is achieved. A typical PDB sampler consists of a low-density polyethylene, 'lay-flat' tube closed at both ends and filled with deionised water before deployment. The sampler is positioned at the target horizon of the well by attachment to a weighted line or fixed pipe. The sampler is used to obtain concentrations of volatile organic compounds (VOCs) in groundwater from wells or at interfaces of groundwater and surface water. The molecular size and shape and hydrophobic nature of a compound influence its ability to diffuse through the polyethylene membrane and thus PDB samplers are not appropriate for all VOCs. PDBs are not suitable for assessment of inorganic species.

3.3. SnapSampler™

The SnapSampler relies on passive flow-through of water through well screens. It is simple to use, and there is minimal field pre-preparation. The sample bottles are open at both ends to the well environment during the deployment period, and contaminants do not have to diffuse through a membrane. Less equilibration time is needed before sampler retrieval and contaminants such as MTBE, 1,2,4-trimethylbenzene and acetone are not selectively inhibited from entering the sampler. The no-pour aspect of the device is unique among common sampling protocols, whether traditional purge or passive. Historic and recent research on VOC sampling techniques indicates that minimising sample transfer steps also minimises VOC losses. This method eliminates all transfer steps outside the laboratory analytical equipment.

3.4. HydraSleeve™

The reported advantages of the HydraSleeve are similar to the PDB. Sladky and Roberts (2002) tested the HydraSleeve for its applicability to sample semi-volatile organic compounds (SVOCs) by comparison with low-flow purging and bottom-loading bailer samplers. Results showed that wells containing LNAPLs were not suited to be sampled using this method and that SVOC concentrations were on average 17% higher than low-flow samples and 150% higher than samples collected using a bottom-loading bailer. Parker and Clark (2002) compared the HydraSleeve to four other discrete interval samplers and found that it yielded representative samples

Figure 3.1   Stainless steel bailer being deployed down a 50 mm monitoring well.

of pesticides, explosives and metals, though the authors recommend that it is used only in low-turbidity wells.

3.5. Inertial pumps

The inertial pump is described for use in groundwater monitoring wells by Rannie and Nadon (1988). The pump consists of just two components: a foot valve, which can be manufactured in high density polyethylene (HDPE), Delrin (an acetal resin thermoplastic that has high strength, rigidity, durability and chemical resistance), Teflon or stainless steel; and a riser tube, also made from HDPE or Teflon. The operating principle of the pump is based on the inertia of a column of water within the riser tubing. Once installed, the water level in the tubing is equal to that of the well. A rapid upstroke closes the foot valve and lifts the water column in the tubing a distance equal to the stroke length. The water column continues to move up the riser tubing due to its inertia. A down stroke opens the foot valve, forcing water to flow into the tubing thus raising the water level. Rapid up and down movement of the tubing causes the water to rise up the tubing to the surface. Flow rates between 1.8 and 6.5 litres per minute were reported depending on the foot valve size, and depths of 40 m (manual operation) and 60 m (motor driven) were achieved. As with bailers, this device fulfils most of the requirements of the 'ideal' sampling device as described above; however, manual operation is labour intensive and the use of a motorised device requires a compressor or generator, thus reducing the portability of the system.

3.6. Peristaltic pumps

Suction lift pumps operate by applying a vacuum to a sample line or tube causing water to be drawn up to the surface. This limits these devices to sampling groundwaters at depths no >7.6 m (Nielsen and Yeates, 1985). Peristaltic pumps operate by creating a low vacuum by the squeezing action of rollers on flexible tubing, as the rollers rotate around the tubing, suction is created drawing sample up from the well. These pumps have an advantage over centrifugal pumps (see below) in that the sample comes into contact only with the tubing. However, silicone tubing is most commonly used, to provide the required flexibility, and this is not suitable for sampling organics due to its propensity to absorb organic compounds (Barcelona *et al.*, 1985; Pearsall and Eckhardt, 1987). The loss of volatiles using peristaltic pumps has been investigated by a number of workers; Barker and Dickhout (1988) found concentrations of volatile halocarbons 23%–33% lower than other pumps tested and concluded that degassing was a problem with this pump. Both pumps require a power source such as a generator or 12 V batteries.

3.7. Gas-operated bladder pumps

There are several designs of bladder pumps; however, the basic design is one of a long rigid casing (often stainless steel) housing a flexible membrane. The bladder, which has a perforated tube inside, is attached to a screened intake check valve and a discharge valve attached to a discharge tube to take sample to the surface. The annulus between the bladder and the housing is pressurised before insertion in the well causing the bladder to collapse. When a sample is to be taken, the annulus pressure is reduced, as the water pressure at the intake valve exceeds that of the annulus, and water flows into the bladder through the perforated tube. When the bladder is full, pressure is increased in the annulus causing the bladder to collapse, closing the intake valve and forcing the water up through the discharge valve and into the discharge tube. The annular gas is then vented to the surface, allowing the bladder to be re-filled and the cycle repeated. Regulating the frequency of the applied and released pressure allows the operator to adjust the flow rate and maintain a steady flow.

Bladder pumps are regarded in the literature as one of the best groundwater sampling devices for a number of reasons: compressed air is used as the driving gas since it does not come into contact with the sample; almost the entire assembly can be made of inert material; depths in excess of 60 m can be sampled and many bladder pumps are designed to sample 50 mm wells; ease of disassembly allows cleaning and repair in the field; variable pumping rates are possible, allowing well purging and low-flow-rate sampling (Nielsen and Yeates, 1985). However, these pumps are often expensive, the need for an air compressor or compressed air tanks and a pump control make these pumps less portable than others and waters with high-suspended solids may block the check valves necessitating removal of the pump. In addition, lifting sample from deep wells requires large amounts of gas.

3.8. Electric submersible pumps

These pumps consist of a sealed electric motor that drives a two-stage centrifugal pump with radial impellers. A partial vacuum is created as the impeller rotates and forces water up a discharge line by centrifugal force, water is then drawn into the impeller housing continuing the pumping action. A screened intake inhibits large particles from entering and blocking the pump and a built in thermal switch turns off the pump if a maximum-operating temperature is exceeded. The most commonly used pump of this type is the Grundfos MP1, which is designed to be used in 2 in. diameter wells or greater. This pump is powered by a 220 V generator and run via an adjustable frequency converter that allows a high range speed adjustment of the pump. High-flow-rate sampling with pumps such as this has been shown to mobilise more colloidal particles than bladder pumps, thus increasing sample turbidity (Puls *et al.*, 1992). Puls *et al.* (1992) also draw attention to the disturbance caused to the water column from the insertion and removal of these pumps and recommend the use of dedicated samplers in each borehole.

Rosen *et al.* (1992) compared downhole sampling using a helical rotor type of centrifugal pump with an absorption cartridge for volatile organic compounds. They found that this method of bringing water to the surface could be reliable in many circumstances but that care was needed with adsorption, cross-contamination and out gassing.

Submersible centrifugal pumps are now available with integral packers that allow the screened interval of the well to be isolated from the stagnant column above, thus negating purging requirements and reducing sampling times and costs.

3.9. Common materials used in sampling devices

It is important that the equipment used for sample collection and the subsequent handling and storage of samples do not contribute to the contaminant load of the sample, for example, by leaching organics. The following are common materials used in sampling equipment construction (after Canter *et al.*, 1990):

- Teflon—advantages include: it is inert, has poor sorptive qualities and low-leaching potential and can be rigid or flexible; disadvantages include: cost
- Stainless steel—easy to clean but expensive
- PVC—good chemical resistance (except chlorinated solvents, ketones and aldehydes) and good for inorganics, but may bias some organic compounds
- High density polyethylene—commonly used in sample containers but can absorb trace metals if the sample is not acidified
- Glass—essential for samples with organic contaminants

4. AVOIDING CROSS-CONTAMINATION

If it is impractical to dedicate a sampling device to a single well or disposable samplers are not suitable, it is imperative that the sampling device is properly cleaned between deployments to avoid cross-contamination of samples and boreholes. Decontamination is essential for microbiological, organic and pesticide sampling and recommended, but not as critical, for major ion analysis. The use of proprietary cleaners is recommended, and the operator should clean inside and outside of the sampling device and the associated hose. All cleaning operations should be undertaken away from boreholes and preferably somewhere where the waste water can drain to a foul sewer. To verify that no residual contamination remains in the sampler, a sample of deionised water should be passed through after the cleaning operation and analysed for the contaminants of concern. Wells that are known to be highly contaminated should have dedicated sampling devices installed whenever possible. At the very least, it is useful to have separate sampling devices to differentiate between 'background' and 'contaminated' samples. Where known, 'background' samples and those that are least contaminated should be collected before those that are more likely to be heavily contaminated.

5. WATER-LEVEL MEASUREMENTS

The level of the water in a well is measured to provide data for groundwater flow direction calculations and to calculate purge volumes required for a particular well. If a low-flow sampling protocol is used, the water level is also measured during pumping to ensure that there is minimal drawdown of the stagnant water column.

The static water level (SWL) should be measured before purging, sampling or inserting any other device into the well. It is also good practice to collect water-level measurements from all site wells within a reasonably short time, that is, by measuring all water levels at the start of the day. This mitigates against measurements not being comparable to one another due to a heavy rainfall event overnight. Measurements should always be taken to a permanent reference point, for example, the casing top. To minimise cross-contamination, the least contaminated wells should be measured first and the measuring device should be thoroughly cleaned before deployment in the next well.

Water-level measurements should be compared to previous data to verify that previously calculated purge volumes remain valid before purging and sampling.

There are level measurement devices (dip metres) available as small as 6.4 mm in diameter to fit narrow diameter wells. In general, the tape is marked every millimetre (although smaller diameter tapes are less frequently marked) and come in lengths of 30–600 m. A high-pitched alarm sounds when water is reached and the distance read from the tape. Dip metres can also be used to measure the total depth of a well and interface metres are available that measure immiscible product levels on the top of the water column.

Automated water-level measurement can be achieved by deploying pressure transducers that can be set to log data as frequently as every second if required. Loggers also record temperature and are able to compensate for altitude, water density, temperature and barometric pressure. These are routinely used when undertaking hydraulic tests on a well but are also useful to provide accurate, regular, long-term water level data and are invaluable for groundwater flow direction calculations.

6. WELL PURGING TECHNIQUES

Groundwater samples representative of *in situ* conditions are difficult to obtain because of complex physical, chemical, geological and bacterial processes. Monitoring wells are often completed with screen lengths that are shorter than the column of water in the well. Water sitting above the screened interval is liable to stagnation as freshwater is unable to flow in this section of the well. In order to collect a sample that is representative of the water flowing through the geological formation, it is important to either remove the stagnant water or sample from the screened section of the well at a rate of flow equal to that flowing through it. Interaction of air in the well with the top of the column of water can change the chemical composition of the water. For example, the dissolved gas content (dissolved oxygen and carbon dioxide) of the water will equilibrate with the air column causing oxidation and precipitation of some metals (e.g., iron and manganese) out of solution. Biological activity, interaction with well casing material and material falling into the well will also affect the water quality. The most common way to mitigate against this is to remove the stagnant water before collecting the sample.

6.1. Specified number of well volumes

Traditionally, the removal of stagnant water from a monitoring well is accomplished by purging a fixed number of well volumes, generally between 3 and 5 volumes, before sample collection. A well volume is defined as the amount of water in the well casing and screened portion of the well at static water-level conditions and is calculated using the formula: Volume $= \pi\, r^2\, h$, where r is the radius of the borehole and h the height of the water column. However, this method does not take into account the annulus between the well casing and the true diameter of the drilled borehole, which is often backfilled with gravel. This annulus should also be taken into account when calculating well volumes and is calculated as the drilled borehole diameter minus the well casing volume multiplied by the porosity of the fill material. Submersible pumps are generally used when purging as high-flow rates are required, although inertial pumps can be used, particularly where smaller purge volumes are required. One technique is to locate the pump above the screen, so it can be lifted to the top of the water column to remove the stagnant water.

The 3–5 well volume purge approach has a number of disadvantages:

1. Large volumes of potentially contaminated waste water requiring off-site disposal are often produced.

2. Turbidity is increased because of the high rate of pumping required to purge large volumes of water in a reasonable timeframe.
3. Water from different zones in the aquifer is mixed, giving an averaged concentration of contaminants in the well. Zones of interest will not be identified.
4. Volatile compounds may be lost because of agitation of the sample.
5. It is relatively easy to pump wells dry mistakenly, allowing aeration of sample that had previously existed in an anaerobic environment.

Another problem with this approach is the lack of scientific evidence to determine when a sample should be collected. This can be alleviated by monitoring water-quality parameters such as pH, temperature (t), electrical conductivity (EC), oxidation–reduction (or redox) potential (ORP, otherwise referred to as Eh) and dissolved oxygen (DO_2) throughout the purge to ascertain when the stagnant water has been removed. Plotting the stabilisation times of these parameters can inform the sampler when a sample can be collected. For subsequent sampling rounds, purge times can be related to parameter stabilisation times, hence reducing the purge time required. Many sampling practitioners use EC as the basis for stabilisation, but Puls and Powell (1992) report that pH, t and EC are the least sensitive indicators of aquifer equilibration. Redox potential, DO_2 and contaminant concentration were reported to be more sensitive, and turbidity the most sensitive indictor of equilibrated conditions. Therefore, the general order of stabilisation is pH, t, EC, Eh, DO_2 and turbidity, and samples should not be collected until at least dissolved oxygen concentration has stabilised. In-line flow cells are recommended to continually monitor these parameters to avoid contact of the sample with the atmosphere.

A study by Gibs et al. (1990) found that monitoring field water-quality parameters for stability is not a reliable indicator of when to collect a representative sample for purgeable organic compounds (POCs), such as chlorinated alkanes, alkenes and aromatics. For these types of compounds, the researchers suggest direct monitoring of POC concentrations.

Re-evaluation of sampling techniques in recent years has led to researchers developing a new technique that reduces the amount of purge water by negating the need to remove the stagnant water column in a well. Controlled pumping of the well at low-flow rates (<500 ml/min) while monitoring indicator parameters to the point of stabilisation has been shown to produce samples comparable with those collected using a full purge technique.

6.2. Low-flow purging

Water flowing through the screened interval of a monitoring well is representative of the formation water and chemically distinct from the overlying stagnant column (Robin and Gillham, 1987). The theory of low-flow purging (sometimes referred to as micro-purging) takes advantage of this and aims to minimise the drawdown in a well during pumping by placing the pump across or just above the screened interval and pumping at a very slow rate. Low-flow refers to the velocity of the formation water entering the well screen. Low-flow purging effectively isolates the stagnant column of water as groundwater flows through the well screen at a sufficiently low velocity such that water is only taken from the screened interval, thus leaving the stagnant water undisturbed.

Significant research into the flow patterns that occur in monitoring wells has been undertaken over the last 10–15 years and is summarised here.

Observations of colloidal movement under natural conditions and during pumping were conducted at several field sites by Kearl *et al.* (1992). Results indicated that the installation of dedicated sampling devices, limited purging of the well before sampling, sampling at a flow rate of 100 ml/min, and not filtering samples may collectively improve the representativeness and cost effectiveness of obtaining groundwater samples for assessing the total mobile contaminant load.

Puls and Powell (1992) recommended the use of low-flow rates during both purging and sampling, placement of the sampling intake at the desired sampling point, minimal disturbance of the stagnant water column above the screened interval, monitoring of water-quality indicators during purging, minimisation of atmospheric contact with samples and collection of unfiltered samples for metal analyses to estimate total contaminant loading in the system. While additional time is often required to purge using low-flow rates, the authors state that this is compensated for by eliminating the need for filtration, decreased volume of contaminated purge water and less re-sampling to address inconsistent data results. The use of low-flow rate purging and sampling consistently produced filtered and unfiltered samples that showed no significant differences in concentrations.

A comparison of micro-purging and traditional groundwater sampling was reported by Kearl *et al.* (1994). To compare methods, duplicate groundwater samples were collected at two field sites using traditional and micro-purge methods. Samples were analysed for selected organic and inorganic constituents, and the results were compared statistically. Analysis of the data using the nonparametric sign test indicated that there was no significant difference (at 95% confidence interval) between the two methods for the site contaminants and the majority of analytes. These analytical results were supported by visual observations using a colloid borescope, which demonstrated impacts on the flow system in the well when using traditional sampling methods. Under selected circumstances, the results suggest replacing traditional sampling with micro-purging based on reliability, cost and waste minimisation. The authors recommend that samples should be collected with dedicated sampling devices such as bladder or submersible pumps. Bailers should not be used, pump intakes should be located in the centre of screens unless depth-specific samples are required and samples should be collected 24 h after pump installation. In addition, only the pump and tubing should be purged, as they believe it is not necessary to purge the well casing and the screen.

Greacen and Slivia (1994) conducted a comparison of low-flow versus high-flow (borehole purge) sampling methodologies on groundwater metal concentrations. They found that although a submersible sampling pump provided an efficient means of collecting groundwater samples, the use of the low-flow sampling methodology did not provide data that could not be obtained by sampling the well at higher flow rates.

Research by Barcelona *et al.* (1994) confirmed that low-flow-rate purging (i.e., ~1 litre per minute) is a valid technique for 2″ (5 cm) diameter monitoring wells with short-screened intervals. They reported that the use of low-flow, and dedicated pumping devices for purging and sampling minimises both the disturbance of stagnant water in the well casing and the potential for mobilisation of particulate or colloidal matter that can lead to sampling artefacts. In addition, these techniques allow the use of purging indicator parameters (e.g., dissolved oxygen and electrical conductivity) to determine when to collect a sample for VOC determinations. The suggested procedure includes documenting purging indicator parameters while purging with dedicated devices at low-flow rates with minimal drawdown. This sampling method is less time-consuming and reduces the need to handle large volumes of purge water since VOC concentrations, DO_2 and specific conductance values stabilised consistently in less than one borehole volume.

An investigation of contaminant migration by low-flow rate sampling techniques by Bangsund et al. (1994) used low-flow rates (10–100 ml/min) to minimise drawdown and colloidal migration. Unfiltered samples collected using these methods are believed to be more representative of actual groundwater quality. Because of the nature of the contaminants, an all-stainless steel positive-displacement gas drive pump was used. Drawdown was monitored continuously during sampling. Redox potential, pH, temperature, electrical conductivity, dissolved oxygen and turbidity were monitored in the field to establish water-quality stabilisation. Water quality in the wells stabilised after removal of one or two well volumes. Turbidity and ORP were found to be the best field indicators of water-quality stabilisation during sampling. Results indicate that the unfiltered samples collected using the low-flow sampling method have concentrations of inorganic parameters roughly equal to historic filtered sample results, and as much as three orders of magnitude lower than unfiltered sample results from standard sampling methods.

In 1995, Puls and Paul (1995) conducted a field study to assess purging requirements for dedicated sampling systems in conventional monitoring wells and for pumps encased in short screens and buried within a shallow sandy aquifer. Low-flow purging methods were used, and wells were purged until water-quality indicator parameters (dissolved oxygen, electrical conductance and turbidity) and contaminant concentrations (chromate, trichloroethylene and dichloroethlyene) reached equilibrium. The data show that purge volumes were independent of well depth or casing volumes. Contaminant concentrations equilibrated in <7.5 litres of purge volume in all wells. Initial contaminant concentration values were generally within 20% of final values. Water-quality parameters equilibrated in <10 litres in all wells and were conservative measures for indicating the presence of adjacent formation water. Water-quality parameters equilibrated faster in dedicated sampling systems than in portable systems and initial turbidity levels were lower.

In 1996, Puls and Barcelona (1996) published a US EPA—Ground Water Issue to provide background information on the development of low-flow sampling procedures and its application under a variety of hydrogeological settings.

Various physical and chemical properties were monitored sequentially by Gibs et al. (2000) during well purging as indicators of stabilisation of the water in the well. Turbidity was correlated with the concentrations of Fe, Al and Mn in oxic groundwater, but appeared to be independent of conductivity, pH, temperature or dissolved oxygen. Pb and Cu were related to the sum of the Fe, Al and Mn.

Stabilisation of turbidity was found to be a good indicator of stable unfiltered trace element concentrations at all wells monitored and for some filtered trace element concentrations.

6.3. 'No purge' sampling

'No purge' sampling (see earlier sections on passive diffusion bags, snap sampler and Hydrasleeve) is becoming an increasingly accepted method of collecting representative groundwater samples for some determinands, in particular, VOCs and some metals using diffusion methods. These sampling systems are deployed in the well and left to equilibrate with the formation water, thus, negating the need for purging. Double-ended bailers can be classed as grab samplers but can also be used to purge shallow wells that have low volumes of water.

6.4. Dedicated pump versus portable sample collection

Passing a sampling device through a stagnant column of water can cause mixing with the screened interval, disturbance to the suspended sediment at the bottom of the casing and the displacement of water into the formation immediately adjacent to the well screen. This can be avoided by using dedicated pumps. This approach also minimises the potential for cross-contamination between boreholes when a common pump is used. In addition, it has been shown that water-quality parameters equilibrate faster in dedicated sampling systems than in portable systems and initial turbidity levels tend to be lower (Puls and Powell, 1992).

7. On-Site Water-Quality Measurements

On-site water-quality measurements are carried out predominantly to monitor effective purging of water at the sampling point before sample collection and to measure unstable parameters that cannot be subsequently reliably determined in the laboratory. On-site measurements can also be used to provide a check on a subsequent laboratory analysis. For example, provided that the on-site SEC is measured accurately, it can be compared with the SEC estimated from the laboratory chemical analysis by one of a number of geochemical programmes. This check can be useful for spotting major errors, such as dilution or typographical errors, as well as systematic errors in analytical methodology.

A flow-through cell should be used for taking on-site chemistry measurements from a pumped sample. This is advised as it produces an airtight environment that isolates the flowing water to be sampled from the atmosphere. All air bubbles must be expelled from the cell to prevent anomalous readings and a constant flow must be maintained. All measurements need to be accurately recorded using a field data sheet or a notebook. Readings should be taken at regular intervals to monitor stabilisation before the final reported measurement is made.

7.1. Temperature

The rate of many biological and chemical reactions is affected by temperature, though deep groundwater temperatures are less susceptible to seasonal temperature fluctuations. Temperature should be recorded to 0.1 °C. Care is needed when making measurements since groundwater temperature can change quickly on exposure to ambient conditions, and it is important to take the measurement as close as possible to the outlet and to exclude direct sunlight. Reporting of other on-site parameters should be corrected to the appropriate temperature.

7.2. pH

pH is a measure of the hydrogen ion concentration in solution and is also referred to as the degree of acidity or alkalinity. As a sample's pH changes, many precipitation, co-precipitation and sorption processes can occur that alter the sample's chemical composition and reaction rates. Biological processes of a sample are also influenced by its pH. Changes in the dissolved gas content of a sample can alter the pH. Groundwater is generally in equilibrium with CO_2 at a partial pressure several times that of the atmosphere. On exposure to the atmosphere, this CO_2 escapes and the pH rises. It is therefore important that pH is measured on-site. The pH is generally measured using a combination electrode calibrated using standards of pH 4, 7 and 10. In addition, it is recommended that a dilute acid solution of known pH is used as a control check and that additional control standards are also checked if the pH is outside the range 4–10.

7.3. Specific electrical conductance

Specific electrical conductance is the measure of a solution's ability to conduct or carry an electric current and depends on the presence of charged ion species such as calcium, sodium, magnesium chloride, etc. Conductivity measurements are approximately related to total dissolved solids (TDS) in a sample, but since different ions carry different amounts of charge and move at different speeds, their individual contribution to the overall SEC varies.

SEC is measured with a conductivity meter, which normally consists of an AC bridge and a conductivity cell or electrodes. The conductance is measured between two electrodes. Two solutions of known conductivity should be used, one to calibrate the metre and the other to check the slope. It is important to correct all data for water temperature, either by calculation or by automatically using the metre's auto-temperature correction mode, since SEC is highly dependent on temperature. SEC increases by about 2% per degree centigrade rise in temperature due principally to an increase in water viscosity.

7.4. Alkalinity

Alkalinity is a measure of the acid-neutralising capacity of water and is usually determined by titration against sulphuric acid to the endpoint of the acid–base reaction. In groundwaters, the carbonate species predominate and an endpoint of about pH 4.5 marks the consumption of bicarbonate in solution. The endpoint can be determined using an indicator dye, such as bromocresol green, or the pH can be monitored and the inflection of the titration curve identified. In relatively uncommon water where the pH is high, the titration curve may also indicate an inflection due to the presence of carbonate. In some waters, other weak acids, such as borate, silicate and organic acids, may also contribute to the alkalinity.

Alkalinity measurements are often made on-site but are sometimes not considered to be as critical as pH, since the loss of CO_2 does not in itself change the alkalinity. However, Shaver (1993) and Fritz (1994) found that significant errors were introduced by the use of laboratory rather than *in situ* data. That said, it is possible for inexperienced samplers to mistakenly identify the endpoint of the titration in the field, and it is always recommended that alkalinity is also determined back in the laboratory. During storage, there may be biological activity that changes the alkalinity or extreme loss of CO_2 may lead to precipitation of $CaCO_3$ in the sample bottle. In carbonate terrains, it may be necessary to filter the alkalinity sample as there may be suspended particles of calcite leading to an overestimate of the value (Fritz, 1994). For some highly contaminated samples, a stable endpoint may not be obtained if there is equilibration with a solid phase.

7.5. Dissolved oxygen

Dissolved oxygen levels in water depend, in part, on the chemical, physical and biochemical activities occurring in the water. Oxygen has a limited solubility in water directly related to atmospheric pressure and inversely related to water temperature and salinity. Low-dissolved oxygen levels can limit the bacterial metabolism of certain organic compounds.

On-site, dissolved oxygen is commonly measured using a membrane electrode of the polarographic type in a flow-through cell. The zero is commonly set using a saturated solution of sodium sulphite and the 100% saturated environment by holding the probe close to the surface of clean water. Below 1 ppm, electrodes provide only a qualitative measure of DO_2 due to slow electrode response (Wilkin *et al.*, 2001).

7.6. Oxidation–reduction potential

Reduction–oxidation reactions are mediated by micro-organisms and involve the transfer of electrons between reactants and products. Free electrons do not exist in solution, so an oxidation reaction (loss of electrons) must be balanced by a reduction reaction (gain of electrons). Redox potential is defined by the Nernst equation and is the energy gained in the transfer of 1 mol of electrons from an oxidant to H_2.

Redox reactions control the mobility of metal ions in solution by changing the valence state, which in turn changes the solubility of metals causing them to dissolve into or precipitate out of solution; a common example is the reduction/ oxidation of iron:

$$Fe^{3+} + e^- = Fe^{2+} \text{ reduction(dissolved in solution)}$$

$$Fe^{2+} = Fe^{3+} + e^- \text{ oxidation(precipitates out of solution)}$$

ORP is measured on-site by monitoring the potential developed at a platinum surface under *in situ* conditions in a flow-through cell. In the presence of oxygen, the electrode behaves as a Pt-O electrode and responds predominantly to pH; consequently, the measurement of ORP in oxygenated waters does not yield much useful information. In reducing waters, the measured potential will be a mixture of potentials developed from a range of redox reactions often involving iron, sulphur and nitrogen species. The redox electrode cannot be calibrated, but correct function can be checked by measuring temperature and the millivolt obtained from Zobell's solution and comparing the temperature-corrected reading with tabulated results (see manufacturer's information and Nordstrom, 1977).

ORP measurements are effective in delineating oxic from anoxic groundwater, but ORP measurements cannot distinguish between nitrate-reducing, Fe (III)-reducing, sulphate-reducing or methanogenic zones in an aquifer (Fig. 3.2).

8. PRESERVATION AND HANDLING OF SAMPLES

Waters are susceptible to change by differing extents as a result of physical, chemical or biological reactions that take place between the time of sampling and analysis. If suitable precautions are not taken before and during transport as well as the time spent in the laboratory, then the nature and rate of these reactions are often such that concentrations determined will be different from those existing at the time of sampling. The causes of variation may include

Figure 3.2 On-site water quality measurements using a flow-through cell.

- Consumption or modification of constituents by micro-organisms
- Photodegradation
- Oxidation by dissolved oxygen
- Precipitation
- Volatilisation
- Degassing
- Absorption of CO_2 from the air
- Sorption of dissolved or colloidal phases to container walls

A summary of the preservation options and recommended maximum storage periods for various determinands is given in Table 3.1. The individual options are described below.

8.1. Filtration

The process of drilling, constructing, purging and sampling a well can mobilise colloids (particles ranging in size from 1 to 1000 nm) and particulates that are not moving through the groundwater under natural flow conditions, resulting in artificially high concentrations of inorganic constituents. These can be removed by filtration, which should be carried out on-site as soon as possible after collection. Standard filters for groundwater investigations are 0.45 µm cellulose membranes. These do not remove all particulates from the sample or colloidal material of biological and non-biological origin in the 0.1–0.001 µm range. Filtration should not be used if the filter is likely to retain one or more of the constituents to be analysed. Different brands may not be identical in performance (Hall *et al.*, 1996).

High-turbidity samples can clog filters reducing the effective pore size. To avoid this, glass fibre pre-filters should be used and clogged filters changed regularly.

It is essential that the filter is not a cause of contamination and, if applicable, filters should carefully be washed before use. If possible a portion of sample should be flushed through the filter before sample collection. In general, samples for most organic determinands should NOT be filtered; highly turbid samples and all samples collected for dissolved metals SHOULD be filtered.

8.2. Addition of preservatives

The most commonly used preservatives for groundwater samples are

- Acids
- Bases
- Biocides
- Specialised reagents, for example, for mercury or sulphides

Acidification to below pH2 is particularly suitable for trace metals since it minimises adsorption of metals to container walls and also reduces biological activity. Acidification before filtration will release metals bound to particulates, giving a false reading only if dissolved metals are required. Samples for anion analysis are generally not acidified. Biocides such as sodium azide are commonly added to samples

Table 3.1 Techniques for the preservation of samples for analysis (simplified from BSI, 1996)

Group	Examples of determinands	Container	Filtration	Preservation	Cool to 2–5 °C for storage	Maximum storage time
Total metals	Al, Cr, Co, Cu, Fe, Pb, Mn, Ni, Ag, U, Zn	Plastic	–	Acidify to <pH2 with HNO_3	√	1 month
Cations and dissolved metals	Al, Cd, Co, Cu, Fe, Pb, Li, Mn, Ni, Ag, Ca, Mg, Na, K, U, Zn, Sb, B, etc.	Plastic	0.45 μm cellulose	Acidify to <pH2 with HNO_3	√	1 month
Redox sensitive cations	Fe^{2+}, NH_4^+	Plastic	–	Acidify to <pH2 and exclude oxygen for reduced species	√	24 hours
Anions	Br, Cl, F‡, NO_3, o-PO_4, P, SO_4^{2-}*, I, B SiO_4^*,	Plastic	0.45 μm cellulose	–	√	1 month
Redox sensitive anions	NO_2^-, Cr^{6+} species, CN^-	Plastic	–	Exclude oxygen for reduced species	√	24 hours
Sulphur species	S^{2-}, SO_2^{2-},	Plastic	–	Fix by alkalization or EDTA as required	√	24 hours
Anionic metals	As, Se, Sn	Plastic	–	Acidify to <pH1 with HCl	√	1 month
Total anions	o-PO_4, P, N	Plastic	–	Acidify to <pH2 with H_2SO_4	√	24 hours
Volatile inorganics	Hg	Borosilicate glass		Acidify to <pH2 with HNO_3 and add $K_2Cr_2O_7$ to 0.05%	√	

(continued)

Table 3.1 (Continued)

Group	Examples of determinands	Container	Filtration	Preservation	Cool to 2–5 °C for storage	Maximum storage time
Oxygen demand	BOD, COD	Glass	–	Exclude light	√	24 hours
Stable isotopes	^2H, ^{18}O, ^{13}C	Glass	–	–	–	
Dissolved gases	N_2, Ar, CH_4	Glass or steel pressure vessel			–	
Fluorocarbons	CFCs, SF_6	Specialised container			–	
Volatile organics	Chlorinated solvents, light hydrocarbons	Glass with septum cap	–	Exclude light	√	7 days
Non-volatile organics	Grease, oil, pesticides, phenols, PAH	Glass	–	Addition of biocide for some determinands, Exclude light	√	7 days
Surfactants,	Anionic, non-ionic cationic	Glass	–	Varies with type	√	
Organic carbon	Total/dissolved	Glass	0.45 μm silver for dissolved	Exclude light	√	7 days
Microbiology		Glass or heat resistant plastic				

⋆ Can be included in cations if total element is required by ICP-OES.
‡ Not PTFE.

for trace organic compounds, such as pesticides. The use of 0.45 μm silver filters for dissolved organic carbon samples both removes particulates and introduces silver into solution, which acts as a biocide. Recommended techniques for the preservation of waters are given in Table 3.1. It should be noted, however, that preservation techniques are often governed by the analytical method to be used; hence, advice should always be sought from the laboratory on the type of bottle and preservation method required for each determinand.

8.3. Solvent extraction

Solvent extraction used for organic analysis is not a feasible field operation, and therefore samples should be stored at 4 °C and transported to the laboratory as soon as possible. The use of opaque or brown glass containers can reduce the photosensitivity of the sample to a considerable extent.

8.4. Cooling or freezing

The sample should be kept at a temperature lower than that at which it was collected. Cooling is effective only if it is applied immediately after the collection of the sample. This normally requires the use of a cool box containing ice or a refrigerator in the vehicle. Cooling is particularly important for minimising microbial activity. In some cases samples can be frozen. The freezing and thawing must be controlled in order to return the sample to its initial equilibrium after thawing. Glass containers are not suitable for samples that are to be frozen.

8.5. Sample containers

It is essential that the sample container and its cap should not be a cause of contamination, or absorb or react with constituents to be determined in the sample. For many inorganic determinands, modern plastic containers such as LDPE are probably the best option.

It is advised that sample bottles are soaked in a 1 molar solution of the preservative acid and thoroughly rinsed in high-grade deionised or distilled water. For phosphate, silicon, boron and surfactants, detergents should not be used for cleaning purposes. For pesticides and their residues, containers should be cleaned with water and detergent followed by thorough rinsing with high-grade water, oven drying at 105 °C and rinsing with the solvent to be used during the analysis. For TOC/DOC, the use of carefully cleaned containers using chromic acid or a specialised surfactant is essential. For microbiological analyses, the container must be able to withstand sterilisation procedures.

In general, it is advisable to fill the container as completely as possible to minimise interaction with the gas phase and the consequent changes in carbon dioxide content and pH. Where samples are to be acidified, sufficient space must be allowed for in the container. For trace volatile organics that are not to be extracted on-site, it is absolutely essential to fill the container completely so that no air bubbles are seen when the container is inverted to minimise partitioning of the volatiles into the gas phase. For microbiological examination, an air space should be left, so the sample container should not be filled to the brim. This aids in mixing before examination and the avoidance of accidental contamination.

Where sufficient volume is available, all containers and caps should be rinsed several times with sample before collecting the sample.

9. QUALITY ASSURANCE AND QUALITY-CONTROL PROCEDURES

It is important that sample integrity be maintained and guaranteed throughout the collection, transport and analytical processes; this can be achieved by including strict QA/QC procedures in the sampling protocol. The areas that should be considered are described in the later sections.

9.1. Blank samples

Blanks consist of deionised water that is carried through all or part of the sampling and analytical processes to provide an indication of contamination. Types of blank sample include both laboratory and field blanks, listed below in ascending order of cumulative potential contamination.

9.2. Laboratory blanks

Instrument blank —a blank analysed with field samples to assess the presence or absence of instrument contamination.

Method blank—an analytical control consisting of all reagents, internal standards and surrogate standards, which is carried through the entire analytical procedure. The method blank is used to define the level of laboratory background and reagent contamination.

9.3. Field blanks

Trip blanks—a clean sample that is sent from the laboratory with the empty sampling bottles and remains with other samples throughout the sampling trip without being opened. This is typically required only for volatiles and assesses contamination during shipping and field handling.

Field blanks—field blanks are performed by passing a volume of contaminant-free water through all processing equipment that an environmental sample would contact, including filtration, addition of preservative, transfer to the sample container in the field and shipping to the laboratory with field samples. The results of field blanks can be used to assess contamination issues associated with processing and transporting the sample.

Equipment blanks—a sample of contaminant-free water poured through decontaminated field sampling equipment before the collection of field samples. For pump blanks, two blanks may be taken: one before the pump is cleaned and one after.

A sample of the water used to pass through the pump should also be collected. These samples should be taken at the beginning and end of each day's sampling. These blanks assess the adequacy of the decontamination process and also assess the total contamination from sampling sample preparation and measurement processes.

9.4. Replicate samples

Replicate samples should be collected at the same time (preferably a split of one sample rather than by collecting two or more concurrent samples in the field) and undergo the same filtration, preservation and storage. These replicate samples measure the variability of the processing techniques and the laboratory precision, but exclude field-sampling variability.

9.5. Spiked samples

Field matrix spikes can be carried out by adding a known amount of a spike solution with a known concentration to a replicate sample. Spike recoveries can be used to identify which compounds are consistently under or over reported or which compounds are variable in their recoveries. This is particularly relevant to the analysis of organic contaminants.

9.6. Labelling

Containers should be labelled in a clear and durable manner to permit identification without ambiguity in the laboratory. Sample labels should adhere firmly to bottles. Sample lists should be provided to the laboratory.

9.7. Transport

It is important that the samples are transported quickly, safely and securely from field to laboratory. In particular, care must be taken to ensure that bottles are protected from breakage (particularly glass bottles) and loss of sample and that they are held at the required temperature.

9.8. Laboratory reception

On arrival at the laboratory, the samples should be preserved under conditions that minimise any contamination of the outside of the containers and that prevent any change in their content.

9.9. Chain of custody

Chain-of-custody paperwork should be completed and copies retained. A chain of custody is a set of procedures used to provide an accurate written record that can be used to trace the possession of a sample from the moment of its collection through its introduction into a data set. Sample identity is maintained by proper labelling. Each person involved in the chain of possession must sign a chain-of-custody form when sample custody is relinquished or received.

A chain-of-custody form is a document used to record the transfer, possession and custody of samples and to ensure the integrity of samples from the time of collection through data reporting. The chain-of-custody form should, at a minimum, contain the following information:

- Contact name and address of sampler
- Signature of sampler
- Order/batch number
- Sample id
- Sample location
- Date
- Time
- Sample type
- Number of containers
- Details of analysis required
- Dispatcher signature and date/time
- Courier signature and date/time
- Laboratory receipt signature and date/time

10. Data Validation

There are a number of relatively simple tests that can be employed to evaluate the analytical data and to check for possible transcription or dilution errors, changes during storage or unusual or unlikely values. A discussion of these can be found in Hem (1985) and Cook *et al.* (1989), among others.

10.1. Comparison of field and laboratory values

The comparison of field and laboratory determined results for parameters such as alkalinity and SEC can be indicative of:

- Sample confusion, for example, errors arising from mislabelling
- Sample storage problems

10.2. Comparison with other samples from the same source

A simple screening procedure for evaluating analyses from the same or similar sources is to compare the results with one another. Transcription or dilution errors become readily evident.

10.3. Comparison with other samples from the area

A table of minimum and maximum values is helpful for identifying unusually low or high values. The data should be evaluated for a consistent pattern of highs and lows and anticipated correlations.

10.4. Comparison of SEC and TDS

An approximate accuracy check is possible using the SEC and TDS determinations. The TDS (in milligram per litre) should be between 0.55 and 0.75 times the SEC (in microsecond per centimetre) for most waters up to a TDS of a few thousand milligrams per litre. Water in which anions are mostly dominated by bicarbonate and chloride should have a factor near the lower end of this range, whereas waters high in sulphate may reach or even exceed the upper end. For repeated analyses from the same area, a well-defined relationship can often be established.

10.5. Evaluation of charge balance errors

The quality of chemical analyses can be checked on the basis of an ionic charge balance. This check should be carried out as soon as possible, while the chemical analysis can still be repeated. The ion balance can be calculated only for samples that have complete chemical analyses. The main purpose is to detect obvious errors and bias in the analysis, but it will only detect these in the major species. Some errors may cancel each other out and this check does not provide confidence in major uncharged species, such as silica, or trace elements. The effect on the balance of minor components, for example, phosphates and organic acids, which are not always included in the analysis, is usually negligible. Bicarbonate ions dissociate into carbonate ions, but this is negligible below $pH = 8$. For neutral groundwater, the charge balance (CB) is calculated as the total cation charge minus the total anion charge divided by the total charge in solution all expressed in micro-equivalents per litre:

$$CB = \frac{100(Na + K + Ca + Mg) - (HCO_3 + Cl + SO_4 + NO_3)}{(Na + K + Ca + Mg + HCO_3 + Cl + SO_4 + NO_3)}$$

For very alkaline waters, it may be necessary to include other species such as carbonate or silicate ions. Acid water will contain H^+ ions, but water with a pH of <4.5 may not provide a usable acidity due to interference from other species such as some iron hydroxides.

As a general guideline, based on the difference and the sum of cation and anion concentrations, the percentage ionic charge balance should be lower than $\pm 5\%$, except for samples with low TDS.

There are some instances where the charge balance may not detect errors:

- Waters with a TDS of >1000 mg/litre tend to have large concentrations of a few constituents, and the charge balance does not adequately evaluate the accuracy of the values of the minor constituents.
- Solutions that are strongly coloured may contain organic anions at sufficiently large concentrations to prevent a satisfactory balance being obtained.
- Waters of low-ionic strength (generally cation or anion totals <1 mg/litre) in which determinands may be close to or less than the limit of quantification.

10.6. Comparison of measured and calculated TDS (or measured and calculated SEC)

Assuming that the TDS or SEC is measured accurately, it can provide a check on the subsequent chemical analysis. This check is useful for spotting major errors, such as dilution or typographical errors as well as systematic errors in analytical methodology. It does not provide a check on any minor species or, in the case of SEC, uncharged species such as silica. Exactly which major species are present in any given water sample will depend on the type of water considered; Table 3.2 provides some guidance for calculating SEC.

$$SEC = \sum C.F$$

where C is concentration in equivalents and F is ionic conductance.

10.7. Apparent anomalies and impossibilities

Species reported should be correct with regard to the original pH of the sample. At a neutral pH, carbonate species will be almost all HCO_3 and high CO_3 cannot exist.

Table 3.2 Ionic conductance at infinite solution and 25 °C for different aqueous species (MacInnes, 1939)

Species	Equivalent ionic conductance (S.cm^2/eq)	Use
Acetate	40.9	Polluted waters
Ba^{2+}	63.6	Where significant
Br^-	78.4	Where significant
Ca^{2+}	59.5	Routine
Cl^-	76.3	Routine
F^-	55.4	Where significant
H^+	349.7	Acid rain and waters below pH5
HCO_3^-	44.5	Routine
I^-	76.8	Where significant
K^+	73.5	Routine
Li^+	38.7	Where significant
Mg^{2+}	53.0	Routine
Na^+	50.1	Routine
$NH_4^+ - N$	73.5	Polluted waters
NO_3^-N	71.4	Routine
OH^-	198.0	Waters above pH12
Other organic anions	Varies	Some brown coloured waters
$SO_4^2 - S$	79.8	Routine
Sr^{2+}	59.5	Where significant

High Fe concentrations can be a problem to interpret since under oxidising conditions all Fe would be expected to be present as a highly insoluble Fe oxide. If high concentrations of iron are found, then either the water must be reducing or some iron has passed the filter; this also applies for Al. Other possible anomalies to monitor data for are:

- Incompatible combinations of species, for example, nitrate in the presence of Fe^{2+} or the absence of dissolved oxygen.
- Totals of any variable less than the sum of the component parts, for example, total iron less than dissolved iron.
- Reported results not in the range of the technique or not theoretically possible, for example, pH>14.
- Apparent zero concentrations for major ions, such as Na or Ca. A zero concentration is rare if these elements have actually been determined.
- Unusual parameter ratios, for example, Ca/Mg or Na/Cl. For example, groups of analyses where all magnesium concentrations are similar but calcium concentrations have a wide range may indicate that calcium and bicarbonate were lost during sampling or storage.

11. HEALTH AND SAFETY IN FIELDWORK

In all aspects of site investigations, health and safety considerations should take priority. Staff should be suitably trained and adequately supervised. Particular care should be taken with monitoring locations that pose particular difficulties for access or that are unsafe in any other way. Risk assessments should be prepared in advance and reviewed at regular intervals. Suitable personal protective equipment and emergency equipment and protocols should be available.

REFERENCES

Bangsund, W. J., Peng, C. G., and Mattsfield, W. R. (1994). Investigation of contaminant migration by low-flow rate sampling techniques. *In* "The Eighth National Outdoor Action Conference and Exposition, Minneapolis Convention Center," Minneapolis, Minnesota.

Barcelona, M. J., Helfrich, J. A., Garske, E. E., and Gibb, J. P. (1984). A laboratory evaluation of ground water sampling mechanisms. *Ground Water Monit. Rev.* **4**(2), 32–41.

Barcelona, M. J., Helfrich, J. A., and Garske, E. E. (1985). Sampling tubing on groundwater samples. *Anal. Chem.* **57**, 460–464.

Barcelona, M. J., Wehrmann, H. A., and Varljen, M. D. (1994). Reproducible well-purging procedures and VOC stabilization criteria for ground-water sampling. *Ground Water* **32**(1), 12–22.

Barker, J. F., and Dickhout, R. (1988). Evaluation of some systems for sampling gas-charged ground water for volatile organic analysis. *Ground Water Monit. Rev.* **8**(4), 112–120.

BSI (1996). Water quality. Sampling Part 3. Guidance on the preservation and handling of samples. BS EN ISO 5667–3.

Canter, L. W., Knox, R. C., and Fairchild, D. M. (1990). "Ground Water Quality Protection." CRC Press, ISBN 0873710185, Florida, USA.

Cook, J. M., Edmunds, W. M., Kinniburgh, D. K., and Lloyd, B. (1989). Field techniques in groundwater quality investigations. *Br. Geol. Surv. Tech. Rep.* WD/89/56, 139 pp.

Fritz, S. J. (1994). A survey of charge-balance errors on published analyses of potable ground and surface waters. *Ground Water* **32**(4), 539–546.

Gibs, J., Imbrigiotta, T. E., and Turner, K. (1990). Bibliography on sampling ground water for organic compounds. *USGS Open File Report,* 90–564.

Gibs, J., Szabo, Z., Ivahnenko, T., and Wilde, F. D. (2000). Change in field turbidity and trace element concentrations during well purging. *Ground Water* **38**(4), 577–588.

Greacen, J., and Slivia, K. (1994). A comparison of low flow vs high flow sampling methodologies on groundwater metals concentrations. *In* "The Eighth National Outdoor Action Conference and Exposition, Minneapolis Convention Center," Minneapolis, Minnesota.

Hall, G. E. M., Bonham-Carter, G. F., Horowitz, A. J., Lum, K., Lemieux, C., Quemerais, B., and Garbarino, J. R. (1996). The effect of using different 0.45 μm filter membranes on 'dissolved' element concentrations in natural waters. *Appl. Geochem.* **11,** 243–249.

Hem, J. D. (1985). Study and interpretation of the chemical characteristics of natural water. *United States Geological Survey Water-Supply Paper* 2254.

Kearl, P. M., Korte, N. E., and Cronk, T. A. (1992). Suggested modifications to ground water sampling procedure based on observations from the colloidal borescope. *Ground Water Monit. Rev.* **12**(2), 155–161.

Kearl, P. M., Korte, N. E., Stites, M., and Baker, J. (1994). Field comparison micropurging vs. traditional ground water sampling. *Ground Water Monit. Rem.* **14**(4), 183–190.

MacInnes, D. (1939). "Principles of Electrochemistry." Reinhold Publishing Corp, New York.

Nielsen, D. M., and Yeates, G. L. (1985). A comparison of sampling mechanisms available for small-diameter ground water monitoring wells. *Ground Water Monit. Rev.* **5**(2), 83–99.

Nordstrom, D. K. (1977). Thermodynamic redox equilibria of Zobell's solution. *Geochim. Cosmochim. Acta* **41**(12), 1835–1841.

Parker, L. V. (1994). The effects of ground water sampling devices on water quality: A literature review. *Ground Water Monit. Rem.* **14**(2), 130–141.

Parker, L. V., and Clark, C. H. (2002). Study of five discrete interval type groundwater sampling devices. *US Army Corps of Engineers Technical Report,* ERDC/CRREL TR-02-12.

Pearsall, K. A., and Eckhardt, D. A. V. (1987). Effects of selected sampling equipment and procedures on the concentrations of trichloroethylene and related compounds in ground water samples. *Ground Water Monit. Rev.* **7**(2), 64–73.

Pohlmann, K. F., and Hess, J. W. (1988). Generalised ground water sampling device matrix. *Ground Water Monit. Rev.* **8**(4), 82–84.

Puls, R. W., and Barcelona, M. J. (1996). Low-flow (minimal drawdown) ground-water sampling procedures. US EPA - Ground Water Issue.

Puls, R. W., and Paul, C. J. (1995). Low-flow purging and sampling of ground water monitoring wells with dedicated systems. *Ground Water Monit. Rem.* **15**(1), 116–123.

Puls, R. W., and Powell, R. M. (1992). Acquisition of representative ground water quality samples for metals. *Ground Water Monit. Rev.* **12**(3), 167–176.

Puls, R. W., Clark, D. A., and Bledsoe, B. (1992). Metals in ground water: Sampling artifacts and reproducibility. *Hazard. Waste Hazard. Mater.* **9**(2), 149–162.

Rannie, E. H., and Nadon, R. L. (1988). Inexpensive, multi-use, dedicated pump for groundwater monitoring wells. *Ground Water Monit. Rev.* **8**(4), 100–107.

Robin, M. J. L., and Gillham, R. W. (1987). Field evaluation of well purging procedures. *Ground Water Monit. Rev.* **7**(4), 85–93.

Rosen, M. E., Pankow, J. F., Gibs, J., and Imbrigiotta, T. E. (1992). Comparison of downhole and surface sampling for the determination of volatile organic compounds (VOCs) in groundwater. *Ground Water Monit. Rev.* **12**(1), 126–133.

Shaver, R. B. (1993). Field vs lab alkalinity and pH: Effects on ion balance and calcite saturation. *Ground Water Monit. Rem.* **13**(2), 104–112.

Schuller, R., Gibb, J. P., and Griffin, R. (1981). Recommended sampling procedures for monitoring wells. *Ground Water Monit. Rev.* **1**(2), 42–46.

Sladky, B., and Roberts, P. G. (2002). Zero-purge groundwater sampling for semivolatile organic compounds. ITRC—Diffusion Sampler Database. Available from http://www.diffusionsampler.org/

Stuart, A. (1984). Borehole sampling techniques in groundwater pollution studies. *British Geological Survey Technical Report,* WE/FL/84/15.

Wilkin, R. T., McNeil, M. S., Adair, C. J., and Wilson, J. T. (2001). Field measurement of dissolved oxygen: A comparison of methods. *Ground Water Monit. Rem.* **21**(4), 124–132.

CHAPTER FOUR

THE COLLECTION OF DRAINAGE SAMPLES FOR ENVIRONMENTAL ANALYSES FROM ACTIVE STREAM CHANNELS

Christopher C. Johnson,* Deirdre M. A. Flight,* Louise E. Ander,*
Robert T. Lister,* Neil Breward,* Fiona M. Fordyce,[†] *and* Sarah E. Nice*

Contents

* British Geological Survey, Keyworth, Nottingham, NG12 5GG, UK
† British Geological Survey, Murchison House, Edinburgh, EH9 3LA, UK

Environmental Geochemistry
DOI: 10.1016/B978-0-444-53159-9.00004-8

© 2008 Elsevier B.V.
All rights reserved.

Abstract

The collection of drainage samples from active stream channels for geochemical mapping is now a well-established procedure that has readily been adapted for environmental studies. This account details the sampling methods used by the British Geological Survey in order to establish a geochemical baseline for the land area of Great Britain. This involves the collection of stream sediments, waters and panned heavy mineral concentrates for inorganic chemical analysis. The methods have been adapted and used in many different environments around the world. Detailed sampling protocols are given, and sampling strategy, equipment and quality control are discussed.

1. INTRODUCTION

Throughout human history, rivers have played a vital role in human sustenance, settlement and transportation and of all the earth's systems it is in the drainage environment where man probably has had the greatest impact. From a geological perspective, the flow of water is an important process in the shaping of the earth's surface, and early civilisations knew how to use drainage channels to trace valuable metals and minerals dispersed in alluvium back to their source. Theophrastus (300 BC) describes how the ancient Greeks searched for ore deposits by tracing detrital anomalies in rivers (Caley and Richards, 1956) and copper for pre-Bronze Age cultures was located using alluvial dispersion trains (Wertime, 1973). The art of panning detrital material for precious minerals both from active and defunct drainage channels is a long-established skill and an activity that supplements the income of farmers in many parts of the developing world.

The collection and chemical analysis of various sample media from the drainage system is a well-established exploration tool having its origins in the Soviet Union (Fersman, 1939) and was subsequently applied outside the USSR, for example, Lovering *et al.* (1950) and Hawkes and Bloom (1955). Although the majority of early surveys were concerned with mineral exploration, there are examples of regional geochemical mapping based on drainage samples being applied to environmental problems (e.g., Plant and Moore, 1979; Thornton and Webb, 1979). The anthropogenic effect on drainage sediment geochemistry was reviewed by Cooper and Thornton (1994).

Webb and Howarth (1979) wrote: 'Although it was apparent more than 20 years ago that geochemical atlases would eventually become a national cartographic requirement, regional geochemical mapping is still in the experimental stage. This trend is now evident in activity in a number of countries. The methods being employed, however, are so diverse that there is an urgent need for international collaboration aimed at securing data that are as mutually compatible as possible.' This need was finally addressed by the International Geological Correlation Programme (IGCP) Project 259 'A global geochemical database for environmental and resource management' (Darnley *et al.*, 1995). This states that wherever the landscape permits, drainage samples, specifically stream sediments, have been the preferred sampling medium for reconnaissance exploration surveys.

Current output of publications on environmental applications of drainage surveys is far greater than for exploration investigations (e.g., Ettler *et al.*, 2006; Gonçalves *et al.*, 2004; Lee *et al.*, 2003; Schreck *et al.*, 2005). There is also a greater emphasis on multi-purpose environmental baseline data generated from multi-media sampling as demonstrated by the Geochemical Atlas of Europe (Salminen *et al.*, 2005) and the environment and resource evaluation of the Tellus geochemical mapping project in Northern Ireland (Young and Smyth, 2005).

A variety of sample media can be classified as drainage samples. These include stream, lake, overbank and floodplain sediments; stream and lake water; suspended/colloidal sediment in streams or lakes; and precipitates or coatings on stones and boulders from the stream bed. The use of drainage sample media in geochemistry was comprehensively reviewed by Hale and Plant (1994). The account presented here is specifically concerned with procedures for collecting samples from the active drainage channel (stream sediment, suspended sediment, stream water and panned heavy mineral concentrates) and lake, overbank and floodplain samples are not discussed further. The procedural section in this account is based on the methods of the British Geological Survey (BGS) G-BASE (Geochemical Baseline Survey of the Environment) project. This is a long-established programme to make geochemical maps of inorganic elements of the United Kingdom land area based mainly on high-density sampling of stream sediments (Johnson *et al.*, 2005).

2. DRAINAGE BASINS

The study of streams is part of the scientific discipline known as hydrology. Some of the terminology associated with the drainage system is summarised in Table 4.1. A stream can be defined as being a body of water confined by the land surface as its base (known as the stream 'bed') and laterally by banks. It is a term often used to cover all naturally flowing water; 'river' is a term generally applied to a large natural stream. Stream water may or may not flow all year but the channel or course along which it flows is a feature that, with or without water, can be sampled throughout the year. Intermittent streams, which flow for only part of the year, are usually found in regions where there is a marked rainfall contrast between seasons or seasonal temperature variations result in the melting of ice or snow. An intermittent stream in an arid area is also sometimes referred to by the Arabic term 'wadi'. These streams are often associated with hazardous events such as flash floods where sudden heavy rain or rapidly melting snow upstream can result in a torrent of debris and water further downstream in areas where there has been no recent precipitation. These events cause significant redistribution of detrital material and are a very important consideration in the health and safety of sampling teams. Streams that form only for a short period during times of precipitation are referred to as ephemeral streams.

Streams are part of a drainage basin system that has inputs, flows, stores and outputs of both water and detrital material. The drainage basin concept is illustrated

Table 4.1 Explanation of terms used in drainage sampling

Term	Definition
Alluvium	a recent deposit of sand, mud, etc., formed by flowing water
Confluence	where two streams merge. If the streams are of equal size then the confluence is often referred to as a 'fork'
Dispersion train	is a feature of variable length found extending downstream from a point and defined by the decreasing presence of a mineral or chemical element
Drainage basin	a region of land where water from precipitation (or snowmelt) drains downhill into a body of water such as a stream. Drainage basins are divided from each other by topographic barriers called a watershed. Also referred to as drainage catchment, water basin or drainage area. Drainage basins can be nested together to form larger basins that can be described by the rank of the largest river, for example, the fourth-order drainage basin
Ephemeral	a term used to describe a stream that forms only during or immediately after precipitation
Floodplain sediment	alluvium accumulated adjacent to high-order stream
Headwater	the section of the stream closest to its source where a discernible stream bed can be identified
Heavy mineral concentrate	in this context is a sub-sample of the stream sediment that has been created by separating out the heavier minerals present in the stream sediment. The most common way of doing this is by panning the sediment to give a panned concentrate
Hydrology	the science dealing with the occurrence, circulation, distribution, and properties of the waters of the earth and its atmosphere
Intermittent	a term used to describe a stream that only flows for part or parts of the year
Mouth	the point at which the stream discharges into a 'static' body of water such as a lake or ocean
Overbank sediment	alluvium accumulated adjacent to low-order stream
Perennial	a term used to describe a stream that flows all year
River	a large natural stream
Source	a spring from which the stream emerges or other point of origin
Spring	the point at which a stream emerges at the land surface
Stream	a body of water with a detectable current confined within a bed and banks. A general term applied to all flowing natural waters regardless of size. Regional names may be used such as beck, bourne, brook, burn, kill, creek or run
Stream bed	the base of a stream
Stream order	hierarchical system of ranking streams whereby low-order streams are small and high-order streams big
Stream sediment	represents a composite sample of detrital material derived from the drainage basin upstream of the sample site. Typically

Table 4.1 (*Continued*)

Term	Definition
	composed of weathered bedrock and material derived from overburden and is generally applied to the <2 mm fraction. Anything bigger than this is generally referred to as clasts, pebbles, stones or boulders
Thalweg	is the stream's longitudinal section, that is, a line joining the deepest point in the channel at each stage from source to mouth
Tributary	a stream joining another stream. When streams of similar size join they are referred to as a 'branches'
Watershed	is a topographic feature dividing drainage basins though American terminology would actually refer to the whole area enclosed by the topographic feature as the watershed

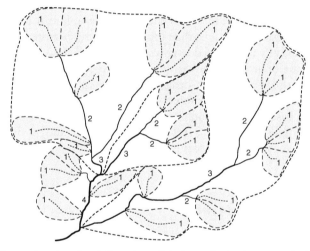

Figure 4.1 A figure showing the delineation of drainage basins by watershed and the Strahler (1957) system for determining stream order. *Note*: The shaded areas define the watersheds for the drainage basins of first-order streams. Note that the first-order basins are components of a much large drainage basins, here the third-order drainage basin is defined by the bold dashed line. According to the Strahler system of stream ordering the end tributaries are designated as the first-order streams. Two first-order streams merge to form a second-order stream segment; two second-order streams join, forming a third-order and so on. It takes at least two streams of any given order joining to form a stream of the next higher order.

in Fig. 4.1 that shows basins defined by a topographic feature known as the watershed. Drainage basins, sometimes also referred to as drainage catchments, can be broken down into smaller or larger 'nested' basins that can be defined in terms of the stream rank or order. A hierarchy of streams can be set up, the more objective and straightforward system is known as the Strahler stream order (Strahler, 1957) which is also illustrated in Fig. 4.1.

High-order drainage basins often define administrative boundaries for the management of resources, for example, the Hydrometric Areas of the United Kingdom (Institute of Hydrology, 2003). In Earth and Environmental Sciences, the drainage basin is an important means of defining an area equivalent in many respects to post or zip codes that have such widespread application in Geographical Information Systems (GIS). The drainage basin area is not only important as a means of delineating a data set within a GIS but it is also fundamental in the sampling and data interpretation processes. In remote underdeveloped areas rivers generally are the only means of ground access into an area, so sampling and mapping is very much based along the river system. In areas of high relief, the drainage channels represent the most likely place to find rock outcrops as a result of downwards erosion by the stream system. The BGS mapping of northern Sumatra, Indonesia (1975–1980), was entirely based on drainage basin areas, and mapping teams were assigned basins to sample during seasonal field campaigns. Drainage basin reports formed the basis for later regional compilations.

Interpreting results from a drainage sample requires an understanding of its derivation and the behaviour of chemical elements in the drainage environment. A drainage sample is a composite of material derived upstream of the sampling site limited by the watershed boundary. In this way, a sample can be collected that is representative of a much larger area than just the point represented by the sample site (its true representativity is discussed later in this account). The area covered depends on the order of the drainage basin with first-order basin generally covering areas of <10 km^2 and the third- and fourth-order basins areas of >100 km^2, though this will very much depend on the geomorphology of the landscape. The drainage sampling described here is based on the sampling of low-order streams generally first to third order, so the sample represents a relatively small area and is not a generalised composite of too many drainage basins.

3. DRAINAGE SAMPLING

Drainage sampling has the significant advantage over other types of media because a wide variety of different sample media can be collected from a single site. This represents a very economical and effective way of generating geochemical information and such a multi-media survey has great value in interpreting the distribution and behaviour of determinands in the surface environment. The distribution of As in stream sediments and stream waters over Mesozoic sedimentary ironstone outcrops in eastern England shows the benefits of being able to compare the geochemistry of different sample media in a regional geochemical survey (Fig. 4.2). In this case, very high levels of As are present in stream sediments over the ironstone outcrops, but the very low mobility of As when bound to sedimentary Fe oxides is illustrated by the absence of corresponding elevated As levels in the stream waters.

From a logistical point of view, the ability to collect a single sample that represents a large area greatly reduces the number of samples needing to be collected.

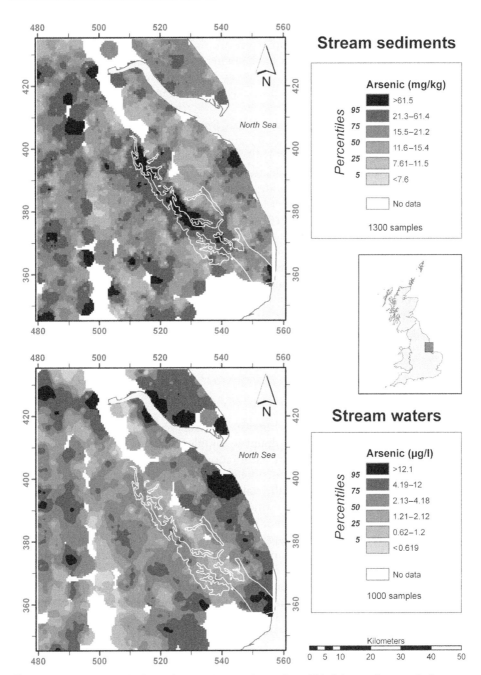

Figure 4.2 An example from the G-BASE project of a gridded image for arsenic in stream sediments (top) and stream waters (bottom) from eastern England. *Note*: The maps show how stream water and stream sediment maps can be used in combination to explain the distribution and behaviour of elements in the surface environment. The Mesozoic sedimentary ironstone referred to in the text is shown by the white outline.

The fact that one can collect drainage samples from adjacent tributaries from sites that may only be tens of metres apart greatly reduces the amount of time spent walking between sites. Anyone that has been involved in sampling will testify to the time and effort saved by not having to walk up hills and mountains to collect samples from near watersheds as the stream effectively brings the material down the valleys to the sampler.

Although there are standard procedures for the collection of environmental samples for very specific purposes (e.g., guidance in collecting bottom sediments in association with water quality, ISO, 1995), there are no nationally or internationally prescribed standards for the collection of drainage samples. The recommendations of Darnley *et al.* (1995) provide a top level of guidance to those planning sampling programmes and, at project level, there should always be detailed instructions on how samples should be collected, for example, Salminen *et al.* (1998) (FOREGS project geochemical mapping field manual) and Johnson (2005) (BGS G-BASE field procedures manual). Such detailed documentation is essential to give quality assurance to results and provide necessary information for the data users to establish whether it is fit for its intended purpose. In any environmental survey, the sampling procedure used should be recorded as a coded field within the resulting database. The G-BASE project now includes a 'sampling protocol' code as part of the field form completed at each sampling site (Fig. 4.3).

4. SAMPLING STRATEGY

The broad range of procedural options available to an environmental scientist planning a sampling project using drainage samples is discussed in detail by Hale (1994). Although this is aimed at the exploration geochemist, the sample campaign planning decisions faced by environmental scientists are just the same. The sampling plan must firstly take into account the stakeholders of the project, that is, organisations or individuals that have an investment or interest in the project and will have a planned use for the data. In this sense, planning the project strategy needs to work backwards from the point of deciding what products need to be delivered and what they have to be used for. At this point, it can be decided which sampling media can best meet the objectives of the user. The environmental scientist or geochemist needs to communicate scientific output in terms that are clear to stakeholders that may not understand the science but need information to satisfy their requirement. For example, determining if stream water concentrations of heavy metals exceed statutory levels.

Whether or not objectives of the stakeholders can be achieved greatly depends on our knowledge as to the usefulness of the various sampling media and the uncertainty that can be associated with a set of results when delivering an objective. This requires the skills and experience of a geochemist who must be able to justify decisions made during the planning phase. The choice of sample media, and the fraction of that sample media to be used, will determine what element concentrations and distributions will be best defined. Analytical methodologies employed will

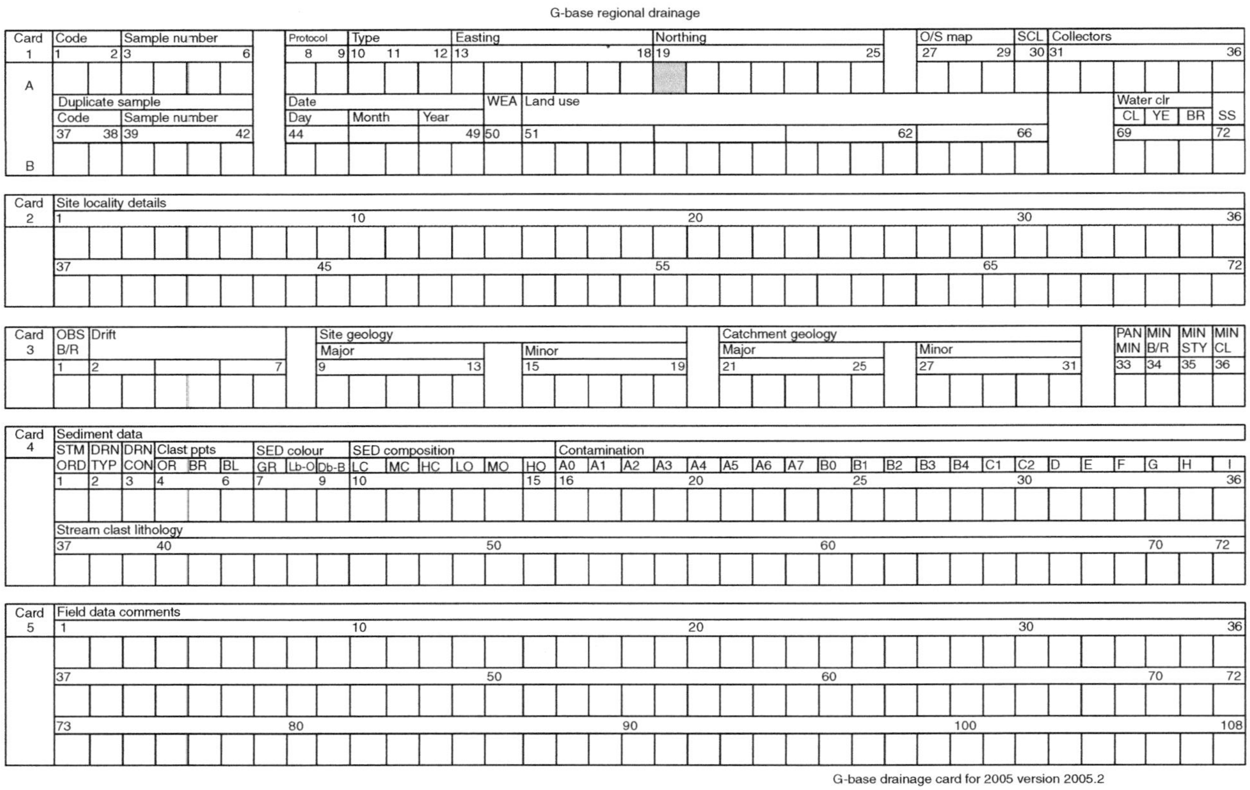

Figure 4.3 Example of field form used for recording information at a drainage sample site.

decide whether element concentrations reflect total levels or merely a partial extraction. This is particularly important in environmental health studies where the availability of essential or toxic elements rather than an understanding of the total concentration determines the level of risk. Multi-media surveys with a wide range of analytical methods employed are a good way of satisfying a wide range of stakeholders (an example of this is the Tellus Project in Northern Ireland, Young and Smyth, 2005).

However, it is rare that surveys have the luxury of a budget to achieve all that is desirable and often compromises have to be made in the sampling and analytical strategy. An example of this could be in the fraction size of sediment collected. Fine sediment may be good at discriminating chemical anomalies for elements dispersed by hydromorphic dispersion, but it may miss coarser grains transported predominantly by mechanical dispersion and not transported in the lighter and finer sediment fractions. With industrial slag, for example, coarser mineral grains that do not readily breakdown would be missed if only very fine sediment was collected. It is in such instances where the multi-media opportunities afforded at a single drainage site make drainage sampling an attractive strategy, especially as a first-stage environmental geochemical sampling tool.

In addition to being able to satisfy the stakeholder objectives within an available budget there are logistical constraints to any project. These would include ease of access to a sampling area and climatic conditions (particularly in relation to seasonal variations). For the latter, consideration of seasonal stream flow in drainage catchments is an important criterion both from a scientific and from safety view point. Furthermore, climate may also be associated with the ease of access with heavy rain or snow making roads and tracks impassable and river crossings difficult. The ease of access more often than not relates to one of the biggest expenditures in the sampling budget, namely the transport of personnel and samples to and from sample sites.

In inaccessible areas, a choice has to be made between high-cost helicopters or collecting on foot. The latter may only require modest expenditure but over a considerably longer period, so overall expenditure could be the same as a helicopterborne survey. In a commercial project, time and money will be of prime concern. For national surveys where there is a large component of institutional building and involvement of local people to be considered, maximising the benefit to the local community by carrying out a ground based rather than helicopter survey will be a more acceptable strategy.

Questions of sampling strategy are reviewed well elsewhere (see Darnley *et al.*, 1995) and include discussion of decisions to be made on the optimum fraction size of sediment to be employed, sampling density and methods of chemical analysis. A knowledge of the behaviour of chemical elements in different environments and an appreciation of element associations (see Tables 4.2 and 4.3) greatly help in the planning and interpretative phases of the project. If it is determined that certain elements are of great importance to the data users (e.g., mercury), then sampling strategy may have to be modified to collect the sample without compromising the results for the most important elements.

Once decisions have been made it is always a good strategy to carry out an orientation survey to test out sampling plans. Examples of such orientations can be

Table 4.2 Table summarising some typical element associations found in stream sediments (elements in italics are not of primary importance in the association)

Geological or environmental features	Principal associated elements and element ratios
Carbonate rocks (limestones, dolomites, calc-schist)	CaO Sr MgO
Argillaceous and pelitic source rocks	Li B Ga
Argillaceous Red Beds (e.g., Mercia Mudstone) with evaporites	K_2O MgO Sr *Se*
Black shales and graphitic schists	Ba Mo V U *Cu Ni Ag Se Cd*
Sedimentary ironstones	Fe_2O_3 As P_2O_5 U
Basic igneous rocks in unmineralised areas	MgO TiO_2 Ni Cu V/Cr
Ultrabasic rocks and derived sediments	Cr MgO Ni Cr/V
Evolved granites	Be Li U Sn Rb/K_2O *Y La Mo*
'Normal' granites	Be K_2O Rb U Li *Sr*
Granodiorites and some intermediate igneous rocks	Be Sr Ca K_2O
Resistate elements for sediment provenance variation, especially in greywackes and arenites	La Y Zr TiO_2 Th Ce Nb—and ratios of these
Generalised urban—industrial contamination	Sn Pb Cu Sb Cd Zn
Industrial contamination—heavy engineering	Sn Pb Cu Sb Cd Zn Cr Ni V Mn
Secondary hydrous oxide formation in stream sediments	Mn Co As Al Fe_2O_3
Mineralisation (vein type sulphide)	Pb Zn Ba Cu Cd *Sb Bi As*
Mineralisation (red-bed type)	Ba Cu *Bi Ag*
Mineralisation (porphyry type)	Mo Cu Sb
Gold mineralisation ('pathfinder' elements)	As Sb Bi

found described in Plant (1971) (drainage sampling in Great Britain), and Ranasinghe *et al.* (2002) (drainage sampling in Sri Lanka).

5. PROCEDURES

The following procedures are derived from the G-BASE field procedures manual (Johnson, 2005). They are presented here as a generic template for multi-purpose drainage sampling and are methods that have been developed and refined over a period of more than 30 years. The basic procedures have changed little since reported by Plant and Moore (1979) and have been adapted and applied to many major geochemical mapping projects around the world spanning many climatic zones (e.g., Sumatra, Indonesia, Muchsin *et al.*, 1997; Stephenson *et al.*, 1982; Ecuador, Williams *et al.*, 2000; and Morocco, Johnson *et al.*, 2001). Such adaptations to different climatic zones and stakeholder requirements are discussed at the end of this section.

Table 4.3 Relative mobility of elements in the different surface environments (taken from Plant and Raiswell, 1994 based on that of Andrews-Jones, 1968)

Relative mobilities	Environmental conditions			
	Oxidising	Acid	Neutral to alkaline	Reducing
Very high	Cl, I, Br S, B	Cl, I, Br S, B	Cl, I, Br S, B Mo, V, U, Se, Re	Cl, I, Br
High	Mo, V, U, Se, Re Ca, Na, Mg, F, Sr, Ra Zn	Mo, V, U, Se, Re Ca, Na, Mg, F, Sr, Ra Zn Cu, Co, Ni, Hg, Ag, Au	Ca, Na, Mg, F, Sr, Ra	Ca, Na, Mg, F, Sr, Ra
Medium	Cu, Co, Ni, Hg, Ag, Au As, Cd	As, Cd	As, Cd	
Low	Si, P, K Pb, Li, Rb, Ba, Be Bi, Sb, Ge, Cs, Tl	Si, P, K Pb, Li, Rb, Ba, Be Bi, Sb, Ge, Cs, Tl Fe, Mn	Si, P, K Pb, Li, Rb, Ba, Be Bi, Sb, Ge, Cs, Tl Fe, Mn	Si, P, K Fe, Mn
Very low to immobile	Fe, Mn Al, Ti, Sn, Te, W Nb, Ta, Pt, Cr, Zr Th, Rare Earths	Al, Ti, Sn, Te, W Nb, Ta, Pt, Cr, Zr Th, Rare Earths	Al, Ti, Sn, Te, W Nb, Ta, Pt, Cr, Zr Th, Rare Earths Zn Cu, Co, Ni, Hg, Ag, Au	Al, Ti, Sn, Te, W Nb, Ta, Pt, Cr, Zr Th, Rare Earths Zn Cu, Co, Ni, Hg, Ag, Au S, B Mo, V, U, Se, Re As, Cd Pb, Li, Rb, Ba, Be Bi, Sb, Ge, Cs, Tl

5.1. Generic sampling considerations

There are some fundamental issues that need to be addressed in order to understand important steps incorporated into the procedures described in the following subsections.

5.1.1. Reliability of sampling teams

The quality of an environmental survey is only as good as the most poorly controlled part of a survey and this will frequently be the sampling. Procedures must be in place to ensure that samplers are reliable and are strictly following protocols, particularly when the work involves the use of many different samplers. They need to be motivated and understand why they are collecting the samples in the prescribed way. Good team leaders are essential in this process. Rotation of individuals in different sampling pairs is important to maintain a check on procedural discrepancies. The more widespread use of GPS has greatly improved the reliability in sample site positioning, and this gives a confirmation that samplers have actually visited sites as scheduled.

5.1.2. Avoiding contamination

Environmental studies involve determining trace levels of elements or in the case of water samples ultra-trace levels and procedures have to be designed so as to avoid introducing contamination through the sampling tools, sample handling and the storage containers. Any orientation studies should be designed to include steps to test for contamination such as subjecting samples of pure silica sand or de-ionised water to all the steps in the sampling procedure. In the following subsections, trademarked products such as Kraft^{TM} paper bags, polyethylene Nalgene^{TM} water containers, Millex^{TM} filter papers or Swinnex^{TM} filter holders are referred to. These are items that the G-BASE project has found suitable for its survey, other surveys may use other trademarked products—the important point is that every piece of equipment used in the sampling must be investigated as a potential source of contamination.

5.2. Sampling team and responsibilities

Sampling is a team effort best led by personnel that have had practical experience of sampling themselves. The organisation of a typical sampling team is shown in Fig. 4.4 and consists of a Project Manager, a Field Team Leader and an Assistant and a number of sampling pairs. The Project Manager is responsible overall for the sampling strategy and logistics, the team's health and safety, managing the samplers' contracts, sampling budget and delivering samples to the laboratory for analysis. As part of the quality assurance, the Project Manager should have regular inspection of the field work. The Field Team Leader is responsible for the day-to-day operations of the sampling team, including the allocation of daily sampling quotas and checking samples and field forms at the end of each sampling period (usually daily). The leader will need to be an experienced sampler and a good team worker in order to maintain the motivation and confidence of the samplers. The leader's assistant will

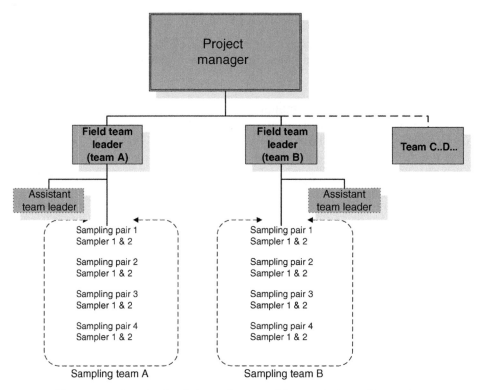

Figure 4.4 Organisational chart of personnel for a typical sampling project.

be less experienced in managing a team but someone that bridges the gap between Team Leader and samplers, with an expectation that they can both share the responsibilities of the leader as well as being actively involved in the sampling.

The key personnel are the samplers who should always work in pairs but not always with the same partner. This satisfies health and safety requirements and by varying the make up of each pair ensures a consistent implementation of sampling procedures throughout the team. Ideally, the samplers should be undergraduates or recent graduates of earth or environmental sciences seeking to gain practical experience in sampling. They generally do not earn a wage but are given a daily subsistence allowance and the knowledge that their work experience is an invaluable addition to their *curriculum vitae*. This model of employment has been used by the G-BASE project for more than 35 years and several BGS geochemical mapping projects internationally (e.g., Morocco and Madagascar). When working in more remote or difficult areas the sampling pair may consist of a graduate and a labourer. If sampling areas of special difficulty, urban environments and remote jungle being examples at two extremes, samplers may need to be accompanied by personnel with specialist knowledge of the area. For example, G-BASE while carrying out drainage sampling in the Glasgow area of Scotland, sampling pairs were assisted by municipal workers.

Any sampling campaign will usually commence with a training day for the samplers, so they have a good knowledge of the sampling procedures and how to complete field forms that record information about the sampling site and the sample itself. Ideally, at the commencement of a sampling campaign the sampling team will have a number of experienced samplers who can be paired off with the inexperienced ones in the first few days of the sampling. The number of sampling teams and pairs depends very much on the logistics of the operation. The use of too many samplers may result in collecting samples faster than they can be collated or analysed and will, if vehicles are being used, require larger or more vehicles for transport. Too few samplers will lead to unacceptably slow-sampling rates and reduces the experience pool, resulting in overdue dependence on a small number of samplers. It is also good practice to ensure Team Leaders are regularly brought together for sampling training to ensure procedures are being correctly followed with no deviations.

All personnel should be asked to contribute to a field campaign report at the end of sampling. In this, any deviations from standard procedure can be documented and suggested improvements to the methodology can be recommended. A review of these recommendations on an annual basis leads to improvements in procedures that can be included in sampling protocol revisions.

5.3. Field sampling equipment

A list of equipment required for drainage sampling is given in Table 4.4. A great advantage of geochemical methods over geophysical methods is that the field equipment is relatively cheap and generally readily available. The equipment can be divided into three parts; these are the items necessary for (i) collecting the sample; (ii) describing and marking the sampling site; and (iii) carrying and transporting the equipment and samples. Such equipment can be adapted to suit the aims and objectives of a survey particularly where local alternatives may be readily more available (see discussion).

5.4. Sampling procedures

The mantra of sampling is to 'avoid introducing contamination' during the sampling process. Samplers should not wear hand jewellery, and all sampling equipment must be free of contaminants and cleaned thoroughly between sample sites. Water sample analysis, in particular, involves determinations to very low concentrations and care should be taken to wash hands in the stream water first to remove any sweat or lotions (such as sun cream), and handling sample bottles must be done in such a way that the inside of bottles or lids are not touched by the hand.

5.5. Site selection

Sites will have been pre-plotted on a topographic map, and samplers should make their way to the designated site. However, the precise sampling site requires the samplers to understand what constitutes a good sampling site. This will to

Table 4.4 List of sampling equipment for collecting stream sediments, stream waters and panned concentrates as used by the G-BASE project

Item	Comments
General	
Topographic maps of field area	Scale as required by project. Need to have clean copies to be used as 'master plots' and working copies for daily use by samplers
Geological and other maps	Maps such as geology, land-use and soil type to help provide samplers with supplementary information about sites and catchments
Binocular microscope and lamp	Used to assist identification of panned concentrate minerals/contaminants
Field forms and folders	Field forms as illustrated in Fig. 4.3 and a folder to keep them dry and clean. Ideally could be replaced by hand-held computer devices
Stationery	Permanent ink markers, biro, pencils, elastic bands, etc.
Communication devices	Mobile phones or short-wave radio if no mobile phone coverage
Sample number checklist	Used for allocating numbers and collating samples
ID passes and letters	Sampling will attract local attention and a sampler's ID and their mission needs to be clearly stated and permitted
First aid kit	To include survival aids such as whistle and survival bags if needed
High visibility jackets	Required in all working environments
Rucksacks	Each sampling pair requires an equipment rucksack and a sample rucksack
Geological hammer and hand lens	Goggles to be used when hammering. Hand lens to look at rocks and minerals
GPS and compass	Used for locating sites and navigating
Knox Protractor	Used for measuring and plotting points on maps
Portable computer	If practical in field location a PC can be used to database field data or with GIS can be used to plan sampling
Stream sediment	
Sieve set	See Fig. 4.5. Ideally made of wood with nylon mesh fixed together with nylon bolts plus fibre glass pan for collecting sediment
Plastic funnel	To be used to help pour sample into sample bag
Rubber gloves	To be used to help rub sediment through sieve mesh
Sample bags	Kraft strong paper $4'' \times 8''$ (10×20 cm) sample bag stuck with waterproof glue. Paper allows sediment to dry.
Plastic bags and containers for transporting samples	Miscellaneous plastic bags are required to place samples for transport to prevent leakages. Also rigid plastic containers are useful to protect samples in the rucksack

Table 4.4   *(Continued)*

Item	Comments
Trenching tool/shovel	Wooden, polyethylene/polypropylene, stainless steel with any paint stripped off
Stream water	
Plastic syringe	25 ml syringes
Millex sealed filters	Pre-loaded with 0.45 μm millipore cellulose filters
Plastic bags	Miscellaneous self-seal plastic bags for keeping dirty and clean equipment apart
30 ml polythene bottles	For collecting water sample for pH determination in field lab
250 ml Nalgene polythene bottle	For collecting water sample to be determined for alkalinity and conductivity in the field lab
30 ml Nalgene polythene bottle	For collecting filtered unacidified sample (for major anions/NPOC)
60 ml Nalgene polythene bottle	For collecting filtered acidified sample (for ICP-MS and -AES)
Conc. HNO_3 and dropping pipette	Used to acidify water samples in field base
Panned concentrate	
Pan	Variety of types of pan available
Sample bags	Kraft strong paper $3'' \times 5''$ (8×13 cm) sample bag
Suspended load	
Self-seal plastic bag	$3'' \times 5''$ (8×13 cm) bag with white panel for storing 'dirty' filter from water sample collection

some extent depend on the nature of the survey. For the G–BASE baseline mapping the instructions are:

- Find a site with flowing water and a good supply of sediment in the stream bed
- Always sample upstream of tracks and roads
- Avoid places where there is obvious bank-fall or soil material in the stream
- Avoid sampling where animals congregate in the stream to drink
- Sample above any obvious contamination to the natural system (e.g., a land drain outlet, mine waste or a dumped car)

One sampler should walk 50–100 m upstream (along the bank) of the intended site to check for any localised contamination, before initiating sample collection. Sometimes, it is impossible to fulfil all the above criteria. For example, during dry periods the stream may have no water or the entire length of the stream may have been dredged and straightened by agricultural activity. If no alternative is available, less suitable sites will have to be used with adequate comments and descriptions recorded on the field

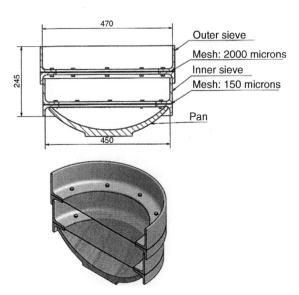

Figure 4.5 Diagram of a nested sieve set used for stream sediment sampling. Top: cross-section plan view (dimensions in millimetre). Bottom: cut-away 3-D visualisation [from engineering drawing by Humphrey Wallis, BGS for ABS (acetyl butyl styrene) polymer plastic sieve sets].

form. As the sampling methodology is based on low-order streams, the depth of water should generally be no more than waist deep.

The selection of the exact spot for sampling is a skill that develops with experience. Figure 4.6 shows a typical cross section of a stream channel for a first- or second-order stream and shows a typical site where sediment should be collected. Samplers should seek to sample typical flow regimes rather than sites which favour depositional sorting, for example, centre of streams away from banks and not behind obstructions. In surveys where heavy or larger detrital minerals are being sought, the higher energy part of the channel (i.e., where the water is flowing fastest) should be sampled or downstream of boulders where an obstruction to stream flow has resulted in the deposition of the heaviest part of the rivers mechanically transported load.

On arrival at site, samples should be collected in the order of stream water, stream sediment and panned heavy mineral concentrate. A flow chart of procedures is shown in Fig. 4.7 which demonstrates the team effort of a sampling pair to efficiently sample a site in a strictly prescribed order. Time spent at each site will vary according to the experience of the samplers and the ease of collecting and sieving fine sediment. On average a drainage site should take 30–40 min to sample.

5.6. Collecting a stream water

1. *Sampler 1* unpacks all the sampling equipment. *Sampler 2* labels the water sample bottles (and the sediment and panned concentrate bags). Four bottles are used for various analytical methods. The pre-allocated site number (taken from the field card) is written on the sample containers, using the black permanent ink marker. The site number becomes the sample number by including a sample type code as

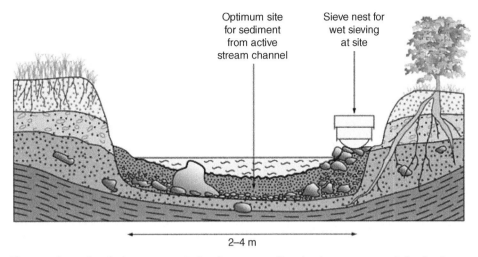

Optimum site
for sediment
from active
stream channel

Sieve nest for
wet sieving
at site

2–4 m

Figure 4.6 A sketch showing a typical sediment sampling site (temperate zone) for the first- or second-order stream.

a suffix, in the case of waters this is 'W'. The number should be written along the side, starting from the cap end. The following sample-type codes must also be written on each container: F/UA—filtered unacidified (30 ml Nalgene bottle); F/A—filtered and acidified (60 ml Nalgene bottle); pH (30 ml polythene bottle); and alkalinity/conductivity—(250 ml Nalgene polyethylene bottle). *Sampler 2* commences by collecting the filtered samples first.

2. Water samples are collected from the mid-stream flow, on the upstream side of the sediment sample location. *Sampler 2* should stand facing upstream, and sediment must not be disturbed. The filter should be removed from the self-seal bag without contaminating the connector, it should only be handled along its sides. Flush the syringe three times with stream water before connecting a clean Millex filter; filters should never be used at more than one site. Flush the filter with 5–10 ml of stream water. Carefully rinse the Nalgene bottles and caps with filtered stream water (minimum 10 ml). Special care must be taken to ensure that the sample containers and lids remain uncontaminated; the inside of lids and containers should not be handled and must not be allowed to come into contact with hands, soil, vegetation or unfiltered water. If they need to be put down while open they must be placed on a clean polythene bag. Fill the 60 ml bottle to the neck and completely fill the 30 ml bottle. Apply caps tightly, ensuring that no leakage occurs. Place the filtered samples into a clean 15 × 43 cm polythene bag tied with a knot then transported inside a self-seal polythene bag. If the bottle, cap or filter are dropped, or otherwise contaminated, a replacement must be used and the process restarted. In situations where filtration is difficult (i.e., very turbid waters), the F/A sample should be collected first. An additional 250 ml Nalgene bottle may be filled with unfiltered water, marked with the sample number and the relevant sample type(s), and taken to the field base for filtration. It is important to try and filter the sample for trace element analysis at site, as an

unquantifiable rate of cation adsorption onto container walls and suspended sediment may occur from the larger bottle before filtration. In areas where the water tends to have a particularly high proportion of suspended material 'pre-filters' are included in the sampling kit. This should be done only where absolutely necessary in order to avoid confusion and incorrect filtration; the sample teams should be reminded of the procedure to use. The pre-filtering process uses a coarser 25 μm pre-filter mounted in a Swinnex filter holder. The sample is first passed through the coarse pre-filter then through the 0.45 μm Millex disposable filter.

3. *Sampler 2* collects the pH and conductivity/alkalinity samples immediately after the filtered water samples. Like the filtered samples, they should be collected from the mid-stream flow, on the upstream side of the sampler. Thoroughly rinse the sample containers and caps with stream water *three times*. Submerge the containers in the stream to fill; then seal underwater, ensuring that all air has been expelled. Place the unfiltered samples into a clean, self-seal polythene bag along with the knotted bag containing the filtered water samples.

4. *Sampler 2*, in order to complete the water colour and suspended solids part of the field form, needs to determine these by filling a polythene bag (15 × 43 cm) with stream water and holding it up against the sky as a background to observe the water colour and opacity.

The 'dirty filter' with the >0.45 μm suspended sediment can be used as an additional sample media from this site and can be bagged and labelled for future analysis. Note that the F/A samples are acidified using Aristar-grade concentrated nitric acid when samples are returned to the field base each day. This is because it is undesirable to have the samplers carrying concentrated acid while in the field. Samples are acidified to 1% v/v (i.e., 0.6 ml conc. acid per 60 ml of sample). An additional precaution against contamination, one that is not employed by G-BASE, is the use of disposable vinyl gloves (Salminen *et al.*, 1998). Many surveys collecting stream waters will routinely determine pH, alkalinity and conductivity at site. The G-BASE project does such measurements back at the field base always within 24 h of sample collection. This gives better reproducibility than on-site field measurements.

5.7. Collecting stream sediment

1. *Sampler 1* downstream of the water sampling will wash the trenching tool (shovel), sieve nest, both pans, the plastic funnel and both sets of thick black protective rubber gloves with stream water. The sieve nest (Fig. 4.5) is assembled on the top of the glass-fibre pan, in a stable position, as close to the sediment collection point as possible. The collection pan and sieves must be clean and free from any particulate matter before the commencement of sampling.

2. The sediment collection position should be an active area of the stream bed (Fig. 4.6), and should ideally be centrally placed in the stream, to minimise contamination from any bank fall material. *Sampler 2, firstly, removes the uppermost (10–20 cm) heavily oxidised sediment with the shovel* and then loads the top sieve

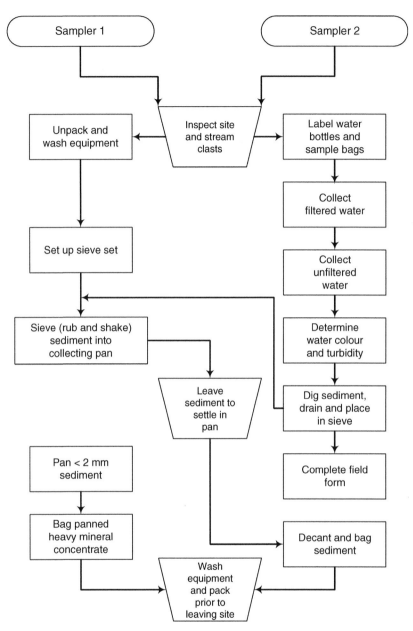

Figure 4.7 A flow chart summarising the procedures at a drainage site for the collection of stream water, sediment and heavy mineral concentrate. *Note*: Certain procedures are repeated when duplicate samples are collected.

with coarsely sorted sediment from beneath the oxidised layer, taking care to drain off excess water and remove any large clasts before placing the material into the top sieve. If the sediment lies on a base of peat or clayey till, take care to ensure that the sediment is sampled without digging into the underlying fixed material. It will normally be necessary to dig 15–25 kg (wet weight) of material to provide a sufficient final sample weight. If there is abundant sediment in the stream bed this can normally be achieved from a single 'hole', if sediment is scarce then several holes may need to be dug over a length of the stream bed of no more than 5 m.

3. *Sampler 1* as loading the sieve proceeds, rubs the stream sediment through the top sieve, providing sufficient (normally 2–3 kg) <2000 μm material in the lower sieve to produce adequate <150 μm material. During this process look out for any contaminant material in the sediment, which should be removed from the sieve and the details noted on the field data card. Before the upper sieve becomes too full and heavy it should be removed and shaken to allow more <2000 μm material to fall through into the bottom sieve. It is very important to make sure the outside of the upper sieve is free of any sediment otherwise coarse grains could fall into the fine sediment when shaking. The upper sieve material can then be discarded and this material is often worth observing for stream clast lithologies, which are noted on the field data card. Several cycles of filling, rubbing, shaking and discarding of the top sieve material may be required to provide enough material in the lower sieve. This is dependant on the physical nature of the stream sediment material.

4. Once there is sufficient material in the lower sieve *Sampler 1* should mix around and rub the <150 μm sieve to help material to pass through into the collecting pan. If the lower sieve material is very dry and sandy it is often necessary to sprinkle a small amount of water into the lower sieve while mixing and rubbing the material. *Care must be taken not to flood the collecting pan with too much water otherwise there is a danger of fine material being washed away.*

5. When the lower sieve material has been well mixed and rubbed through, *Sampler 1* rinses the rubber gloves and then uses the funnel to rinse any particulate material off of the top rim and outer sides of the lower sieve, ensuring that the volume of water that goes into the sieve is kept to a minimum. The lower sieve should then be picked up carefully, without disturbing the collecting pan, and gently shaken to allow additional <150 μm material to fall through into the collecting pan. If there appears to be insufficient material in the collecting pan, the lower sieve may be replaced and the material re-mixed and rubbed while sprinkling with a small volume of water (<100 ml). The gloves and sieve top and outer sides should then be re-rinsed and the sieve carefully lifted and shaken as before. Take particular care at this stage to avoid biasing the sediment sample by incorporating oversize material. Once there is enough sediment in the collecting pan, remove the lower sieve and retain the <2000 μm material which it contains. *Leave the pan containing the <150 μm sample undisturbed for about 20 min to allow the settling out of suspended material.*

6. While *Sampler 1* is panning (see below) *Sampler 2* completes the field form (Fig. 4.3) with site and sample descriptions.

7. When the fine material has settled in the collecting pan, *Sampler 2* decants the fine sediment slurry into a sample bag. *Sampler 2* puts on a pair of rubber gloves, cleans them in the stream water, then slowly decants excess water from the surface of the sediment collecting pan. The sediment sample is then homogenised by firmly,

but carefully, shaking the pan to mix the dense, particulate material with the fine colloidal fraction. This is important as if there is an excess of material, any portion discarded must be the same as the portion which is retained (final sample volume should be 200–250 ml). At this stage, the sediment details (colour, clay, organic and colloidal content) should be noted. Next, *Sampler 2* thoroughly rinses clean the polypropylene funnel with stream water then transfers the sample, via the funnel, to the appropriate, numbered Kraft paper bag 10×20 cm (4×8 in). See the water sampling procedure, step 1, for sample number allocation—for stream sediments the G-BASE project uses the sample type code 'C'. The Kraft bag is sealed by folding the tab over three times and bending the wire fixings over the ends of the envelope. Place the sealed Kraft bag in a 15×43 cm (6×17 in) polythene bag and tie a loose knot in the polythene bag to prevent loss or contamination during transport. The sealed bags are placed into a plastic box, and then into a rucksack, taking care to ensure that the sample bag is upright.

5.8. Collecting panned concentrate

1. *Sampler 1* commences collecting the panned concentrate while the sediment is settling in the collecting pan. The <2000 µm material retained from the sediment collection process is tipped into the wooden Malaysian 'dulang' style pan, using water from the funnel to wash all the material from the sieve.
2. Further wet sieve <2000 µm sediment from as deep as possible within the stream bed, using the top sieve placed directly on the wooden pan. Copious amounts of water may be used to aid sieving at this stage. Once the wooden pan is almost full of <2000 µm material the panned heavy mineral concentrate is then collected using the following three stages:

   (a) Removal of clay and organic material that binds grains together by repeated washing and stirring of the material in the pan. The pan should not be submerged during this procedure, but clean water should be continually added and dirty water poured out. Once the grains feel well separated and the water being poured out looks relatively clean, proceed to stage (b).

   (b) Formation of heavy-mineral bed by vigorous shaking of pan with ample water for a minimum duration of 2 min. This allows density separation in the pan material and is extremely important before proceeding to stage (c).

   (c) Selective removal of the less dense fraction by circulating the pan on the surface of the water in an elliptical fashion to yield 20–40 g of heavy mineral (density >2.9 g/cm^3) concentrate. This process is best demonstrated by an experienced sampler and it is important during stage (c) to regularly stop circulating and re-shake the material to maintain density separation.

3. *Sampler 1* inspects the final concentrate with a hand lens and notes the presence and relative abundance of heavy minerals on the field form. The funnel is used to transfer the concentrate material to a numbered, 8×13 cm ($3'' \times 5''$) Kraft sample bag using sufficient water to ensure complete recovery of all grains. The sample type code for a panned concentrate is 'P'.

5.9. Completion of sampling

Before leaving the site, as was performed on arrival, all equipment should be thoroughly rinsed to remove traces of particulate material to avoid between site contamination. The field form should be checked to ensure that all observations have been noted. If any field observations are not applicable at a site, for example, there is no contamination, the relevant box should be struck through so it is clear that the observation was investigated but there was nothing to record. Finally, on departure, the site should be clear with all of the samples and field equipment packed in the rucksacks ready to be taken to the next site.

On return to field base, stream sediment samples require careful handling. At the end of each sampling day, the field team leader and assistant will go through a procedure of checking samples in by marking checklists confirming the sample numbers are readable and correct. The paper sample bags should be hung up to air dry. The Kraft paper sample bags will allow water to seep out but not the fine sediment. However, the samples should not be dried until they are rock hard, the paper bag should be dry and the sediment having a plastic consistency. Wet paper bags left unattended will rot in a very short time and samples will be lost.

5.10. Collecting duplicate samples

One sample in every hundred samples collected by the G-BASE project is a duplicate sample. A predefined sample and duplicate site number is allocated to samplers who must collect a duplicate stream sediment and stream water sample from a single site. A duplicated panned sample is not generally collected because such samples are not routinely submitted for chemical analysis. It is stipulated that a stream sediment and water duplicate should be collected within 25 m of the original sample.

5.11. Dry sites

During periods of prolonged dryness, small first-order streams in the UK may dry out so collecting a water sample is not possible. However, the stream sediment and panned concentrate can still be sampled. The 2000 μm sieve can be used to collect ∼5 kg of material and stored in labelled self-seal polythene bags. If the material is predominantly clay, less material needs to be collected and if the sample is still moist then dry sieving will not be appropriate and larger stones and fragments are removed by hand. The bulk sample can be carried to the next wet site or base camp for the subsequent wet-sieving and panning.

5.12. Control samples

An important part of the sampling methodology relates to the use of control samples in order that the data can be quality controlled and quality assured (see Johnson *et al.*, 2008). While it is only duplicates that are created during the sampling process, these, along with replicates, blanks (for waters) and secondary reference materials need to be assigned sample numbers so they are included as part of the routine sample submission and are 'blind' to the analyst. Duplicate samples are collected from a single site according to the procedures described earlier. A pair of duplicate samples (i.e., a normal sample and an additional sample collected at the same site) help to define the within site sampling uncertainty. Each sample of the duplicate pair can be split in the laboratory (again following a strictly defined procedure) to give replicates. Replicates help to define any laboratory or within sample uncertainty. For waters, a blank sample is created from de-ionised water and acidified in the same manner as normal water samples. This acts as a check to see if any contaminants are added during the process of sample acidification and sample storage. Nested ANOVA analysis of the duplicate and replicate samples (Sinclair, 1983) can be used to assign the source of variability between sample sites, within sites and between sites. The point of such an analysis is to demonstrate the validity of the sampling strategy. It is obviously highly desirable to have the maximum variance attributed to between sites variability. If the combined sampling and analytical variance (i.e., within site and within sample variance) exceeds 10%, then there are issues with the sampling strategy that need to be explained. A common cause of a high-analytical variance is when element results are at or near the lower limit of detection.

The G-BASE project collects samples in random number order (Plant, 1973), as this helps identify any correctable systematic errors introduced during sample preparation and analysis, processes in which the samples are handled in numeric order. For every block of one hundred numbers, five numbers are reserved for control samples so when they are submitted within a batch of samples they are 'blind' to the analyst. The control samples inserted are one duplicate sample, two replicate samples, two blanks, and two secondary reference materials (SRM) used to monitor accuracy and precision as well as to level data between different field campaigns (see Johnson et al., 2008). Along with the original sample of the duplicate pair, this means 8% of samples submitted are control samples, a point not to be overlooked in setting the budget for analyses.

These blind control samples are in addition to any primary reference materials (PRM) that the laboratory may also analyse. For the G-BASE project, the BGS laboratories usually insert a PRM at the beginning and end of each batch of 500 samples. As G-BASE generally collects and analyses 2000–3000 samples each field campaign 8% of the samples is more than adequate to carry out quality control procedures. However, if sample numbers are <500, then it is recommended that the number of duplicates and replicates per hundred samples should be doubled.

5.13. Health and safety

Health and safety considerations should be an important part of planning a drainage sampling campaign, and instruction on health and safety issues must be given to sampling teams before they commence work. Apart from the obvious duty of care a project manager has for their sampling team and the samplers have for each other, serious accidents to personnel can seriously disrupt or terminate a sampling programme. It is surprising how often samplers or labourers employed to collect and carry samples, often involving crossing or working near deep water, have never been asked such a fundamental question as to whether they could swim. A risk assessment should always be part of a sampling plan whether or not it is a statutory requirement of the country of work. A summary of the G-BASE project, health and safety concerns and mitigating actions are given in Table 4.5. In urban areas, the health and safety aspects of working in heavily polluted culverted streams make it essential that the field geochemists have the assistance and involvement of municipal authorities during the sampling.

 ## 6. DISCUSSION

6.1. Sampling equipment

For collecting drainage samples, the fundamental piece of equipment is the sieve set and this is shown diagrammatically in Fig. 4.5. This is the wooden sieve set with nylon mesh that has been used throughout the history of the BGS G-BASE project and has a proven history of robustness. The use of a nylon sieve mesh and wooden sieve relate to a desire to minimise any contamination that may be introduced through sampling equipment. Plastic polymer sieve sets are currently undergoing trials on several BGS projects. Steel equipment can introduce elements such as V, Cr, Mn and W and plastics can contain high levels of Cd and Sb. Similarly, sample bags and water containers have to be free of sources of contamination. Water containers can be particularly problematical if they, for example, contain metal foil-inserts in the cap or are made of plastic from which trace elements can readily be leached by the acidic water sample.

The sieve set can readily be adapted to suit the working environment. For the BGS reconnaissance geochemical mapping in the tropical climate of Sumatra (at a sampling density of ~1 sediment per 15 km² —Muchsin *et al.*, 1997; Stephenson *et al.*, 1982), the 2000 μm mesh sieve was used but not the 150 μm sieve—the fine sediment sieving was carried out in the laboratory. In the semi-arid Anti-Atlas Mountains (geochemical exploration with drainage sediments at one sample per kilometre square), the sieve sets were reduced in diameter to 25 cm to lighten the load samplers had to carry in the hot climate. These were equally efficient at sieving dry sediment. Orientation work in this Moroccan environment, where the streams were predominantly dry for more than 10 months in the year, showed that a coarser stream sediment fraction was more suitable for defining anomalies associated with mineralization (British Geological Survey (BGS), 1999a). BGS's ongoing geochemical sampling in Madagascar has not used sieve sets and is instead using just flat wooden framed sieve screens sieving directly into a collecting pan. This greatly reduces the amount of equipment samplers have to carry but will increase sampling error due to the greater risk of unsieved material entering the pan. The G-BASE project has a preference for wet sieving as fine particles are more likely to be disaggregated than in dried sediment where the fine fraction of the sediment may form concretions of larger particle size.

Information collected at site is generally recorded onto a field form then later entered into a database on a PC. BGS has tested hand-held computers, with GPS attachment, for inputting data directly into a digital field database (Scheib, 2005). Although these offer the potential for greater efficiency in creating the field database, they still have an unproven reliability in areas of difficult terrain and extremes

Table 4.5   A table summarising the main health and safety issues for the G-BASE project in the UK and suggested mitigating actions

High/medium risk activity	Summary of measures to reduce risk
Driving in field area	• receive appropriate vehicle driving training • use vehicle appropriate for the type of fieldwork
Transporting heavy loads and equipment by vehicle	• do not overload vehicles • secure equipment and samples • transport acid in special anti-spill containers
Lifting heavy loads/loading and unloading samples	• receive manual handling training • use appropriate storage crates for sample transportation • don't overload storage crates • do not load/unload heavy items alone
Carrying heavy loads in the field	• use good quality rucksacks offering high level of support and adjusted appropriately for the carrier • share the load between the two samplers • sensible handling of load whilst negotiating obstacles (e.g., pass load across a wall rather than climbing over the wall with rucksack still on)
Sampling drainage samples	• attend sampling training day • dress appropriately with good footwear and always take waterproof clothing • stick to recognised paths. Do not take risks crossing barbed wire fences/stone walls or rivers/streams for the sake of making a shortcut
Walking on roads used by frequent traffic	• always use Hi-vis jackets and rucksacs with Hi-vis strips • seek alternative footpaths if available • where no footway, walk in direction of oncoming traffic except when approaching the brow of a hill
Remote working	• always sample in pairs • inform team leaders of proposed route • carry emergency telephone contact numbers
Adverse weather	• pay attention to weather forecasts • do not sample areas in times of flood • take appropriate measures against exposure to the sun

(continued)

Table 4.5 *(Continued)*

High/medium risk activity	Summary of measures to reduce risk
	• during thunderstorms follow standard procedures to avoid lightening strikes and in particular don't carry a metal equipment
Attack by animals	• avoid potentially dangerous animals (e.g., bulls and guard dogs) where possible by choosing an alternative route
Military, shooting area and other hazardous land use	• always have permission to enter such areas first • team leaders to advise samplers of such potential areas on their map • team leaders plan daily sampling areas so hazards such as large rivers or railways do not have to be crossed • always wear Hi-vis jackets
Exposure to infection, agrochemicals and pesticides	• samplers to be advised of dangers on training day • avoid contaminated sites or fields being sprayed • observe agricultural exclusion notices when encountered in the field
Exposure to substances used by the field team	• receive training and H&S procedures for handling conc. acids • glue sediment and pan bags in a well-ventilated area

of climate, and are not suitable for use by untrained samplers. GPS has greatly assisted in the accurate positioning of sample sites but where reliable topographic maps are available grid references from these are still the prime locational reference. It should be borne in mind that in steep-sided stream gorges or wooded areas GPS units can report inaccurate coordinates.

6.2. Representative nature of drainage sediments

A lot has been written on the subject of just what a drainage sediment represents, particularly with respect to trying to trace anomaly trains from mineralization (e.g., Hawkes, 1976; Ottesen and Theobold, 1994). Potentially, drainage sampling has a great advantage over other types of sample media (such as soil or rock sampling) which have a much reduced area of representativity and are far more inhomogeneous. Soil sampling presents considerable problems for regional geochemical mapping because of the variation in soil types, the variable nature of horizons and the depths at which they occur, the limited cover in upland areas, and the wide variation in pH and Eh in soils which critically affects the solubility and concentration of metals (Plant and Moore, 1979). However, Plant and Moore (1979) do concede that soils may be the optimum medium in agricultural areas of lowland England particularly for larger scale geochemical maps.

The nature of the processes that combine to produce the water or detritus at a drainage sampling site means that it is unlikely that the sample is truly representative of the entire drainage basin upstream from the site (Bolviken *et al.*, 2004; Ottesen *et al.*, 1989; Peh *et al.*, 2006). While the fine sediment of overbank and flood plain alluvium in certain environments can be more representative of the drainage catchment, they have more limited landscape of applicability and generally are a composite from a very broad area on account of the order of the drainage basin with which they are associated. The long-established and widely employed method of stream sediment sampling (see listing in Plant *et al.*, 1988, 1997) testifies to the fact that this method is the preferred method of regional mapping that, providing the sampling procedure has been strictly followed, consistently produces satisfactory regional geochemical maps. There are specific parts of the procedure (e.g., the settling of the fine sediment) that are designed to address some of the issues of representativity (see discussion at the end of Plant and Moore, 1979) and this further emphasises the importance of following instructions in a precise manner.

There are circumstances where the drainage sample does not represent material derived from the basin upstream of the site, and these should be dealt with at the orientation phase of the project. Such circumstances would include wind-blown material collecting in the drainage channel or exotic materials introduced to the drainage basin through anthropogenic activity.

Anthropogenic activity has significantly impacted on drainage basin systems throughout the world (Owens *et al.*, 2005). Even in the more remote inaccessible regions of the world such as the Amazon basin or the jungles of SE Asia logging activities have significantly changed rainwater run-off and percolation and resulted in huge volumes of soil entering the stream system more rapidly than under natural undisturbed conditions (Fletcher and Muda, 2005). Humans have dammed, straightened, dredged and redirected rivers throughout the history. Although most nations of the world are now more appreciative of the importance of good environmental management of river systems, there are still many economically less-developed communities that would see the river system as the principal means of removing waste and rubbish. If the strategy of the sampling is to determine the presence of anthropogenic contamination, then sampling sites should not seek to avoid these effects. If the survey is concerned with the natural baseline, then anthropogenic effects should be actively avoided by the appropriate selection of sampling sites, such as avoiding old mine dumps or always sampling upstream of urban areas.

Urban environments provide a particularly challenging environment for drainage sampling. For example in 2003 BGS adapted its standard regional drainage sampling methods to carry out a drainage survey of the city of Glasgow, Scotland (Fordyce *et al.*, 2004). Access to sampling sites was difficult in the city environment, and the project was greatly helped by the involvement of Glasgow City Council in the work. The adaptability of the drainage sampling method for use in both rural and urban environments enabled a direct comparison between rural and urban sediments and waters. A quantitative assessment of the impact of urbanisation and industry can then be made by comparing the rural and urban areas with similar geological settings. For the first time, this Glasgow urban survey involved the G-BASE project in the analysis of drainage samples for organic analyses (Total Petroleum Hydrocarbons (TPH), Polycyclic Aromatic Hydrocarbon (PAH), Poly-Chlorinated Biphenyls (PCB) and Organo-Tin). This demonstrated the many difficulties and procedural changes that were needed in the collection and storage of drainage sediment samples to be analysed for organic compounds.

The introduction of wind-blown sediment from outside the drainage basin is not a problem in the United Kingdom, but it was an issue that BGS had to address during orientation studies for reconnaissance geochemical mapping in the semi-arid Anti-Atlas Mountains of Morocco. Located to the north of the Sahara Desert there was more than ample evidence of wind-blown deposits and a sampling strategy had to be devised that would minimise any dilution by wind-blown sediment. The orientation was carried out by the chemical analysis of many fraction sizes of the stream sediment along stream channels down from mineralised and unmineralised areas. This work suggested that a coarser fraction of sediment (-250 μm) was more appropriate for this type of climatic zone (British Geological Survey (BGS), 1999a). Evidence from Ti/Zr ratios suggested that some aeolian dilution was present in the finest -63 μm fraction (Dickson and Scott, 1998).

A further frequently raised issue concerning representativity relates to the as yet undiscussed dimension of time. Concerns of temporal variability are often aired when geochemical maps based on drainage samples collected over many years are compiled. This is particularly the case for stream waters where even daily let alone seasonal variations in the water system might be expected to cause considerable variability in element concentrations. Geochemical maps produced by BGS have consistently shown that this is not the case and is a tribute to the robustness of the sampling method. In the stream water geochemical atlas of Wales (British Geological Survey (BGS), 1999b), in spite of the documented influence that temporal variations are known to have on stream water composition, it is spatial controls that predominate at a regional scale (Hutchins *et al.*, 1999). Indeed, analytical uncertainty in the geochemical maps is probably more significant than any temporal variations. The drainage sampling of the Tellus Project in Northern Ireland in 2004 extended the stream sediment and water sampling carried out by the G-BASE project in 1994–1996. Again, the compilation of the two data sets derived from samples collected more than 10 years apart shows the dominance of spatial controls over temporal ones.

6.3. Other drainage site media

This account has been specifically concerned with stream sediments, stream waters and panned heavy mineral concentrates, all collected from the same site. For the G-BASE project, the stream sediments and stream waters are submitted for inorganic chemical analyses, water samples benefiting by improvements to detection limits in the past decade enabling ultra-low element concentrations to now be reported.

The panned heavy mineral concentrates have not generally been analysed unless follow-up work has been carried out. However, all are inspected at site and observed minerals and contaminants are recorded. The samples are an excellent resource for identifying drainage catchment mineralisation and lithologies as well as anthropogenic contamination (Fig. 4.8). Indeed, all the G-BASE excess samples are stored at the National Geological Data Centre, Keyworth, UK, and are available for further study. The value of excess sample powders in research should not be underestimated.

The availability of other sample media at a drainage site has been briefly mentioned previously. The dirty water filters with the >45 μm suspended sediment captured on the filter discs could be very useful in studies of the suspended-sediment load. The Fe and Mn coatings on pebbles and other chemical precipitates (e.g., insoluble Fe hydroxides) could also be collected as these are known to have good scavenging properties for certain trace elements.

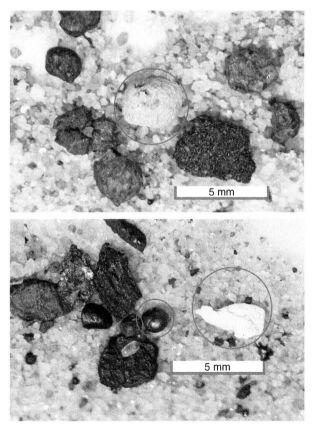

Figure 4.8 Photographs showing anthropogenic contamination observed in some G-BASE panned heavy mineral concentrates. Top: Sample 307349P showing oxidised lead shot. Bottom: Sample 307364P showing steel shot and a flake of paint.

7. CONCLUSIONS

1. Drainage samples, specifically stream waters and sediments in this account, have a long and well-established use in geochemical and environmental studies and, given the right climatic and geomorphological conditions, should be the sampling media of choice.
2. A single drainage site enables the environmental scientists to study several different media at one location, each of which can assist in interpreting the chemical behaviour and distribution of an element in the surface environment.
3. There are issues relating to how representative of the drainage catchment results from drainage samples are. Geochemical maps from many areas of the world have repeatedly proven the accurate representation of drainage basin geochemistry using stream sediments.
4. Representativity can be addressed in an orientation phase of a project and can be satisfied by following strict well-documented procedures. Such procedures will address issues of health and safety as well as quality control.
5. There is surprisingly little temporal control on the spatial patterns on water or stream sediment geochemical maps where samples may have been collected over periods of many years. Analytical method variability over a period is generally a bigger problem than any temporal effects when creating seamless geochemical maps for drainage samples collected over a long period.

ACKNOWLEDGMENTS

This chapter is published with the permission of the Director of the British Geological Survey (NERC). Figures 4.1 and 4.5 were prepared by Lauren Noakes and Henry Holbrook, respectively (both BGS).

REFERENCES

Andrews-Jones, D. A. (1968). The application of geochemical techniques to mineral exploration. Colorado School of Mines, Mineral Indus. Bull., 11, No 6.

Bolviken, B., Bogen, J., Jartun, M., Langedal, M., Ottesen, R. T., and Volden, T. (2004). Overbank sediments: A natural bed blending sampling medium for large-scale geochemical mapping. *Chemom. Intell. Lab. Syst.* **74,** 183–199.

British Geological Survey. (1999a). Résultats de l'étude d'orientation et analyses chimiques des "Stream Sediments" dans le domaine de l' Anti-Atlas (Maroc). British Geological Survey Report prepared for the Moroccan Ministry of Mines and Energy, Rabat, Morocco. British Geological Survey, Keyworth, UK.

British Geological Survey (1999b). "Regional Geochemistry of Wales and Part of West-Central England: Stream Water." British Geological Survey, Keyworth, Nottingham. ISBN 0 85272 363 6.

Caley, E. R., and Richards, J. F. C. (1956). "Theophrastus on Stones: Introduction Greek Text, English Translation, and Commentary." The Ohio State University, Columbus, Ohio, USA.

Cooper, D. C., and Thornton, I. (1994). Drainage geochemistry in contaminated terrains. 447–497 in *Drainage Geochemistry. In* "Handbook of Exploration Geochemistry" (M. Hale and J. A. Plant, eds.), Vol. 6. Elsevier, Amsterdam.

Darnley, A. G., Bjorklund, A., Bolviken, B., Gustavsson, N., Koval, P. V., Plant, J. A., Steenfelt, A., Tauchid, M., and Xuejing, X. (1995). A global geochemical database for environmental and resource management. *UNESCO publishing*, 19.

Dickson, B. L., and Scott, K. M. (1998). Recognition of aeolian soils of the Blayney district, NSW: Implications for mineral exploration. *J. Geochem. Explor.* **63**, 237–251.

Ettler, V., Mihaljevic, M., Sebek, O., Molek, M., Grygar, T., and Zeman, J. (2006). Geochemical and Pb isotopic evidence for sources and dispersal of metal contamination in stream sediments from the mining and smelting district of Pribram, Czech Republic. *Environ. Pollut.* **142**, 409–417.

Fersman, A. E. (1939). Geochemical and mineralogical methods of prospecting for useful minerals. *in U.S. Geol. Surv. Circ. 127, 1952*, 37pp. (US Geological Survey.)

Fletcher, W. K., and Muda, J. (2005). Dispersion of gold in stream sediments in the Sungai Kuli region, Sabah, Malaysia. *Geochemi.: Explor. Environ. Anal.* **5**, 211–214.

Fordyce, F. M., Dochartaigh, B. É. Ó., Lister, T. L., Cooper, R., Kim, A. W., Harrison, I., Vane, C. H., and Brown, S. E. (2004). Clyde tributaries: Report of urban stream sediment and surface water geochemistry for Glasgow. *British Geological Survey, Keyworth, UK*, Commissioned Report No. CR/04/037.

Gonçalves, M. A., Nogueira, J. M. F., Figueiras, J., and Putnis, C. V. (2004). Base-metals and organic content in stream sediments in the vicinity of a landfill. *Appl. Geochem.* **19**, 137–151.

Hale, M. (1994). Strategic choices in drainage geochemistry. 111–144 in *Drainage Geochemistry. In* "Handbook of Exploration Geochemistry" (M. Hale and J. A. Plant, eds.), Vol. 6. Elsevier, Amsterdam.

Hale, M., and Plant, J. A. (1994). Drainage Geochemistry. *In* "Handbook of Exploration Geochemistry," Vol. 6. Elsevier, Amsterdam.

Hawkes, H. E. (1976). The downstream dilution of stream sediment anomalies. *J. Geochem. Explor.* **6**, 345–358.

Hawkes, H. E., and Bloom, H. (1955). Heavy metals in stream sediment used as exploration guides. *Min. Eng.* **8**, 1121–1126.

Hutchins, M. G., Smith, B., Rawlins, B. G., and Lister, T. R. (1999). Temporal and spatial variability of stream waters in Wales, the Welsh borders and part of the West Midlands, UK. 1. Major ion concentrations. *Water Res.* **33**, 3479–3491.

Institute of Hydrology (2003). "Hydrological data UK. Hydrometric Register and Statistics 1996– 2000." Institute of Hydrology, NERC, Wallingford, England.

ISO (1995). Water quality-Sampling-Part 12: Guidance on sampling of bottom sediments. International Organization for Standardization. ISO 5667–12.

Johnson, C. C. (2005). 2005 G-BASE Field Procedures Manual. *British Geological Survey, Keyworth, UK*, Internal Report No. IR/05/097.

Johnson, C. C., Flight, D. M. A., Lister, T. R., and Strutt, M. H. (2001). La rapport final pour les travaux de recherches géologique pour la realisation de cinq cartes géochimique au 1/100 000 dans le domaine de l'Anti-Atlas (Maroc). *British Geological Survey Confidential Internal Report prepared for the Moroccan Ministry of Mines and Energy*, Commissioned Report Series, No.CR/01/031.

Johnson, C. C., Breward, N., Ander, E. L., and Ault, L. (2005). G-BASE: Baseline geochemical mapping of Great Britain and Northern Ireland. *Geochem.: Explor. Environ. Anal.* **5**, 347–357.

Johnson, C. C., Ander, E. L., Lister, T. R., and Flight, D. M. A. (2008). Data conditioning of environmental geochemical data: Quality control procedures used in the British Geological Survey's regional geochemical mapping project. *In* "Environmental Geochemistry: Site Characterization, Data Analysis and Case Histories" (B. de Vivo, H. E. Belkin, and A. Lima, eds.), Chapter 5, 93–118.

Lee, S., Moon, J. W., and Moon, H. S. (2003). Heavy metals in the bed and suspended sediments of Anyang River, Korea: Implications for water quality. *Environ. Geochem. Health* **25**, 433–452.

Lovering, T. S., Huff, L. C., and Almond, H. (1950). Dispersion of copper from the San Manuel copper deposit, Pinal County, Arizona. *Econ. Geol.* **45**, 493–514.

Muchsin, M., Johnson, C. C., Crow, M. J., Djumsari, A., and Sumartono. (1997). Atlas Geokimia Daerah Sumatera Bagian Selatan/Geochemical Atlas of Southern Sumatra. "Regional Geochemical Atlas Series of Indonesia." No. 2. Directorate of Mineral Resources, Bandung, Indonesia and British Geological Survey, Keyworth, UK.

Ottesen, R. T., and Theobold, P. K. (1994). Stream sediments in mineral exploration. 147–184 in *Drainage Geochemistry. In* "Handbook of Exploration Geochemistry" (M. Hale and J. A. Plant, eds.), Vol. 6. Elsevier, Amersterdam.

Ottesen, R. T., Bogen, J., Bölviken, B., and Volden, T. (1989). Overbank sediment: A representative sample medium for regional geochemical mapping. *J. Geochem. Explor.* **32**, 257–277.

Owens, P. N., Batall, R. J., Collins, A. J., Gomez, B., Hicks, D. M., Horowitz, A. J., Kondolf, G. M., Marden, M., Page, M. J., Peacock, D. H., Petticrew, E. L., Salomons, W., *et al.* (2005). Fine-grained sediment in river systems: Environmental significance and management issues. *River Res. Appl.* **21**, 693–717.

Peh, Z., Miko, S., and Mileusnic, M. (2006). Areal versus linear evaluation of relationship between drainage basin lithology and geochemistry of stream and overbank sediments in low-order mountainous drainage basins. *Environ. Geol.* **49**, 1102–1115.

Plant, J. A. (1971). Orientation studies on stream sediment sampling for a regional geochemical survey in northern Scotland. *Trans. Inst. Min. Metall.* **80**, 323–346.

Plant, J. A. (1973). A random numbering system for geological samples. *Trans. Inst. Min. Metall.* **82**, 63–66.

Plant, J. A., and Moore, P. J. (1979). Geochemical mapping and interpretation in Britain. *Philos. Trans. R. Soc.* **B288**, 95–112.

Plant, J. A., and Raiswell, R. W. (1994). Modifications to the geochemical signatures of ore deposits and their associated rock in different surface environments. 73–109 in *Drainage Geochemistry. In* "Handbook of Exploration Geochemistry" (M. Hale and J. A. Plant, eds.), Vol. 6. Elsevier, Amsterdam.

Plant, J. A., Hale, M., and Ridgeway, J. (1988). Developments in regional geochemistry for mineral exploration. *Trans. Inst. Min. Metall. B. Appl. Earth Sci.* **97**, 116–140.

Plant, J. A., Klaver, G., Locutura, J., Salminen, R., Vrana, K., and Fordyce, F. M. (1997). The Forum of European Geological Surveys Geochemistry Task Group: Geochemical inventory. *J. Geochem. Explor.* **59**, 123–146.

Ranasinghe, P. N., Chandrajith, R. L. R., Dissanayake, C. B., and Rupasinghe, M. S. (2002). Importance of grain size factor in the distribution of trace elements in stream sediments of tropical high grade terrains—A case study from Sri Lanka. *Chemie der Erde Geochem.* **62**, 243–253.

Salminen, R., Tarvainen, T., Demetriades, A., Duris, M., Fordyce, F. M., Gregorauskiene, V., Kahelin, H., Kivisilla, J., Klaver, G., Klein, P., Larson, J. O., Lis, J., *et al.* (1998). FOREGS geochemical mapping field manual. *Geol. Surv. Finland, Guide No. 47.* pp. 42.

Salminen, R., Batista, M. J., Bidovec, M., Demetriades, A., De Vivo, B., De Vos, W., Duris, M., Gilucis, A., Gregorauskiene, V., Halamic, J., Heitzmann, P., Lima, A., *et al.* (2005). Geochemical Atlas of Europe. Part 1—Background information, methodology and maps. Geochemical Atlas of Europe. *Geol. Surv. Finland* ISBN 951-690-921-3.

Scheib, A. (2005). G-BASE trials of SIGMA digital field data capture; Feedback and recommendations. *British Geological Survey, Keyworth, UK,* Internal Report No. IR/05/015.

Schreck, P., Schubert, M., Freyer, M., Treutler, H. C., and Weiss, H. (2005). Multi-metal contaminated stream sediment in the Mansfeld mining district: Metal provenance and source detection. *Geochem.-Explor. Environ. Anal.* **5**, 51–57.

Sinclair, A. J. (1983). Univariate Analysis. 57–81 in *Statistics and Data Analysis in Geochemical Prospecting. In* "Handbook of Exploration Geochemistry" (R. J. Howarth, ed.), Vol. 2. Elsevier, Amsterdam.

Stephenson, B., Ghazhali, S. A., and Harwidjaja. (1982). Regional geochemical Atlas of northern Sumatra. "Regional Geochemical Atlas Series of Indonesia. No. 1." Directorate of Mineral Resources, Bandung, Indonesia and British Geological Survey, Keyworth, UK.

Strahler, A. N. (1957). Quantitative analysis of watershed geomorphology. *Trans. Am. Geophys. Union* **38**, 913–920.

Thornton, I., and Webb, J. S. (1979). Geochemistry and health in the United Kingdom. *Philos. Trans. R. Soc.* **B288**, 151–168.

Webb, J. S., and Howarth, R. J. (1979). Regional geochemical mapping. *Philos. Trans. R. Soc.* **B288**, 81–93.

Wertime, T. A. (1973). The beginning of metallurgy: A new look. *Science* **182**, 875–887.

Williams, T. M., Dunkley, P. N., Cruz, E., Actimbay, V., Gaibor, A., Lpez, E., Baez, N., and Aspden, J. A. (2000). Regional geochemical reconnaissance of the Cordillera Occidental of Ecuador: Economic and environmental applications. *Appl. Geochem.* **15**, 531–550.

Young, M. E., and Smyth, D. (2005). New geoscience surveys in Northern Ireland. European Geologist, Vol. No. 19, pp. 24–26. European Federation of Geologists, Brussels, Belgium.

CHAPTER FIVE

DATA CONDITIONING OF ENVIRONMENTAL GEOCHEMICAL DATA: QUALITY CONTROL PROCEDURES USED IN THE BRITISH GEOLOGICAL SURVEY'S REGIONAL GEOCHEMICAL MAPPING PROJECT

Christopher C. Johnson,* Louise E. Ander,* Robert T. Lister,* *and* Deirdre M. A. Flight*

Contents

Abstract

Data conditioning procedures involve the verification, quality control and data levelling processes that are necessary to make data fit for the purpose for which it is to be used. It is something that has to be planned at the outset of any project generating geochemical data. Whether it is in the sampling phase, for example, determining how sites and

* British Geological Survey, Keyworth, Nottingham, NG12 5GG, UK

Environmental Geochemistry
DOI: 10.1016/B978-0-444-53159-9.00005-X

© 2008 Elsevier B.V.
All rights reserved.

samples should have a unique identity, or through to the data presentation phase in which disparate data sets may have to be joined to form a seamless map. This account describes the methods currently used by the British Geological Survey's regional geochemical mapping project that has been generating geochemical data for various sample media for nearly 40 years. It is important that users of the data are given information that will help them ascertain whether the provided environmental data is suitable for the purpose of its intended use.

1. INTRODUCTION

Data conditioning is the process of making data fit for the purpose for which it is to be used. Users of the geochemical data need to have confidence that anomalous results are not an artefact of the sampling or analytical method. The results need to be interpreted in the context of existing data whether it is as a comparison with statutory concentration levels for an element or using the data in a regional context. Failure to condition the data has both time and cost consequences that can easily be avoided if a system of quality assurance is followed. The data conditioning process is a three-stage process.

Initially, data will undergo a series of error checking and verification procedures that relate to the number of samples submitted; the methods used; the elements requested; element ranges; limits of detection; absent, not determined and not detected results; and mis-numbering errors. These procedures are essentially a check of collation errors and that the laboratory has carried out what they were asked to do. This checking and verification phase is generally completed before the laboratory receives payment for their work. This is referred to in subsequent discussions as the 'raw data checking' phase where the term 'raw data' is given to the analytical results as recorded by the laboratory. A second phase of processing involves a series of plots and statistical tests of the data that measure the accuracy and precision of the results and attempts to attribute their sources of variability. Finally, the data may need to be levelled in order to be joined with existing data from earlier batches that may even have been derived using different analytical procedures.

This account of data conditioning describes methods used by the British Geological Survey's (BGS) Geochemical Baseline Survey of the Environment (G-BASE) project (Johnson *et al.*, 2005). This has been a long running high-resolution geochemical mapping programme to establish a geochemical baseline for the land area of Great Britain and Northern Ireland. This is achieved principally by the collection of drainage samples (stream sediments and waters), but other environmental samples such as soils are also collected. Rock samples are not routinely collected but procedures described here would equally apply to a range of sampling media. Quality control throughout the sampling and analytical phases has always been identified as crucial to the production of reliable environmental data and has been an important part of the programme since its inception (Plant, 1973; Plant and Moore, 1979; Plant *et al.*, 1975). These procedures are applicable whether for specific site investigations or regional mapping and are in addition to any internal procedures the analytical laboratory may operate as a condition of its accreditation certification.

Quality control commences in the planning phase of a project and discussion here begins with decisions that need to be made before any samples are collected or analysed.

2. PLANNING QUALITY CONTROL—QUALITY ASSURANCE

Quality assurance is the planned and systematic activities necessary to provide adequate confidence that the service (or product) will satisfy given requirements for quality. Quality control is the combination of operational techniques and activities that are used to fulfil the requirements for quality (Potts, 1997). Such activities must be planned at the outset of any project.

2.1. Appropriate and well-documented procedures

If environmental data are to be fit for the purpose for which they are to be used, then the media and methods of sampling and chemical analysis must be appropriate to the objectives of the project. Additionally, the environmental samples collected and the subsequent analyses are a long-term national asset, particularly when undertaken by public-funded organisations, and every effort should be made to ensure procedures conform to nationally and internationally recognised standards. This is the sentiment expressed by Darnley et al. (1995) in the IGCP (International Geological Correlation Project) 259—'A global geochemical database for environmental and resource management'. Although IGCP 259 was concerned with the production of a global database, the recommendations covering all aspects of sampling through to data management form a useful generic guide that should be referenced when planning any environmental sampling programme. At a project level, there should be procedures manuals that should be referenced in the meta-data attached to the database of results. Users of the environmental data can then assess for themselves whether results are appropriate for their needs. The G-BASE project has a field procedures manual (Johnson, 2005), a field database manual (Lister et al., 2005) and a data conditioning manual (Lister and Johnson, 2005). The Geochemistry Database now includes a sampling protocol code which references the version of the field manual current at the time of sampling.

Environmental site investigations in recent years have largely been legislatively driven. In Europe, this has been in response to European Commission environmental directives (e.g., Water Framework Directive and Sewage Sludge Directive (EC, 1986, 2000)) or national legislation such as the United Kingdom's Environmental Protection Act (1990) Part IIa (DETR, 2000). Such legislation means that laboratories produce results to a certified standard to guarantee the quality of the results such as MCERTS accreditation. MCERTS[1] is the England and Wales Environment Agency's Monitoring Certification Scheme (Environment

---

[1] www.mcerts.net

Agency, 2006). The scheme provides a framework within which environmental measurements can be made in accordance with the Agency's quality requirements and this includes documentation of the sampling and analytical procedures. Internationally, laboratories use the ISO/IEC 17025 Standard (British Standards, 2005) to implement a quality system aimed at improving their ability to produce valid results consistently. As the Standard is about competence, accreditation is simply formal recognition of a demonstration of that competence.

2.2. Sample numbering

A simple but fundamental requirement in any environmental sampling programme is that the sample can be identified with a unique and meaningful identity referred to here as the sample number. Potentially, sample numbering errors present one of the highest risks to the production of high-quality data. A good sample numbering system covering the way in which sample numbers are attributed and labelled is a very important part of ensuring a reliable output of data. When an organisation carries out many discrete projects, the work must conform to a sample numbering system otherwise there is high likelihood of duplicate sample numbers, particularly as the majority of projects will commence sampling using sample number 1.

The BGS uses a combination of field codes to give a sample a unique identification (Fig. 5.1) as prescribed by the rules governing sample numbering for the BGS corporate Geochemistry Database (Coats and Harris, 1995; Harris and Coats, 1992). The complete sample number is composed of (i) a single digit numeric sample numbering system code that distinguishes between the main programme areas of work that generate geochemical results (i.e., mineral exploration, regional geochemical mapping or environmental investigations); (ii) a 1–3 character alphanumeric code known as the project or area code that identifies the area being sampled; (iii) a site number recorded as an integer number from 1 to 99999; and (iv) a single character alphabetic sample type code, for example, 'S' for soil, 'W' for water or 'R' for rock. These field codes can be used to search and retrieve samples from the BGS database that currently holds more than half a million sample records.

The field sampling protocol should ensure that there are quality control measures in place to check that the sample identities attached to samples are both legible and permanent before their dispatch to laboratories. An analyst misreading an ambiguous sample label is a common source of error. Sample-labelling problems can be dealt

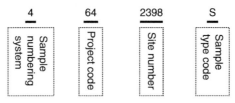

Figure 5.1 Example of a Geochemical Baseline Survey of the Environment (G–BASE) sample number defined by four key fields to give a unique sample identity.

with rapidly if the samples are accompanied by sample lists. Such lists are essential to create an audit trail, so the point at which errors are introduced can be established.

A further quality control measure that needs to be initiated at the planning phase of the project is the use of a random numbering system when collecting the samples. Systematic sample preparation and analysis errors can more easily be identified if the samples are collected in a random number order then prepared and analysed in sequential number order (Plant, 1973). Figure 5.2 illustrates this point showing that when collecting samples in sequential number order it is more difficult to distinguish

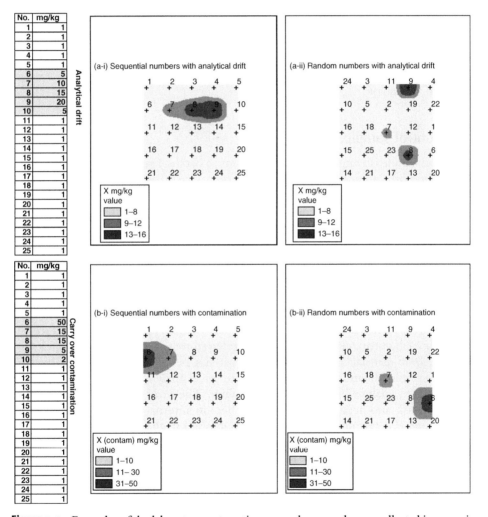

Figure 5.2 Examples of the laboratory systematic errors when samples are collected in numeric and random number order. (a) Shows error due to analytical drift and (b) 'carry over' contamination during sample preparation after the preparation of a 'high' sample. The random numbering of samples makes it easier to identify systematic laboratory errors. If the samples are collected and analysed in sequential order the results look like naturally occurring anomalies while the random numbering method tends to distinguish such errors as isolated highs.

systematic laboratory errors from naturally occurring trends than if the sample numbering was random. When using a random number sampling method it is important that the laboratories are requested to analyse the samples in numerical order.

A key to quality control of environmental analyses is the insertion of 'blind' (hidden) control samples (duplicates, replicates and reference materials) among the routinely collected samples. The control samples need to be allocated sample numbers that make them indistinguishable from the normal samples when submitted for analysis. This can be achieved by the use of sample number list sheets such as that illustrated in Fig. 5.3.

There should also be a field in the field database that indicates which sample numbers are control samples, though this suffix to the sample number should not be given to the analysts. In the G–BASE field database, there is a field called SAMP_STD which is populated with the codes shown in Table 5.1. This field can be used to quickly retrieve all control samples from the data set when data are stored in a database.

2.3. Control samples

A control sample is a sample that is inserted into a batch of samples during the process of sampling or analysis for the purpose of monitoring error, precision and accuracy. There are four main categories of control samples. These are

(i) *Duplicate*: A duplicate sample is collected from the same site as another sample in a manner defined by the project's sampling procedures manual. It is a control sample that can be used to show the amount of variability in results that can be attributed to the process of sampling by collecting two samples from the same location. A duplicate sample collected in the field is also referred to as a 'field duplicate'.

(ii) *Replicate*: This is a control sample created in the laboratory by dividing a sample into two identical parts according to a well-defined protocol. It is used to help define laboratory error. It can also be referred to as a 'sub-sample' (Fig. 5.4).

(iii) *Reference material*: This is a sample that has been prepared and analysed to acceptable documented procedures to give analytical results that through repeated analysis become accepted values. They can be used to indicate the precision and accuracy of results. When an international certified reference sample is determined, results can be published in the context of a sample for which results are recognised internationally. Reference materials should be of a similar composition to the samples being analysed. Reference materials can be subdivided into primary and secondary reference materials (PRMs and SRMs). The PRM is an international reference standard, whereas the SRM will generally be a project created reference material which is submitted at more frequent intervals and blind to the analyst.

(iv) *Blank*: This is a control sample generally only submitted for water analysis where better than 18 MΩ quality water is handled and included with a batch of normal samples. This helps to indicate any trace contamination that may be introduced during the handling, bottling, acidification or storage parts of the

Random number list 1
sediment

Project code.........

Number range......................

18	01	36	29
49	99	70	73
46	03	59	43
41	38	88	82
32	91	66	55
45	67	64	14
94	07	52	87
98	34	79	06
56	89	05	12
15	83	60	92
26	95	08	02
19	21	96	63
39	84	25	31
28	93	47	53
54	40	100	27
62	71	24	30
80	57	77	11
16	61	09	76 Duplicate A
17	48	85	81 Duplicate B
72	35	50	86 Sub sample A
65	13	33	78 Sub sample B
37	51	42	68 Standards
04	97	20	22 Standards
90	74	58	10 Blank waters
69	23	44	75 Blank waters

Figure 5.3 Example of a Geochemical Baseline Survey of the Environment (G-BASE) random number list used for issuing site numbers with 8% of the numbers reserved for duplicates, replicates (sub-samples), standards and blanks. The blank column is used to record details of the date and sampling pair assigned to each number.

water sampling and analysis procedures. G-BASE project submits two blank waters in every hundred water samples, one of which is filtered (as the samples are) and one which is unfiltered blank water.

Table 5.1 Table showing the codes used in the Geochemical Baseline Survey of the Environment (G-BASE) field database to identify control samples

Code	Sample description
DUPA	Duplicate A (original sample)
DUPB	Duplicate B (collected at same site as Dup A)
DUPC	Duplicate C (original sample)
DUPD	Duplicate D (collected at same site as Dup C)
REPA	Sub-sample A (laboratory replicate of DUPA)
REPB	Sub-sample B (laboratory replicate of DUPB)
REPC	Sub-sample C (laboratory replicate of DUPC)
REPD	Sub-sample D (laboratory replicate of DUPD)
STD	Secondary reference material (SRM)
BW	Blank water used only for W

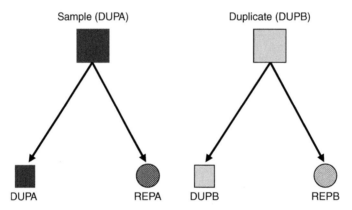

Figure 5.4 Figure showing the relationship between field duplicates and the laboratory replicates.

The number of control samples required for each batch of analyses depends on a number of factors, including the number of samples to be analysed in relation to the number of analytical batches, whether analytical batches are analysed over a continuous short period or intermittently over a longer period and the data quality requirement to which the data is being applied. For the global geochemical database for environmental and resource management, Darnley *et al.* (1995) recommend the inclusion of 3% duplicate samples and 4% SRMs. Eight percent of the samples the G-BASE project submits for analysis are control samples (see Fig. 5.3). This project generally collects 1000s of samples each field campaign and the number of control samples is sufficient to give substantial quality control data, particularly for graphical plots that require a range of concentrations to give meaningful linear regressions. When the project has been involved in smaller sampling projects collecting only tens or hundreds of samples the number of duplicates and replicates is increased to 8 per 100.

 ## 3. Raw Data Checking

3.1. Data checking

The initial phase of data checking consists of simple and obvious procedures that need to be carried out as the first stage of acceptance of the results from the laboratory. This needs to be done systematically and as soon after the results are received as is possible. The original digital file of results as received from the laboratory should always be archived unchanged and the subsequent modifications described in the following sections performed on a copy of the original data file.

(i)   Has the laboratory reported all the analytes it was contracted to do by the methods agreed in the contract?
(ii) Does the number of samples submitted for analysis correspond with the number of data records received and do all the sample numbers correspond to the sample number list submitted?
(iii) Are the results given with the correct units, absent, outside limit or not detected analyses reported correctly? Have elements reported as oxides been reported as percentages? Reporting results as a mixture of percentage element oxide and element milligram per kilogram can be misleading, particularly for non–geochemists interpreting the results.
(iv) Has the laboratory correctly reported control samples that were not blind to the analyst?
(v)   Does the range of values for each element reported look sensible for the area sampled?
(vi) If the samples were collected in random number order but analysed in sequential number order is it possible to identify any analytical drift or carry over contamination from high samples?

3.2. Dealing with missing, semi-quantitative and unreliable data

During the initial data checking phase missing, non-numeric, coded and unreliable results need to be dealt with. To correct bad data you first must find them (Albert and Horwitz, 1995). At this stage, it is simply a case of looking at the data to identify obvious misfits such as results that are way out of expected ranges or control samples that clearly do not conform to expected values. If there is a problem with control samples, then the reliability of the complete data set must be questioned and the matter taken up with the analyst. There is little point proceeding with any data conditioning unless the results can be deemed acceptable. However, the acceptance of the results at this stage is conditional on precision and accuracy tests discussed later in this account. Suspicious analyte determinations can be tagged at this stage with a qualifier or removed altogether until the interpretative phase of the project is complete. The G–BASE project qualifies such analyte determinations with a '*' in the final database so users can see that there is a documented data quality issue associated with the result. Outlying results may have no explanation as to their cause but are still of great value to the data user.

Different laboratories use different codes to represent missing or semi-quantitative data, examples from the BGS XRFS (X-ray fluorescence spectroscopy) laboratory are shown in Table 5.2. The Analytical Methods Committee of the Royal Society of Chemistry makes a distinction between the *recording* of results and the *reporting* of results (AMC, 2001a, 2001b). Recording of results just as they come, including negative and below detection values, is how the laboratory produces the data. The reporting of results, that is, the data passed onto the customer, is a contractual matter between the analyst and the customer. Most types of statistical processing of data sets containing low concentrations of analyte should be undertaken on the unedited data, that is, the data should be reported in the same way as it is recorded.

For a site investigation, where the question might be whether or not a result falls above or below a statutory level, the problems of below detection limit values may seem of little consequence if the method of analysis can satisfactorily determine results to the level of the statutory value. However, if an assessment is to be based on the average value of a data set it is important that no bias is added to the results by substituting below detection limits or over-range values with a constant that may introduce a high or low bias into the censored (that is modified) data. The laboratory must produce a statement of analytical uncertainty that would usually include a value for each element below which results have a high uncertainty, that is, a detection limit. It is important that such information is passed on to the data user. With modern analytical instrumentation delivering digital results via complex analytical software it is difficult to often pin the analyst down to a single limit of detection. For example, some elements determined for the G-BASE project are done by ED (energy dispersive)–XRFS, and this method will record variable detection limits for each element for every sample depending on the compositional matrix of the sample medium making it difficult for the analyst to cite a single detection limit for the entire analytical batch.

The low-detection limits as reported by analysts are generally more conservative than are the actual limits. This is suggested by the slope of the curve on the data

Table 5.2 Table showing the codes used by the British Geological Survey X-Ray fluorescence spectroscopy (BGS XRFS) laboratory to represent absent or semi-quantitative results

Code	Comment
−94	Insufficient sample (e.g., sample collected but not enough to analyse)
−95	Not determined because of high concentration, but exceeds calibration limit
−96	Not determined because of interference; probably of high concentration
−97	Not determined because of interference; probably of low concentration
−98	Not determined because of interference; no estimate
−99	Absent data (e.g., not requested, sample not submitted or sample lost)

shown in Fig. 5.5. This data represents Al concentrations in the same 10,000 samples, determined by two different methods of analysis. The slope of the graph suggests no sharp change in curvature until a concentration well below the reported detection limit. This indicates that there may be some regional pattern which can be usefully obtained from this data, and that the practical detection limit may be lower than has been reported by the laboratory. However, it should be recognised that the actual concentrations have much higher uncertainties associated with them when concentrations are low with respect to the detection limit. There are statistical plotting methods that can be used to formally determine more realistic levels of detection. These include a method that plots duplicate difference against duplicate means (Thompson and Howarth, 1973, 1976, 1978) and determining limits of analytical detection by identifying bottom truncation on cumulative probability plots (Sinclair, 1976). However, these graphical methods do require a good range of element concentrations with at least some values approaching or below detection, such as is shown in Fig. 5.5 where bottom truncation is indicated by the curve becoming a horizontal line.

Historically, the G–BASE project has replaced values recorded as 'detection' by a value one half the detection limit for soils and stream sediments and for consistency with older data. This practice continues. There is no sound basis for such a remedy

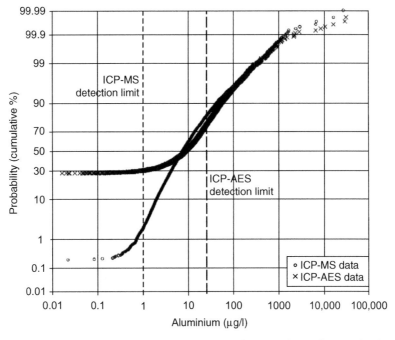

Figure 5.5 Cumulative probability plot indicating true detection limits for samples determined by two different analytical methods. Such plots can demonstrate that the real detection limits are often much lower than those cited by the analyst (the cited detection limits are marked on the plot). Data for 10,000 samples from part of central and eastern England.

except the acceptance that the real value probably lies between the detection limit and zero and the data needs to have numeric rather then semi-quantitative or alpha-numeric values in order to use the data in statistical analyses. For stream waters, for which there is a shorter history of collection and analyses for a wide range of analytes, data are not modified by the application of the detection limits before storage in the geochemistry database. The benefit of this lies in the lack of bias introduced into any statistical tests for which the data are used. Clearly, this does not preclude later censoring of the data where required by the user. Inherent in this approach is the requirement to have a field in the database specifying the detection limit for each analyte in each analytical batch. For instance, the approaches described above to assess the reported detection limit against the data supplied could not possibly be undertaken on censored data.

Current procedures emphasise the importance of storing the laboratory recorded data ('raw data') and the XRFS analyses are now directly transferred via a Laboratory Information Management System (LIMS) to data tables in the BGS corporate geochemistry database. Conditioned data is loaded to different data tables and importantly every analyte result has an accompanying qualifier field that can be provided to the user to explain any data quality issues.

So far, all outside limit discussion has been concerned with the lower rather than the upper detection limit. Similar procedures apply with the data being qualified with a code and the value set to the upper detection limit. Generally, semi-quantitative 'above detection' values are less of a problem than the 'below detection' values. This is because the analysts are better able to find simple remedies for very high results than they are for very low results. For example, when an extractive procedure is used less sample can be used in the analysis, or the final solution being determined can be diluted. With low concentrations increasing the sample weight for analysis or reducing dilution results in increased interference problems. The upper limit of elemental totals should not exceed 100%, though in reality uncertainty in results for abundant oxide values (e.g., silicon) can lead to totals >100%.

4. STATISTICAL ANALYSES AND PLOTTING OF CONTROL SAMPLE DATA

Once the preliminary error checking of the raw data has been done, the control samples should be separated from the normal samples for more detailed examination. This process of separation is greatly aided by the inclusion of the STD_SAMP field in the field database (see earlier) and a comprehensive sample list that identifies control samples and their relationships (Fig. 5.3). Control sample results can then be subjected to a number of statistical and plotting procedures that determine the accuracy and precision of results. These processes give an indication of the levels of uncertainty that are associated with the results, information that is essential to interpret the data and present it in a meaningful manner.

4.1. Control charts

G-BASE uses control charts (Shewhart plots) that are time sequenced quantitative data plotted as a graph that has fixed defining limits (Miller and Miller, 2005). The upper and lower limiting lines are usually defined as the mean (accepted) value for an analyte ± 3 standard deviations. Both PRMs and SRMs are repeatedly analysed over a period by being included in every batch of samples submitted for chemical analysis. The results can be plotted on the control chart to see how close the result falls to an accepted value and whether any analytical drift is recognisable over a sequence of successive analytical batches. The G-BASE project has used a number of software packages to plot control charts, the simplest being MS Excel. Specialist software packages such as the MS Excel add-in SPC XL (Fig. 5.6) and QI Analyst (Fig. 5.7) will flag values that exceed the upper or lower limits or whether there is an upward or downward drift in results. If a reference material result is suspect, that is, it plots outside the defined limits or a significant shift or drift in values is shown, then all the results from the analytical batch should be treated as suspect and, in the absence of any satisfactory explanation, the whole analytical batch should be reanalysed.

4.2. Duplicate–replicate plots

An effective but simple way of graphically illustrating the variability associated with the analytical data is to plot x–y plots of the duplicate and replicate pairs. Most statistical packages will have an option for plotting simple x–y plots. The G-BASE project uses MS Excel running a macro that will automatically plot duplicate-replicate and duplicate-duplicate results. Figure 5.8 shows three examples from the G-BASE East Midlands atlas area duplicate–replicate data for soils. This method gives an immediate visual appreciation of any errors present in an analytical batch and an indication of 'within site' variability, as shown by the duplicate pairs, or the 'within sample' variability, as indicated by the replicate pairs that demonstrate

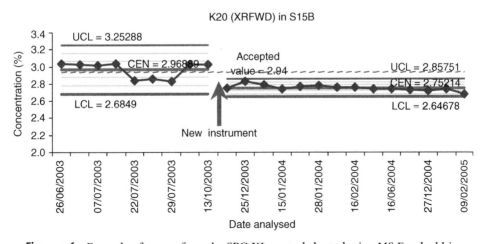

Figure 5.6   Example of output from the SPC XL control chart plotting MS Excel add-in.

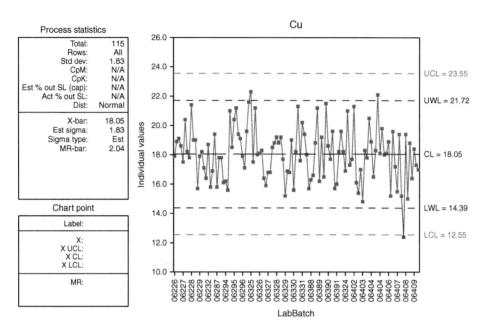

Figure 5.7 Control chart plot using QI Analyst software (Cu in mg/kg on *y*-axis).

analytical uncertainty. In the examples shown in Fig. 5.8, the uranium results show a broad spread about the line of gradient 1 (the lines in Fig. 5.8), particularly around the detection limit, indicating that for this element there is high sampling and analytical variance. The iodine results show generally low analytical variance (shown by the dup-rep plot) but have higher within site variability, particularly at higher concentrations. The copper data shows a good correspondence with the line of gradient 1 suggesting that for this element both the within site and within sample (i.e., analytical variability) are low.

In instances where only a few duplicate–replicate pairs are available, individual results can still be plotted to see where they fall relative to the line of gradient 1. For larger surveys, collecting many duplicate–replicate pairs, a good spread of values can be produced, giving plots such as those shown in Fig. 5.8.

Thompson (1983) and Thompson and Howarth (1978) describe a method of estimating analytical precision using duplicate pairs. This is not a procedure routinely used by the G-BASE project but is a particularly useful way of estimating the analytical precision when no truly representative reference materials are available.

4.3. Hierarchical analysis of variance

In the previous section, the duplicate–replicate control data set was used to give graphical representation of sampling and analytical variability. A statistical procedure referred to as analysis of variance (ANOVA) analysis can be done on the same data set to give a more quantitative statement on variability. Sinclair (1983) describes this method that compares variations that arise from different identifiable sources,

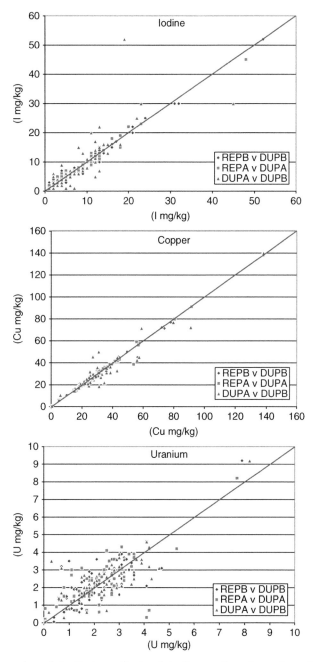

Figure 5.8 Examples of duplicate–replicate plots from soil analyses from the Geochemical Baseline Survey of the Environment (G-BASE) project generated by an MS Excel macro (for the relationship of control samples, see Fig. 5.4). Note that the cited detection limits for Cu, U and I are 1.3, 0.5 and 0.5 mg/kg, respectively.

namely analytical error, sampling error and regional variation. This procedure is therefore an important test of a project's strategy. It would be hoped that the analytical (within sample) variations and the sampling (within site) variations are small relative to regional variations. If not, then the analytical and sampling methodologies need to be reviewed. In many instances, particularly where the natural levels of an element are at or below the lower analytical detection limit, it is not possible to make improvements to the methodology. It is important that users of the data are made aware that for some elements the analytical variance accounts for a high percentage of the total variance in the results. The table that attributes variation generated by the ANOVA (Table 5.3) is an excellent way of disseminating such information.

The G-BASE project has used several statistical packages to perform this nested ANOVA analysis (e.g., Minitab and SAS). It currently uses an MS Excel procedure with a macro based on the equations described by Sinclair (1983) in which the ANOVA is performed on results converted to $\log_{10}$ (Johnson, 2002). Ramsey *et al.* (1992) suggest that the combined analytical and sampling variance should not exceed 20% of the total variance with the analytical variance ideally being $<4\%$.

5. LEVELLING DATA

The final part of the data conditioning procedure is the process of making the data fit with existing data sets. The accuracy of small data sets derived from a single analytical batch, and not used along with other geochemical data, can be put into context simply by publishing the certified international reference materials results. However, if a data set is composed of many analytical batches analysed over a period of months or even years, then the data will need to be levelled between batches to ensure the seamless creation of geochemical images. An example of a geochemical image where results have not been levelled is shown in Fig. 5.9. An annual field campaign boundary can be seen.

Another issue concerning the levelling of results can arise from improvements in the analytical lower detection limits. Such improvements will give much greater resolution in the data at low values which will be absent from older data with a poorer detection limit. In such instances, it is difficult to produce geochemical maps and images when the different data sets have differing degrees of resolution. Indeed, levelling may not be desirable as it would involve degrading the results that have the better resolution. The G-BASE approach to dealing with this is described later in this section. Finally, a problem very common to any project that requires building a national geochemical database is the requirement to combine geochemical data generated by different analytical methods.

A good discussion of the levelling of geochemical data sets using the mathematical process of normalisation is given in Darnley *et al.* (1995). This work describes how the term normalisation is used in a mathematical sense, that is, 'to adjust the representation of a quantity so that this representation lies within a prescribed range (Parker, 1974), or, any process of rescaling a quantity so that a given integral or other functional of the quantity takes on a pre-determined value (Morris, 1991), rather

Table 5.3 Example of nested analysis of variance (ANOVA) of stream sediment control samples from the Geochemical Baseline Survey of the Environment (G-BASE) sampling of east Midlands, UK

Element	Between site %	Between sample %	Within sample %
Na_2O	88.71	6.53	4.76
MgO	97.18	2.43	0.39
Al_2O_3	96.34	3.45	0.21
SiO_2	97.53	2.28	0.19
P_2O_5	91.83	7.98	0.19
K_2O	98.46	1.45	0.10
CaO	96.61	3.16	0.22
TiO_2	97.43	2.45	0.12
MnO	92.79	6.81	0.40
Fe_2O_3	94.25	5.25	0.50
Sc	92.02	2.83	5.15
V	95.55	4.00	0.45
Cr	96.62	2.54	0.84
Co	83.79	7.78	8.44
Ba	96.96	2.71	0.33
Ni	92.73	6.40	0.87
Cu	96.58	3.08	0.34
Zn	94.96	4.88	0.16
Ga	95.83	3.48	0.68
Ge	65.50	8.82	25.68
As	93.86	5.30	0.84
Se	88.84	3.85	7.31
Br	89.50	9.96	0.54
Rb	97.49	1.81	0.70
Sr	96.81	2.84	0.35
Y	93.35	5.96	0.69
Zr	94.97	4.74	0.28
Nb	97.89	1.55	0.56
Mo	76.58	−0.51	23.93
Hf	82.13	9.19	8.68
Ta	29.89	−13.36	83.47
W	58.12	−6.52	48.39
Tl	12.97	6.49	80.54
Pb	97.37	2.04	0.59
Bi	34.61	−12.75	78.14
Th	96.24	1.42	2.35
U	70.81	−0.92	30.11
Ag	70.49	0.29	29.22

(continued)

Table 5.3   (*Continued*)

Element	Between site %	Between sample %	Within sample %
Cd	87.71	5.36	6.94
Sn	88.49	5.62	5.89
Sb	65.07	6.89	28.05
Te	−2.65	0.00	102.65
I	84.27	11.79	3.94
Cs	72.25	−1.37	29.11
La	88.98	4.26	6.77
Ce	94.28	1.08	4.64

Note: The table was generated using an Excel procedure (Johnson, 2002). Elements where <80% of the variance is attributable to regional variations are highlighted, and results for these elements should be treated with caution.

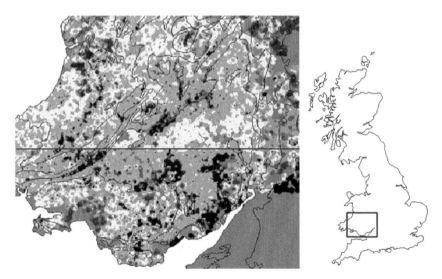

Figure 5.9   Example of a geochemical image with unlevelled data (copper in stream sediments from Wales). The horizontal E–W line is shown just above the southern extent of the 1988 field campaign boundary. A subtle but noticeable linear feature can be seen coincident with the boundary between samples collected in different years. The results in the south had to be reduced by levelling in order to remove this analytical artefact.

than in the statistical sense, where it connotes a transformation of a data set so that it has a mean of zero and a variance of one'. Normalisation of the secondary reference material results gives levelling factors that are applied to the data to give, ultimately, a single discrete national G–BASE data set.

The G–BASE project predominantly uses parametric levelling as described in the following section. When comparing geochemical results analysed by different

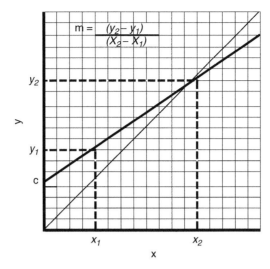

Figure 5.10   Cartesian equation $y = mx + c$ representing a straight line intercepting the y-axis at c with a gradient of m. The line shown passing through the origin has gradient $m = 1$.

geochemical methods, or by the same geochemical method but at a different time one can anticipate that the results will show a proportional relationship between the two results x and y. The simplest model is a directly proportional relationship between x and y represented by a straight line and the Cartesian equation $y = mx + c$ illustrated in Fig. 5.10. Secondary reference materials included in every analytical batch over a long period can be plotted as x–y plots where the 'accepted' value (i.e., a result determined by repeated analysis) is plotted as the y value against the determined value for a particular batch plotted as the x value. G-BASE submits samples in batches of 500 including sufficient SRMs that have a range of elemental values allowing a linear graph to be plotted (Fig. 5.11). From the linear regression modelling of the straight line, the derived Cartesian equation can then be used to determine levelling factors for each element.

5.1. Between batch and between field campaign data levelling

The between batch and field campaign levelling is best illustrated by the use of a real example. A potassium control plot for G-BASE secondary reference material S15B (determined by XRFS) over a period of 19 months is shown in Fig. 5.6. This shows that although all results fell within accepted limits of analytical variability (which is generally high for major oxides determined by XRFS on pellets); there are significant analytical shifts observable during this period. The most noted change came when the XRF machine was replaced, but before that there was also a period when results were consistently lower. The data therefore needs to be divided into three groups: (1) high values/old instrument, (2) low values/old instrument and (3) values from new instrument; and then levelled against the accepted result for K_2O in S15B. Although the results for S15B are shown here, other SRMs were analysed in the same batches and exhibited similar shifts. The data for all reference materials can be

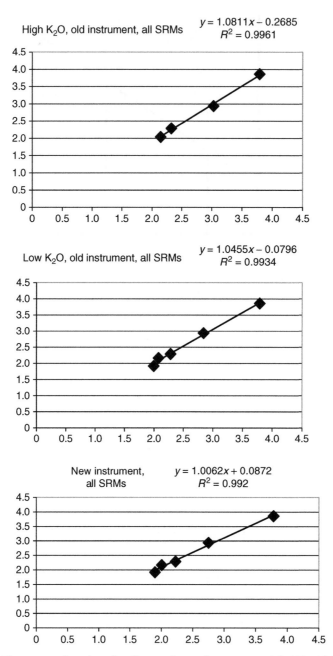

Figure 5.11 Liner regression plots for all secondary reference materials (SRMs) for potassium. The three data subsets relate to shifts indicated in Fig. 5.6. Axis units are % K_2O with the accepted value plotted as y and the batch value plotted as x.

plotted as shown in Fig. 5.11. The regression analysis (equations shown on Fig. 5.11) gives the levelling factors that need to be applied to bring the results to the level of the 'accepted' SRM results. For example, for the batches of high K_2O results on the old instrument (top graph in Fig. 5.11), all K_2O results should be multiplied by 1.0811 and 0.2685 subtracted. These factors can be applied to their respective data subsets and the results 'levelled' as shown in Fig. 5.12.

5.2. Levelling data with differing lower limits of detection

Variations (usually decrease) in detection limits occur with time, affecting both long-running projects, and the comparison of time-series data or adjacent project areas separated in time. It arises as analytical methods improve, and has its greatest impact on the trace elements, where the natural abundance is low in relation to the lowest measurable concentration. Such improvements can significantly increase the number of sample locations with measurable values in comparison to older data. The ability to make use of all the data acquired to its best potential is important, particularly for national mapping programmes. The data can be levelled as described above, if the standards used fall above the detection limit of the older method.

Illustrations of two sets of data with different lower levels of detection can be problematical as one set will show greater resolution of low results than the other. Presentation must be done in such a way as to not downgrade the best data (by applying the highest detection limit to all results), nor giving a false impression (by suggesting that samples below two very different detection limits represent the same data distribution). The example in Fig. 5.5 shows that a substantial amount of useful data occurs below the ICP-AES detection limit of 14 µg/l Al and that the ICP-MS data should not be truncated at that concentration. The graph also shows

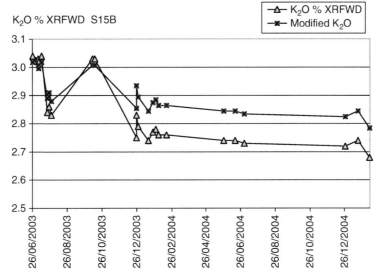

Figure 5.12   A graph showing the effect of levelling K_2O values for secondary reference material (SRM) S15B using the levelling factors determined in Fig. 5.11.

that to represent all the data at <14 µg/l Al in the same way as samples below the ICP-MS detection limit of <1 µg/l Al would be a false representation. This process is best illustrated by the techniques used to combine stream water data for England and Wales, exemplified by a significant lowering of the detection limit (in relation to the natural abundance of Al) caused by a change in analytical method. The earliest data, from Wales, was acquired by Inductively Coupled Plasma Atomic Emission Spectroscopy (ICP-AES). However, the advent of routine Inductively Coupled Plasma Mass Spectrometry (ICP-MS) analysis of stream water samples led to a marked increase in the number of samples with measurable Al, with only ∼2% of data below the detection limit, compared to ∼50% by the ICP-AES. Figure 5.13 illustrates the method used by the G-BASE project to overcome these very different data distributions, using stream water Al data. The ICP-AES data for Wales is plotted using the same percentile scale as is derived for the ICP-MS data in England, until the detection limit of 14 µg/l is reached. Beneath that concentration, data are shown in a dark grey colour; this is to demonstrate that the measurement of the sample has been made in that location, but that the information it provides is not as detailed as occurs in the more recent data.

5.3. Levelling data determined by different analytical method

Over a long period, it is inevitable that analytical methods will improve and change. The G-BASE project made significant changes in the 1990s when Direct Couple Optical Emission Spectroscopy (DCOES) was replaced as the principal method of analysis for sediments and soils by XRFS. This occurred during the regional geochemical mapping of Wales (BGS, 2000). Since this major change in analytical methodology geochemical images of the United Kingdom are plotted with data levelled with reference to the Welsh XRFS data. The levelling procedure removes features on maps that would solely be attributed to different analytical methods.

The levelling was achieved by a combination of techniques. The first of which involved a similar procedure to that described above to level between analytical batches, namely, regression analysis of SRM samples that had been determined by both DCOES and XRFS. Levelling factors for elements were further refined by plotting elemental results, DCOEC versus XRFS for some 3000 samples analysed by both methods. A final check of the levelling was then done by using a percentile regression method (see Darnley *et al.*, 1995) in which the percentile distributions of both methods are compared and adjusted until they match.

6. DISCUSSION

Data conditioning is a collection of processes that makes data suitable for the purpose for which it is to be used. Some or all of the processes described may be applied to the data, depending on the way it is to be used. Its importance is illustrated by geochemical images such as that shown in Fig. 5.13 where geochemical data collected over a long period, and determined by different analytical methods, can be presented as a single geochemical image without additional analytical or temporal artefacts. At a local or site scale, it is important that geochemical results are subjected

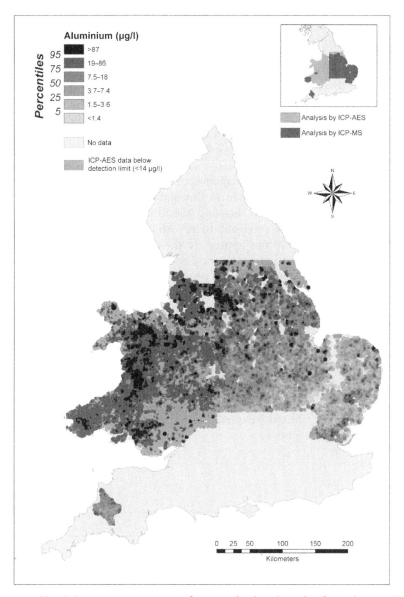

Figure 5.13 Aluminium in stream waters from England and Wales from the Geochemical Baseline Survey of the Environment (G-BASE) project illustrating how data sets with different lower detection limits can be combined.

to a similar level of conditioning, so they can readily be interpreted in the context of a regional setting.

In the preceding sections, the importance of documenting procedures and passing on quality information to the end user has been repeatedly stressed. Two areas not covered by the previous discussions are management responsibilities and user-appreciation of quality control information. The former requires good coordination and communication from the planning phase to data delivery. Lack of understanding on how the customer uses the data may negate all the effort put into making the data fit for purpose.

Worldwide there is an understandable perception that collecting and preparing samples would generally be considered as a 'labouring' job, whereas chemical analysis and data interpretation are more intellectual tasks. This perception must be managed carefully. The reality is that the sampling and sample preparation are the phases at which data quality is at the greatest risk. The efforts and the value of the samplers and sample preparation staff must not be underestimated in the final outcome of the project. Good management requires that all those involved in the project are aware of all the procedures throughout the project. A sampling team which, for very good reason and showing good initiative, may decide to collect an additional sample from a site and label them with identical site numbers with an 'A' and 'B' suffix will probably not appreciate the dilemma this produces for a field database that only accepts unique numeric sample numbers. Everyone must be made aware that small, unimportant tweaks to procedures can have serious impacts later in the programme.

Finally, the effort to assess the quality of the data is wasted if the information is not passed on to the data user. Such users may not be a scientist, so effort must be made to present the data quality information in a manner that they can understand. The G-BASE project when distributing data to users makes use of font formatting, colour coding and highlighting in Excel spreadsheets to pass on information on quality, and Fig. 5.14 is a key to such an Excel document. This gives the user data in a completely numeric format with simple and straightforward advice about how it should be used. It is then up to the data user to choose whether to ignore the advice.

Colour/ format	Signifies	User action
56.7	No quality issues	None
25.4	This result is associated with a qualifier	Be aware that the result is qualified
Red	Data of dubious quality with significant issue(s)	Pay careful attention to what the quality issue is and if necessary don't use results
Yellow	Results <= 0	Be aware that the result could give problems in some statistical or plotting packages
Grey	<null> value	Be aware that no result is present though transferring to some software packages could erroneously reset this to 0
25 10 5 0.6	Generally a data issue relating to representation of results above or below limits of detection	Be aware that this result has some quality issue but is unlikely to restrict its use

Figure 5.14   A key to the colour–coded quality information sent out to data users when the data is distributed in an Excel spreadsheet. Note that when produced in colour the colour-fill is used but is shown above by the text red, yellow and grey in the data cells. Similarly italic font in different colours (only shown in black/grey above) are used to indicate different data quality issues.

ACKNOWLEDGMENTS

This chapter is published with the permission of the director of the British Geological Survey (NERC). The authors also wish to acknowledge the efforts of several generations of geology students who have worked as 'voluntary workers' collecting samples for the G-BASE project. Their efforts working to strict procedures have ensured that the G-BASE project output is based on data of high quality and reliability. We appreciate the constructive reviews from C. A. Palmer and an anonymous referee.

REFERENCES

Albert, R. H., and Horwitz, W. (1995). Incomplete data sets: Coping with inadequate databases. *J. Assoc. Anal. Chem. (AOAC) Int.* **78**(6), 1513–1515.

AMC (2001a). What should be done with results below the detection limit? Mentioning the unmentionable. *In* "AMC (Analytical Methods Committee)" Technical Brief No 5. Royal Society of Chemistry, United Kingdom.

AMC (2001b). Measurement of near zero concentration: Recording and reporting results that fall close to or below the detection limit. *Analyst* **126**, 256–259.

BGS (2000). Regional geochemistry of Wales and part of west-central England: Stream sediment and soil. *In* "British Geological Survey." Keyworth, Nottingham, UK. ISBN 0 85272 378 4.

British Standards (2005). General requirements for the competence of testing and calibration laboratories. BS EN ISO/IEC 17025: 2005. ISBN 0 580 46330 3.

Coats, J. S., and Harris, J. R. (1995). Database design in geochemistry: BGS experience. *In* "Geological Data Management" (J. R. A. Giles, ed.), No. 97, pp. 25–32. Geological Society of London Special Publication. London, UK.

Darnley, A. G., Bjorklund, A., Bolviken, B., Gustavsson, N., Koval, P. V., Plant, J. A., Steenfelt, A., Tauchid, M., and Xuejing, X. (1995). A global geochemical database for environmental and resource management UNESCO publishing, 19.

DETR (2000). "Contaminated Land Implementation of Part IIa of the Environment Protection Act." HMSO London.

EC (1986). Council Directive 86/278/EEC of 12 June 1986 on the protection of the environment, and in particular of the soil, when sewage sludge is used in agriculture. (Sewage Sludge Directive). Published in the Official Journal (OJL 181/6), 4 July 1986.

EC (2000). Directive 2000/60/EC of the European Parliament and of the Council establishing a framework for the Community action in the field of water policy. EU Water Framework Directive, 23 October 2000. Published in the Official Journal (OJL 327) 22 December 2000.

Environment Agency (2006). Performance standard for laboratories undertaking chemical testing of soil, March 2006, version 3. Environment Agency for England and Wales.

Harris, J. R., and Coats, J. S. (1992). Geochemistry database: Data analysis and proposed design. British Geological Survey Technical Report, WF/92/5.

Johnson, C. C. (2002). Within site and between site nested analysis of variance (ANOVA) for Geochemical Surveys using MS EXCEL. British Geological Survey, UK, Internal Report No. IR/02/043.

Johnson, C. C. (2005). 2005 G-BASE Field procedures manual. British Geological Survey, Keyworth, UK, Internal Report No. IR/05/097.

Johnson, C. C., Breward, N., Ander, E. L., and Ault, L. (2005). G-BASE: Baseline geochemical mapping of Great Britain and Northern Ireland. *Geochem. Explor. Environ. Anal.* **5**(4), 347–357.

Lister, T. R., and Johnson, C. C. (2005). G-BASE data conditioning procedures for stream sediment and soil chemical analyses. British Geological Survey, BGS Internal Report Number IR/05/150.

Lister, T. R., Flight, D. M. A., Brown, S. E., Johnson, C. C., and Mackenzie, A. C. (2005). The G-BASE field database. British Geological Survey, BGS Internal Report Number IR/05/001.

Miller, J. N., and Miller, J. C. (2005). "Statistics and Chemometrics for Analytical Chemistry," 5th edn. Ellis Horwood, Chichester.

Morris, C. G. (1991). "Academic Press Dictionary of Science and Technology," pp. 2432. Academic Press, San Diego.

Parker, S. B. (1974). "McGraw-Hill Dictionary of Scientific and Technical Terms," pp. 2088. McGraw-Hill, New York.

Plant, J. A. (1973). A random numbering system for geological samples. *Trans. Inst. Min. Metall.* **82**, 64–65.

Plant, J. A., and Moore, P. J. (1979). Geochemical mapping and interpretation in Britain. *Philos. Trans. R. Soc.* **B288**, 95–112.

Plant, J. A., Jeffrey, K., Gill, E., and Fage, C. (1975). The systematic determination of accuracy and precision in geochemical exploration data. *J. Geochem. Explor.* **4**, 467–486.

Potts, P. J. (1997). A glossary of terms and definitions used in analytical chemistry. *Geostand. Newsl.* **21**(1), 157–161.

Ramsey, M. H., Thompson, M., and Hale, M. (1992). Objective evaluation of precision requirements for geochemical analysis using robust analysis of variance. *J. Geochem. Explor.* **44**, 23–36.

Sinclair, A. J. (1976). Applications of probability graphs in mineral exploration. *Assoc. Explor. Geochem.* Special Volume No. 4, pp. 95.

Sinclair, A. J. (1983). Univariate analysis. *In* "Statistics and Data Analysis in Geochemical Prospecting" (R. J. Howarth, ed.). Handbook of Exploration Geochemistry, Vol. 2, pp. 59–81. Elsevier, Amsterdam.

Thompson, M. (1983). Control procedures in geochemical analysis. *In* "Statistics and Data Analysis in Geochemical Prospecting" (R. J. Howarth, ed.), Handbook of Exploration Geochemistry, Vol. 2, pp. 39–58. Elsevier Science, Amsterdam.

Thompson, M., and Howarth, R. J. (1973). The rapid estimation and control of precision by duplicate determinations. *Analyst* **98**, 153–160.

Thompson, M., and Howarth, R. J. (1976). Duplicate analysis in geochemical practice. *Analyst* **101**, 690–709.

Thompson, M., and Howarth, R. J. (1978). A new approach to the estimation of analytical precision. *J. Geochem. Explor.* **9**, 23–30.

CHAPTER SIX

Gas Chromatographic Methods of Chemical Analysis of Organics and Their Quality Control

Christopher Swyngedouw,* David Hope,† *and* Robert Lessard‡

Contents

* Consulting Scientist, Bodycote Testing Group, Calgary, Canada
† CEO, Pacific Rim Laboratories, Surrey, Canada
‡ QA Manager, Bodycote Testing Group, Edmonton, Canada

Environmental Geochemistry

DOI: 10.1016/B978-0-444-53159-9.00006-1

© 2008 Elsevier B.V.

All rights reserved.

Abstract

Analytical organic chemistry employing common gas chromatographic techniques involves dissolving the analyte in organic solvent, removing the interfering coextractives by solid phase extraction and then injecting the purified extract onto a gas chromatograh coupled to a detector. Analytical methods for the analysis of organic chemicals are widely available. This chapter provides procedures to extract, isolate, concentrate, separate, identify, and quantify organic compounds. It also includes some information on the collection, preparation, and storage of samples, as well as specific quality control and reporting criteria. This chapter attempts to summarize the best practices. A performance-based process is described, whereby individual laboratories can adapt methods best suited to their situations.

1. INTRODUCTION

Analytical organic chemistry is broad ranging in its scope, with analytes ranging from highly volatile single carbon molecules like chloroform to large macromolecules like the six ringed dibenzo[a,i]pyrene. The analytes can be aliphatic (e.g., hydrocarbons), aromatic [e.g., polycyclic aromatic hydrocarbons (PAHs)], or a combination thereof (e.g., nonylphenol). They can be halogen substituted [e.g., polychlorinated biphenyls (PCBs) and pesticides] or by carboxylic acid groups (e.g., phthalates, haloacetic acids), hydroxyl (e.g., pentachlorophenol) and amine (e.g., aniline, toluidine) moieties.

All sampling and laboratory activities are aimed at one target: the production of quality data that is reliable and has a minimum of errors. Further, reliable data must be produced consistently. To achieve this, an appropriate program of quality control (QC) is needed. Quality control includes "the operational techniques and activities that are used to satisfy the quality requirements or data quality objectives (DQOs)" (FAO, 1998).

Quality assurance (QA) programs are very important for demonstrating the performances of the analytical methods for the organics within a laboratory and between laboratories. QA requirements for organics analysis are well known and include the use of certified reference materials, field and laboratory blanks, the use of QC charts to monitor long-term laboratory performance, participating in interlaboratory studies and proficiency tests, and the use of guidelines for sampling and analysis.

This chapter includes only gas chromatography and methods of sample preparation used for this technique. Other chromatographic techniques, such as high pressure liquid chromatography (HPLC), size exclusion chromatography (SEC), supercritical fluid chromatography (SFC), or thin layer chromatography (TLC), are not covered, neither are techniques that use unextracted solids, such as pyrolysis gas chromatography.

The organic analytical techniques covered here may be considered those commonly found in routine (commercial) analytical laboratories. Some laboratories that also perform research and method development may use other techniques in addition to the ones mentioned in this chapter. These additional techniques often involve spectroscopy methods like infrared spectroscopy, nuclear magnetic resonance, and X-ray diffraction and microscopy.

2. Sample Preparation—Aqueous Samples

A number of techniques can be used to isolate analytes from water. The technique used will depend on the volatility of the analyte. Volatile compounds (i.e., more volatile than n-C12) can be analyzed using "Purge and Trap" techniques or by Headspace analysis. Semivolatile compounds are extracted using liquid–liquid or solid phase extraction techniques.

2.1. Purge and Trap

Purge and Trap (P&T) involves sparging a small quantity of water—typically 5 ml—with helium and then trapping the analytes on a sorbent phase (Tenax). The sorbent is rapidly heated, the analytes are released and swept with helium onto a GC column. When coupled with a mass spectrometer, the technique is very broad ranging and can be used for the simultaneous analysis of volatile organics, including trihalomethanes, BTEX (benzene, toluene, ethylbenzene, xylene), methyl t-butyl ether (MTBE), carcinogenic halogenated aliphatics, etc., with detection limits in the range of 0.2–2 µg/l. Surrogates (e.g., D_8-Toluene, Bromofluorobenzene) can be added to the sample either before taking a 5 ml aliquot or after transferring to the sparging vessel. Excessive carryover of water will affect the ability of the sorbent to trap compounds; therefore, a water trap should be installed in front of the sorbent trap. Memory effects following the introduction of high-level samples can lead to false positives.

2.2. Headspace

In this technique, an aliquot of sample (10 ml) is placed in a septum vial (20 ml) to a maximum of 50% capacity. The vial is spiked with surrogates and then heated for a moderate period (~30 min) to create an equilibrium for volatile organic compounds between the air phase (headspace) and the water. The headspace is sampled (20–100 µl) with an airtight syringe and injected into a GC. Analytes are similar to

P&T. Headspace does not have problems with water, and memory effects are greatly reduced from those found with P&T. Problems will be encountered if the sample contains matrix modifiers that change the analyte pK_as. In this case, quantification should be based on standard additions methods.

2.3. Liquid/liquid extraction

Samples are transferred to a separatory funnel, surrogates are added, and an immiscible solvent (dichloromethane, hexane, etc.) is added. The liquids are shaken vigorously for a few minutes and then allowed to rest until a separation between the two phases occurs. The solvent is removed and the extraction process is repeated twice more. The extracts are combined, dried over anhydrous sodium sulphate, and processed further (cleanup) as required. Some laboratories have automated this tedious procedure by performing extractions in bottles. In this case, solvent and water are placed in a bottle and rotated (windmill rotators) or shaken (platform shakers) for 1–2 h. The lack of vigorous shaking is replaced by an extended time for extraction. Liquid:liquid extraction is used for all semivolatile analysis (hydrocarbons >C12, PAH, pesticides, PCB, dioxins). By lowering the pH, extraction of phenols (pentachlorophenol) and acidic compounds (2,4-dichlorophenoxyacetic acid—2,4-D) will be enhanced. Increasing the pH will increase extractability of basic (aromatic amines) and neutral compounds (PAH).

2.4. Solid phase extraction

Rather than extracting water with solvent, the water sample is poured through a column or filter containing an absorbent resin. The organics will preferentially adsorb to the resin, which is subsequently desorbed with solvent. This technique has been used for PAHs, pesticides, and PCBs and has been well characterized for drinking water. Laboratories should take proper steps to evaluate the efficiency of this technique for effluent samples or turbid samples and may refer to EPA method 3535A or to guideline documents from SPE suppliers (e.g., Supelco bulletin 910).

3. SAMPLE PREPARATION—SOIL SAMPLES

Organic pollutants (e.g., pesticides, PCB) are not naturally occurring in the environment. They are introduced at specific locations, at specific times, and with different application practices. They may undergo transformation (degradation) with half lives, ranging from days to centuries.

In most cases, the soil sample that arrives in the laboratory cannot be analyzed entirely. Only a small subsample is usually analyzed, and the analyte concentration of the subsample is assumed representative of the sample itself. A subsample cannot be perfectly representative of a heterogeneous sample, and improper subsampling may introduce significant bias into the analytical process. Subsampling can sometimes be improved by procedures such as grinding and homogenizing the original samples (Gerlach *et al.*, 2002).

Maintaining environmental samples in their original wet state is regarded as the most appropriate approach for preparing samples. Homogenized samples should be mixed with a desiccant such as sodium sulfate or celite to bind water.

The appropriately prepared sample can then be extracted by a number of techniques. The main points to consider are to allow adequate time of exposure of the solvent system in the sample matrix and to limit sample handling steps, that is, avoid filtration steps by using soxhlet (sample in a glass thimble), extraction columns (sample matrix eluted after soaking in soxhlet), or semiautomated systems (pressurized liquid extractors).

3.1. Soxhlet extraction

Once a soil is subsampled, extraction of the organic constituents is typically performed via soxhlet. The sample is chemically dried with anhydrous sodium sulfate, ground to a fine powder and then soxhlet extracted for 16 h with either hexane:acetone, dichloromethane, or toluene. Soxhlet extraction is used for a wide range of compounds; however, the extraction efficiency is the greatest with neutral analytes and drops off with increasing acidity.

3.2. Shaker table

Shaker table techniques have been developed with increased efficiency for acidic and basic compounds. The sample is placed in an extraction vessel (Erlenmeyer flask or glass jar) together with an extracting solvent, dichloromethane: methanol: H_2SO_4 (70:29:1) for acidic compounds or dichloromethane: methanol: NaOH (70:29:1) for basic compounds. The vessel is placed on a platform shaker for 1 h and then the liquid is decanted into a separatory funnel-containing water. The separatory funnel is gently shaken, the layers are allowed to separate as for a normal liquid:liquid extraction, and the solvent is collected for further processing.

3.3. Ultrasonic probe

An alternative extraction technique is sonication of the soil with an ultrasonic probe. The technique is relatively quick, but its speed is limited by the sequential nature of sample processing. Samples are placed in a glass extraction vessel and mixed with a drying agent (anhydrous sodium sulphate) until free flowing. Solvent is added and the ultrasonic disruptor is placed in the solvent and above the sample for a period of 3 min. The solvent is decanted, fresh solvent is added to the sample, and the process is repeated twice more. The extracts are combined and concentrated for further processing (cleanup).

3.4. Accelerated solvent extraction

Accelerated solvent extraction (ASE) is becoming more popular because of its rapidity. This however is offset by sequential sample processing. Samples are placed in a stainless steel vessel and subjected to solvent at elevated pressure and temperature. The pH of the extraction can be modified to assist in extracting acidic or basic compounds.

The time and solvent consuming nature of soxhlet extraction (or related techniques involving percolation of a solvent through the sample) is generally thought to be related to the slow diffusion and desorption of the analytes from the sample matrix. Semivolatile compounds (e.g., naphthalenes) can also be lost from the soxhlet apparatus via volatilization. The use of elevated temperatures and pressure (as in ASE) increases the rates of diffusion and desorption and thus speeds up extraction.

 4. CLEANUP TECHNIQUES

Organic analytes must be separated from nonvolatile materials that may affect the performance of GC columns, such as pigments, inorganic sulfur, and triglycerides. Also, there is a need to separate, as much as possible, the analytes from each other before GC analysis in order to limit co-elution problems. Typical techniques used for cleanup are column chromatography or SPE and/or acid/base washes.

Column chromatography involves loading a glass column (1–3 cm i.d.) with granular stationary phase, prerinsing with solvent, adding the sample extract to the top of the column and then eluting the column with solvent(s). The column can be an absorption type where coextractives are left behind (i.e., silica gel) or a fractionating column in which the analytes and coextractives are split into fractions by changing the polarity of the solvents [i.e., PCB/organochlorine pesticides (OCPs) are columned on Florisil with hexane, 2% dichloromethane:hexane, 10% dichloromethane: hexane, etc. to separate different groups of pesticides and PCB] (Table 6.1).

4.1. Adsorption "cleanup" columns

The separation of the analytes from coextractives can be relatively straightforward for low-lipid samples such as soils sediments and vegetation. Generally small silica gel or Florisil columns will suffice. The purpose of this step is to remove coextractive pigments and to separate nonpolar analytes (e.g., PCBs, p,p$'$-dichlorodiphenyldichloroethylene—p,p$'$-DDE) from more polar ones (hexachlorocyclohexane—HCH, chlordane, dieldrin). This is achieved by applying the extract in a small volume of a polar solvent and fractionating by eluting with hexane followed by one or two elutions of increasing polarity. Polar compounds are retained on the column. The effectiveness of these adsorption columns depends on the mass of the adsorbent, polarity of the solvent, and water content of the adsorbent.

Lately, chromatography using laboratory-made open columns for cleanup is being replaced by the increasingly popular SPE, where cartridges ("columns") in ready-to-use format are employed, resulting in an improved batch-to-batch reproducibility.

Table 6.1 Column cleanup materials used for environmental organic extracts

Compound class	Packing	Solvent	Type
Organochlorine pesticides (OCP), acid herbicides, chlorophenols	Florisil	Hexane, dichloromethane	Fractionation
PAH	Silica gel	Hexane, dichloromethane	Adsorption (remove polar materials), Fractionation (split alkanes from PAH)
Dioxins/furans, PCB	Acid/base/ neutral silica gel	Hexane	Adsorption
Dioxins/furans, PCB	Basic alumina	Hexane, dichloromethane	Fractionation
Dioxins/furans	Carbon/celite	Toluene	Fractionation
Dioxins/furans, PCB	Florisil	Hexane, dichloromethane	Fractionation
Lipid removal	Biobeads SX-3	Cyclohexane, dichloromethane	Gel permeation (size exclusion)

4.2. Size-exclusion columns

For high-lipid samples, a lipid removal step must be included. This can be achieved using size exclusion or gel permeation chromatography (GPC). The advantage of GPC is that it is nondestructive and columns can be reused. GPC eluates generally require an adsorption fractionization step on silica or Florisil to remove the remaining low-molecular-weight lipids, waxes, and pigments that are not completely separated from the analytes.

4.3. Lipid destruction

Lipid removal using sulfuric acid washing is also effective but does result in loss of some analytes. The acid-treated extracts are then subjected to an adsorption column fractionization step on silica or Florisil.

4.4. Sulfur removal

Sulfur is coextracted with PCBs and OCPs and presents a problem especially for the GC-ECD analysis of soil or sediment extracts because of strong response in this detector. Sulfur can be removed by GPC but can also be removed using activated copper turnings.

4.5. Evaporation steps

Solvent evaporation is used to 1 °C or another in all organic analyses. Minimizing losses during this step is critical to successful analyte recoveries. Most widely used equipment are rotatory evaporators, Kuderna–Danish apparatus, and TurboVap® equipment. A final evaporation step may use nitrogen gas to gently blow away excess solvent. This can be done using a stream of regulated gas via a disposable glass pipet and heating block or via multi-needle devices (e.g., "N-Evap").

5. INSTRUMENTAL ANALYSIS

Organic extracts are complex mixtures of chemicals and require further separation by means of gas chromatography (GC). GC involves injecting a liquid sample—either a concentrated extract or diluted liquid—into a heated (180 °C–275 °C) injection port. The liquid is flash vaporized and swept into a GC column with carrier gas (helium, hydrogen, argon/methane, nitrogen, etc.). By adjusting oven temperatures and carrier gas flow rates, analytes migrate through the column, moving back and forth between stationary and gas phases. In general, smaller molecules elute first; however, retention time is also affected by the polarity of compound and of the stationary phase.

The detector associated with a GC depends on the analyte of interest and the specificity required.

5.1. Flame ionization detector

Flame ionization detector (FID) requires a carbon–hydrogen bond. Compounds are ionized by a flame as they exit the column, thus making further analysis impossible. It is applicable to all organic compounds, but because of its lack of sensitivity (ppm range) and specificity, it is usually used for hydrocarbon analysis.

5.2. Electron capture detector

Electron capture detector (ECD) uses a radioactive beta particle (electrons) emitter—^{63}Ni foil—and a positively charged anode to capture compounds eluting from the GC that have electron-capturing capacity, such as halogens (Cl, Br, and F), carboxylic acids, and nitrates. It has good sensitivity (ppb range) and linearity (10,000-fold). Its most common uses include organochlorine pesticides, PCB, chlorophenols, chlorobenzenes, and phthalates. GC-ECD suffers from a potential for false positives due to interferences such as sulfur, phthalate esters, and negative peaks generated by hydrocarbons.

5.3. Nitrogen/phosphorus detector

As its name implies, nitrogen/phosphorus detector (NPD) is sensitive to nitrogen and phosphorus compounds. The detector is similar to an FID, but uses a heated rubidium silicate bead in place of a hydrogen/air flame to generate ions. Nitrogen- and phosphorus-containing compounds such as pesticides cause the bead to emit ions that are then collected on a collector to produce a current. Although not as sensitive as an ECD, it is extremely selective, thereby removing many interferences that hamper other detectors.

5.4. Photo ionization detector

The photo ionization detector (PID) contains an ultraviolet lamp that emits photons that are absorbed by compounds (aromatic rings, alkynes, and alkenes) in the ionization chamber. The ions created are then collected at electrodes, thus giving a signal. The most common use for this detector is for the analysis of BTEX.

5.5. Mass spectrometry

Mass spectrometry (MS) has become the detector of choice for many laboratories due to its sensitivity, selectivity, and a wide range of applicability to environmental analysis. Most environmental compounds listed for other detectors can also be analyzed on the MS at equal or better sensitivity. Analytes included PAH, aromatic amines, nitrobenzenes, phthalates, and volatiles.

5.6. High-resolution mass spectrometry

High-resolution mass spectrometry (HRMS) is used for the ultimate in sensitivity and specificity. The resolution of a mass spectrometer is defined as mass divided by the change in mass ($m/\Delta m$). A standard quadrapole MS is tuned to unit resolution, meaning it can distinguish between m/e 300 and 301 (resolution of $300/(301-300) = 300$). For HRMS, the resolution is set to $>10,000$. Therefore, if the mass analyzed for was m/e 300, the HRMS would detect all compounds with m/e 299.985–300.015. This is crucial for analyzing ultra trace organic compounds such as polychlorinated dibenzo(p) dioxins and dibenzofurans (PCDD/F) in the part per trillion and part per quadrillion (10^{-15}) range.

6. DATA ANALYSIS

The modern GC data system will produce a report of peaks detected with the retention time, peak area, and peak height. In order to identify the analytes of interest and quantify the data, a series of calibration standards are required to be analyzed followed by samples. The calibration standards will identify retention times for analytes, surrogates, and internal standards. With the exception of MS analysis, compounds are identified in chromatograms based solely on their retention time. Positive confirmation can be done by analyzing the same sample extract on a different type (polarity) of GC column. If the compound is detected at the same concentration from both GC columns, then the data can be reported (e.g., US EPA Method 8081—OC Pesticides—requires analysis on a DB-5 column with confirmatory analysis on a DB-17 column). For MS analysis, multiple ion chromatograms are possible, therefore compounds are detected if they have peaks in the correct ratio in the appropriate ion channel.

To ensure the integrity of the data, the following quality controls should be taken with instrumental analysis:

- Linearity needs to be proven by running a multi-point—usually five—calibration. The range will depend on detector and should cover at least two orders of magnitude.
- The calibration should be confirmed every 12 h or at the end of the run (whichever comes first) by analyzing a midpoint standard made from a secondary source.
- Positive identification should be confirmed by a second column analysis or by confirmatory ions on the MS.
- If analyte peak area exceeds the upper calibration point, sample needs to be diluted.
- Analyte peak area cannot be <10% of the lowest-level standard. In addition, the peak must have a signal-to-noise ratio or at least 3.

Once a compound has been identified in a GC scan, there are three methods of quantifying GC data: external standard, internal standard, and isotope dilution.

6.1. External standard calculations

An external standard method is used when the standard is analyzed on a separate chromatogram from the sample. Quantitation is based on a comparison of the peak area/height (HPLC or GC) of the sample to that of the reference standard for the analyte of interest. The external standard method is more appropriate for samples with a single target analyte and narrow concentration range, where there is a simple sampling procedure, and for the analysis of hydrocarbon fractions. The calculation requires an accurate extract final volume and constant injection size. The peak area of an analyte is compared with that from a standard or standard curve and corrected for volume:

$$C_x = \frac{A_x}{RF_x} \times \frac{FV}{W}$$

where C_x = concentration of analyte x; A_x = area of x; FV = final volume of extract; W = weight or volume of sample; and RF_x = response factor of x (calculated from calibration standard)

$$= \frac{A_x}{C_x}$$

6.2. Internal standard calculations

With an internal standard (IS) method, a compound of known purity that does not cause interference in the analysis is added to the sample mixture. Quantitation is based on the response ratio of the compound of interest to the internal standard versus the response ratio of a similar preparation of the reference standard. The internal standard method is more appropriate for samples that undergo complex sample preparation procedures (e.g., multiple extractions), for low-concentration samples, and for samples with a wide range of expected concentrations.

A relative response factor (RRF) for each analyte relative to the IS is calculated from calibration standards. Therefore, with a known amount of IS added to the sample, the concentration can be calculated as follows:

$$C_x = \frac{A_x}{A_{IS}} \times \frac{C_{IS}}{RRF_x} \times \frac{1}{W}$$

Where C_x = concentration of analyte x; A_x = area of x; A_{IS} = area of internal standard; C_{IS} = amount of internal standard added; W = weight or volume of sample; RRF_x = relative response factor of x calculated from the calibration curve

$$= \frac{A_x}{A_{IS}} \times \frac{C_{IS}}{C_x}$$

6.3. Isotope dilution calculations

In this method, stable isotopes (e.g., ^{13}C or ^{2}H labeled) of the analytes of interest are added to the sample before extraction. The isotopes are used as internal standards and will correct for any analyte losses during sample work up. They do not give an indication of extraction efficiency. By adding a recovery standard before GC analysis, the recovery of the isotopes can be calculated and reported.

7. Quality Control

A laboratory QA/QC program is an essential part of a sound management system. It should be used to prevent, detect, and correct problems in the measurement process and/or demonstrate attainment of statistical control through QC samples. The objective of QA/QC programs is to control analytical measurement errors at levels acceptable to the data user and to assure that the analytical results have a high probability of acceptable quality.

The data quality is ordinarily evaluated on the basis of its uncertainty when compared with end-use requirements. If the data are consistent and the uncertainty is adequate for the intended use, then the data are considered to be of adequate quality. An important consideration in planning for sampling and analysis is the type and number of QC samples to take.

8. Internal QC

Internal QC monitors the laboratory's current performance versus the standards and criteria that have been set, normally at the time of method development or validation. To ensure that quality data are continuously produced during all analyses and to allow eventual review, systematic checks are performed to show that the test results remain reproducible. Such checks also show if the analytical method is measuring the quantity of target analytes in each sample within acceptable limits for bias (Environment Canada, 2002; IUPAC, 1995; CAEAL, 1999).

Analytical QC procedures that determine whether the sample handling procedures and laboratory methods are performing as required are as follows (Table 6.2):

8.1. Method blanks

Purified matrix (water, soil, and tissue) are processed as though they were a sample. This will indicate interferences caused by the reagents used. It will also identify cross-contamination from samples. Frequency should be once per batch of up to 20 samples.

8.2. Method spikes (laboratory control samples)

A known quantity of the analytes of interest are added to blank samples and processed. The recovery of the analytes will give an indication of the performance of the method. Frequency should be once per batch of up to 20 samples.

8.3. Matrix spikes and matrix spike duplicates

Similar to an LCS, but in this case the actual sample analyzed is being spiked with the analytes of interest. Data will give an indication of possible matrix interferences, and relative percent differences (RPD) can be calculated (30%–50% depending on matrix and analyte). Matrix spikes and matrix spike duplicates (MS/MSD) should be run with each new matrix.

8.4. Surrogates

These are compounds that are similar to the analytes of interest and will give an indication of the method performance but do not interfere. Many surrogates are isotopes of the analyses (i.e., D_{10}-phenathrene or $^{13}C_{12}$–2,3,7,8-TCDD). Surrogates do not give an indication of the extraction efficiency of the method. They can only be used to identify losses. All organic methods should have surrogates, and the recovery of those surrogates should be recorded.

8.5. Duplicates

Dulpicate analyses will track the variability in the laboratory to subsample and process a sample. Typical controls have a variability of up to 50% for soils.

Table 6.2 Data verification procedures for QC samples

QC sample	Why	Data verification
Method blanks	To demonstrate the absence of sample contamination.	Normally, only method blanks and specific blanks submitted with the samples will be reported. No blank should have a reportable concentration of any compound of interest above the reporting limit. Any compound that has a concentration above the reporting limit in the blank and is present in any sample must be considered a non-detect up to 5 times the level in the blank.
Method spikes	To verify appropriate analyte recovery (accuracy or bias).	Verify that the recoveries are reasonable. Some typical values for organic analysis: 30%–150% recovery.
Matrix spikes and matrix spike duplicates	To determine possible matrix specific effects affecting recovery.	Verify that the recoveries are reasonable and not greatly different from method spike. Some typical values for organic analysis: 30%–150% recovery.
Surrogates	To determine whether matrix effects may be present.	Surrogates are compounds that are spiked into every (organic) sample. A surrogate is a compound that is not found in nature and is not a "normal" pollutant. Surrogate recoveries should be within a defined range to be acceptable. If the surrogate recovery is low, then positive values are flagged and non-detects are rejected. If it is high, then non-detects are considered acceptable and positive data are flagged as estimated.
Duplicates	To verify appropriate analytical precision.	Verify the relative percent difference of the concentrations X and Y of each duplicate. A criterion for organic analysis is that the RPD $<50\%$ RPD $= |X - Y| \times 200/(X + Y)$
Calibration samples	To optimize quantitation of target analytes.	Quantitation range is defined and reproducable.

8.6. Reference materials

Reference materials give an indication of bias and precision for your method. They can be purchased from a number of vendors (e.g., NIST, Environment Canada) and should be analyzed on a monthly basis. When reference materials are not available, a reference standard can be used. A reference standard is a highly purified compound that is well characterized.

9. EXTERNAL LABORATORY QC

External laboratory QC involves reference help from other laboratories and participation in national or international interlaboratory sample and data exchange programs such as Proficiency Testing (PT). Such programs may involve:

- The exchange of samples with another laboratory. These samples would be prepared by a staff member other than the analyst or by the QC department. Similarly, samples prepared by the QC department could also be used as internal check samples.
- The participation in interlaboratory sample exchange programs (such as round robins and/or PTs).

The necessary components of a complete QA/QC program include internal QC criteria that demonstrate acceptable levels of performance, as determined by a QA review (audit). External review of data and procedures is accomplished by the monitoring activities of accreditation organizations such as the Standards Council of Canada (SCC, 2005). This includes laboratory evaluation samples (PT samples, see above) and a periodic (sometimes every 2 years) on-site assessment of all QA/QC procedures, performed by external assessors from the accrediting organization.

REFERENCES

Canadian Association of Environmental Analytical laboratories (CAEAL) (1999). Template for the design and development of a Quality Manual for Environmental Laboratories [Online] and QC for Environmental laboratories [Online]. Available:http://www.caeal.ca/t_caealpubs.html[07 March 2005].

Environment Canada (2002). Contaminated site remediation section technical assistance bulletins. [Online] available:http://www.on.ec.gc.ca/pollution/ecnpd/contaminassist_e.html[04 March 2005]. This webpage shows a list of Technical Assistance Bulletins. See Section 5: QC samples in Tab#4: Sampling and Analysis of Hydrocarbon Contaminated Soil.

Gerlach, R. W., Dobb, D. E., Raab, G. A., and Nocerino, J. M. (2002). Gy sampling theory in environmental studies.1. Assessing soil splitting protocols. *J. Chemom.* **16,** 321–328. Published online in Wiley InterScience www.interscience.wiley.com DOI: 10.1002/cem.705.

International Union of Pure and Applied Chemistry (IUPAC) (1995). Harmonized guidelines for internal quality control in analytical chemistry laboratories. *Pure Appl. Chem.* **67,** 649–666. http://www.iupac.org/publications/pac/1995/pdf/6704x0649.pdf.

Standards Council of Canada (SCC) (2005). Programs and services—Laboratories. [Online] available: http://www.scc.ca/en/programs/laboratory/index.shtml [07 March 2005].

United Nations – Food and Agriculture Organization (FAO) (1998). Guidelines for quality management in soil and plant laboratories. [Online] available: http://www.fao.org/docrep/W7295E/w7295e00.htm [04 March 2005].

Evaluation of Geochemical Background at Regional and Local Scales by Fractal Filtering Technique: Case Studies in Selected Italian Areas

Annamaria Lima*

Contents

Abstract

Backgrounds represent, in the environmental field, the borderline between concentrations of a chemical element that naturally occurs in a media as compared with the concentrations of the same analytes present as a result of anthropogenic activities. To evaluate background values, there are two basic approaches: statistical frequency analysis and spatial analysis. Statistical frequency analysis uses techniques for characterizing the frequency distribution based on the assumption that point data from different locations may originate from different sources and represent different populations. Spatial analysis refers to methods dealing with spatial distribution of values on a 2-D map where geochemical point data are generally interpolated. In this chapter, the application of GeoDAS^{TM} software to perform multifractal inverse distance weighted (MIDW) interpolation and a fractal filtering technique, named spatial and spectral analysis or simply (S–A) method, will be illustrated to evaluate geochemical background at regional and local scale.

* Dipartimento di Scienze della Terra, Università di Napoli "Federico II", Via Mezzocannone, 8, 80134 Napoli, Italy

Environmental Geochemistry
DOI: 10.1016/B978-0-444-53159-9.00007-3
© 2008 Elsevier B.V.
All rights reserved.

1. INTRODUCTION

The total concentrations of elements in air, water, soil, and sediment may vary depending on natural factors and also for the human impact.

In environmental geochemistry, the term "baseline" indicates the actual content of an element, independently of its origin, in the surficial environment at a given point in time as opposed to the term "background" that indicates the content depending on natural factors like lithology, genesis of the overburden, and climate (Salminen and Gregorauskiene, 2000). The term background in environmental studies does not imply the absence of anomalies as it does when used by exploration geochemists. They use it to differentiate between element concentrations within samples devoid of enrichments and those which show positive anomalies. Backgrounds represent, in the environmental field, the borderline between concentrations of a chemical element and component that naturally occurs in a media as compared with the concentrations of the same analytes present as a result of anthropogenic activities. They are not necessarily equal to low concentrations, but they need to be quantified in order to be used as a reference limit (Matschullat *et al.*, 2000). In addition, the evaluation of "geochemical background" levels also assumes particular relevance as a function of environmental legislation, which fixes intervention limits for elements, such as the harmful ones (e.g., As, Pb, Cd, and Hg). Under Italian legislation, for example, if the intervention limits are exceeded, remediation of the contaminated land becomes mandatory, whereas concentrations above the fixed intervention limits are allowed only in situations where the investigated site is characterized by natural background values higher than the intervention limits. For the above reasons, it is important to discriminate efficiently background concentrations from anthropogenic anomalies.

To process geochemical data and to evaluate background values, there are two basic approaches: statistical frequency analysis and spatial analysis. Spatial analysis refers to methods dealing with spatial distribution of values on a 2-D map where geochemical point data are generally interpolated. Statistical frequency analysis uses techniques for characterizing the frequency distribution based on the assumption that point data from different locations may originate from different sources and represent different populations. Hence, statistically it is possible to distinguish between different geological processes. In exploration geochemistry, assuming that the distribution of geochemical data obeys a certain form of distribution (e.g., normal or lognormal) to separate anomaly from background values, various statistical methods are used, including univariate and multivariate analysis, and cumulative probability graphs (Garrett, 1989; Govett *et al.*, 1975; Miesch, 1981; Sinclair, 1974, 1976, 1991; Stanley, 1988; Stanley and Sinclair, 1987, 1989). In geochemical exploration, statistical methods are the most common techniques utilized to create data subsets in populations and as a consequence distinguish anomaly from background values. However, because of significant population overlap, background values can be erroneously classified as anomalous and vice versa anomalous values can be erroneously classified as background. This happens mostly at a regional scale where significant geo-lithologic variations make it difficult to select one or two general threshold to utilize as reference to identify anomalies (exploration geochemists define the threshold as the upper limit of normal background values). Once anomalies have been identified, also the spatial distribution and the geometry of geochemical anomaly can give important information on natural versus anthropogenic source of the anomalous values.

The use of geochemical maps for environmental purposes is of great relevance because they contain a large amount of helpful information and they also help characterize element background concentration values from anthropogenic anomalies.

The goal of this chapter is to show the application of some new techniques of plotting geochemical data, including multifractal interpolation and fractal filtering (Cheng, 2001) that have been applied successfully to determine background levels of potentially harmful elements (PHEs) in stream sediments at the regional scale, for the Campania region of Italy (Albanese *et al.*, 2006; De Vivo *et al.*, 2006a; Lima *et al.*, 2003, 2005) and for As in European waters (Lima *et al.*, 2008), and at the local scale for the urban area soils of Napoli, Avellino, Caserta, Salerno, and Ischia Island (Albanese *et al.*, 2007; Cicchella *et al.*, 2003, 2005; De Vivo *et al.*, 2006b; Fedele *et al.*, in press; Frattini *et al.*, 2006a, b; Lima *et al.*, 2007).

In this chapter, the application of multifractal inverse distance weighted (MIDW) interpolation method and a fractal filtering technique, named spatial and spectral analysis or simply (S–A) method (Cheng, 2003), will be illustrated to evaluate geochemical background at the regional and local scale. For this purpose, two case studies will be discussed for two different Italian areas:

- The evaluation at the regional scale of Pb and U natural background concentrations for 2389 stream sediment samples of Campania region (13,600 km^2) (Lima *et al.*, 2003).
- The evaluation, at the local scale, of Pb background concentrations for 982 volcanic soil samples from the metropolitan and provincial areas of Napoli (1171 km^2) (Cicchella *et al.*, 2005).

2. MULTIFRACTAL INTERPOLATION AND FRACTAL CONCENTRATION–AREA (C–A) METHOD

GeoDAS^{TM} software (Cheng, 2003) is used for the MIDW interpolation method to generate, starting from dot data, geochemical maps. The latter can be decomposed by a fractal filtering technique, named spatial and spectral analysis or simply (S–A) method, in two additional maps one representing the background/baseline concentrations and the other the anomalies of the area under investigation. The mathematical approach of these methodologies is fully described in Cheng *et al.* (1994a, b, 1996, 2000), Turcotte (1997), Cheng (1999, 2001, 2003), and Agterberg (2001), and only a brief outline is included here.

The MIDW (Cheng, 2003) interpolation method estimates the distribution of chemical values based on fitted power law trends between chemical values and the grid area using the observed data points. It is a technique that fits the source data accurately and preserves local high-concentration values in the interpolation grid. This is useful in distinguishing anomaly from background.

The MIDW interpolation method takes into account both spatial associations and local high values (singularity). Singularity is an index representing scaling dependency from a multifractal point of view, and it characterizes how statistical behavior changes as the scale of geochemical values changes. Spatial association and scaling are two different aspects of the structure of surfaces locally, and both should be taken into account in data interpolation and surface mapping. The multifractal method incorporates both the singularity and spatial association and, compared with other conventional interpolation techniques (e.g., kriging), improves map interpolation significantly (Lima *et al.*, 2008), especially for locally high or low values (with significant singularity). For a detailed discussion on MIDW data interpolation, see Cheng (2003, Appendix C).

The color scale classification of the MIDW interpolated map is based on the concentration–area (C–A) fractal method. This method allows images to be subdivided into components for symbolizing distinct image zones, representing specific features on the ground. For example, the pixels of the image in Fig. 7.1A have been classified by means of the fractal C–A plot showed in Fig. 7.2A. The latter plots the pixel Pb concentration values against the cumulative area on log–log paper. C–A plot characterizes not only the frequency distribution of pixel values but also the spatial and geometrical properties of the features reflected by pixel zones in the image. In C–A plot area concentration [$A(\rho)$] with element concentrations greater than ρ shows a power–law relation. If the geochemical surface is fractal, a single straight-line relationship would occur, whereas for a multifractal surface, values will fall on several straight lines (Fig. 7.2A and B). The breaks between straight-line segments on this plot and the corresponding values of ρ have been used as cutoffs to separate geochemical values into different components, representing different factors, such as lithologic changes and geochemical processes (e.g., mineralizing events, surficial geochemical element concentrations, and surficial weathering) (for more details, see Cheng, 2003, and references there in).

3. BACKGROUND/BASELINE GEOCHEMICAL MAP OBTAINED BY FRACTAL FILTERING (S–A) METHOD

To decompose the MIDW interpolated map to obtain background/baseline and anomaly maps, a fractal filtering technique, named spatial and spectral analysis or simply (S–A) method, is applied.

In theory decomposing the MIDW interpolated map, background and anomaly maps should be obtained but, often happens that the background map shows areas where concentration values are still affected by contaminations and they should hence be interpreted as baseline values.

The S–A method is based on a Fourier spectral analysis (Cheng, 2001; 1999). It uses both frequency and spatial information for geochemical map and image

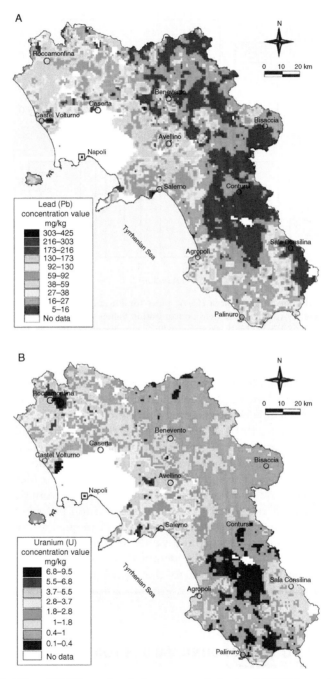

Figure 7.1 Pb (A) and U (B) geochemical maps compiled using MIDW interpolation method. Search distance is 3.5 km and map resolution 1 km. Pixel values have been reclassified by fractal C–A plot based on the frequency distribution of pixel values (Fig. 7.2A and B).

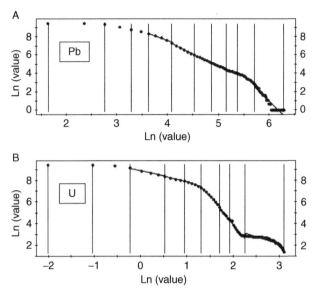

Figure 7.2 Fractal concentration–area (C–A) plots for Pb and U. The vertical axis represents cumulative pixel areas A(ρ), with element concentration values greater than ρ, and the horizontal axis the pixel values (ρ). Breaks between straight-line segments and corresponding values of ρ have been used as cutoffs to reclassify pixel values in the MIDW interpolated map (Fig. 7.1A and B).

processing. The basic geological assumption for S–A method is that a geochemical field or image generated by specific geological processes may be described in terms of its fractal properties. These properties can be measured in both the frequency and the spatial domain (Turcotte, 1997). The fractal filter to be used is defined on the basis of the power–law properties of a power spectrum in the frequency domain. The purpose of this is to divide the power spectrum into components characterized by similar scaling properties. The filter, therefore, can be used to identify anomalies from the background and to extract other meaningful pattern from the original map. Fourier transformation and inverse Fourier transformation provide the foundation for converting field (maps) between the spatial and frequency domains. Fourier transformation can convert geochemical values into a frequency domain in which different patterns of frequencies can be identified. The signals with certain ranges of frequencies can be converted back to the spatial domain by inverse Fourier transformation.

4. Pb AND U BACKGROUND VALUES FOR CAMPANIA REGION STREAM SEDIMENTS

The Campania region covers about 13,600 km^2. Morphologically, Campania is made up, in its eastern sector, by the Apennine mountains oriented roughly NW–SE and in the west, by two coastal plains. Lithologies mostly consist of sedimentary and volcanic rocks, spanning from the Triassic to the Recent age (Fig. 7.3A). Stream sediment samples, numbering 2389, were collected in the region with a density of one sample per 5 km^2 (Lima *et al.*, 2003). In the metropolitan area of Naples (Fig. 7.3B), because of the lack of drainage, top soils (sampled at depths between 0 and 15 cm below the surface) have been collected; the soil data have been studied separately from stream sediment data (Cicchella *et al.*, 2005; De Vivo *et al.*, 2006b).

Univariate statistical analysis has been performed on the 2389 Campania region stream sediment data. Table 7.1 shows Pb and U statistical parameters. The frequency distributions of Pb and U concentrations were evaluated using traditional methods such as histograms and probability plots. Figure 7.4 shows the Pb and U histograms and the plot of percent cumulative frequency versus concentrations (log-transformed values) on probability paper. For both Pb and U, the data show a tendency toward a multimodal lognormal distribution, though the histograms do not help characterize the population distribution, this distribution becomes clearer using cumulative probability curves. Cumulative probability curves for both Pb and U (Fig. 7.4A and B) show the existence of three data populations: A, B, and C. For Pb, the relative proportion of each population can be estimated from inflection points on the curve at 2%, 83%, and 15%, respectively (there is clearly a small 2% population of high lead, very much the urban lead population, a low population of about 15% of the data, and an intermediate population of about 83%), and for U at 5%, 47%, and 48%, respectively. Sinclair (1976) describes a method for partitioning the curve into these component populations (assuming lognormal distributions), and each population can be defined with a mean (defined by the 50 percentile) ±1 standard deviation (defined by the 84 and 16 percentiles). Each population can therefore be assigned a size and a range of values but, as indicated by the histograms in Fig. 7.4, they are not discrete and can overlap or be completely contained within another population making spatial distribution difficult to determine. A rough estimate (without partitioning) of values associated with each population is indicated on Fig. 7.4, using the curve inflection points. The low population is associated with values around <11 and <0.8 mg/kg for Pb and U, respectively, the high population with anomalous values (>94 and >3.8 mg/kg for Pb and U, respectively), and an intermediate data population falling between 11 and 94 mg/kg for Pb and between 0.8 and 3.8 mg/kg for U. However, the selection of thresholds in this way may not be appropriate at a regional scale (e.g., Campania), because the background values are characterized by significant variations due mostly to lithologic variations and the partitioning into just three populations is an oversimplification of the real situation. The value distributions are not sufficient to divide the dataset in different populations, because the occurrence of overlaps, and give no idea of the spatial distribution and geometry of the geochemical anomalies.

The stream sediment geochemical maps (Fig. 7.1A and B) of Campania have been interpolated by MIDW using the GeoDAS program (Cheng, 2003). Search distance and map resolution are 3.5 and 1 km, respectively.

Comparing geological map (Fig. 7.3) of the Campania region with Pb (Fig. 7.1A) and U (Fig. 7.1B) MIDW interpolated maps, it can be seen that the distribution of high Pb values reflects both a lithologic and an anthropogenic control, whereas U reflects mostly a lithologic control.

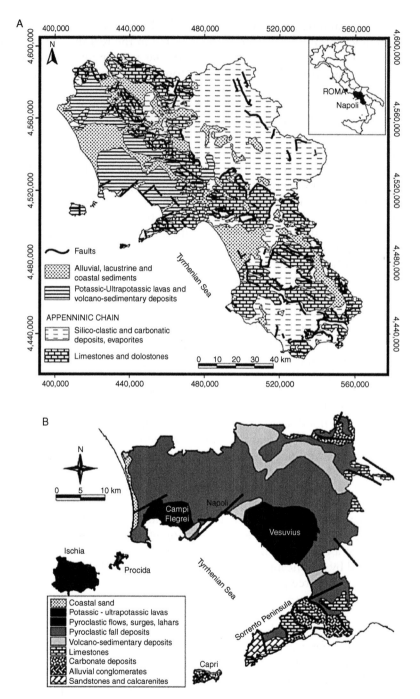

Figure 7.3 Simplified geological map of Campania region (A) and of metropolitan and provincial areas of Napoli (B).

Table 7.1 Pb and U statistical parameters of 2389 stream sediment samples from campania region

mg/kg	Pb	U
Mean	31.7	1.6
Median	22.4	0.9
Geometric mean	24.5	1.1
Minimum	3.3	0.1
Maximum	546	22.5
25th percentile	14.5	0.6
75th percentile	39.5	1.9
Standard deviation	32	1.6

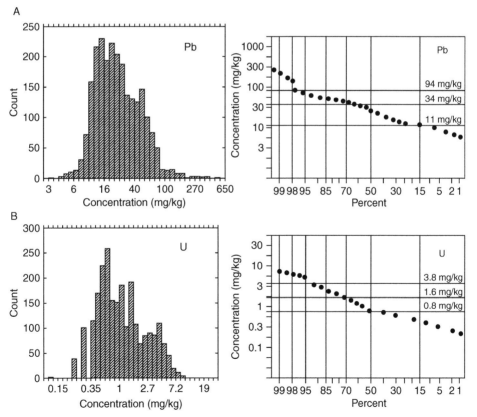

Figure 7.4 Pb and U log-transformed values histograms and cumulative probability curves for the Campania region 2389 stream sediments.

Lead values, between 38 and 92 mg/kg (Fig. 7.1A), are located in correspondence with the potassic–ultrapotassic volcanic rocks (Roccamonfina, Mt Somma-Vesuvius, Ischia) along the western border of the region. In contrast, the highest Pb concentrations (up to 425 mg/kg) are located in urban areas mostly around the big cities (e.g., Avellino, Caserta, Salerno), and where human activities are more intense. This situation is perfectly outlined by the Pb background/baseline map (Fig. 7.5A). The latter has been compiled using spatial and spectral analysis (S–A method). Pb values have been obtained by decomposition of the Pb geochemical map (Fig. 7.1A) into separate components using the fractal filter selected on the S–A plot (Fig. 7.6A). Pb background map is generated only with signals with power spectrum (E) >15,713 (selected cutoff in Fig. 7.6A). Finally, pixel values have been reclassified once again by fractal C–A plot.

High Pb values (>107 mg/kg) occurring in small areas around the big cities and SE of Castel Volturno (Fig. 7.5A) are interpreted as anomaly of anthropogenic origin. In general, Pb values >91 mg/kg, in the north-western volcanic sector of the region, do not reflect natural background values but anthropogenic pollution (they have for this reason, to be considered baseline values). This interpretation is based on the observation that Pb MIDW interpolated values (Fig. 7.1A) in uncontaminated areas, around volcanic complexes, are always <91 mg/kg and as well by the fact that the Pb content in Campanian volcanic rocks has an average value of 64 mg/kg (Paone *et al.*, 2001).

Uranium maps (Figs. 7.1B and 7.5B) show a different pattern to the Pb maps. With the exception of the area, SE of Castel Volturno that might represent pollution caused by illegal waste disposal, a close relationship can be observed between the U background values (Fig. 7.5B) and the different lithologies of the Campania region (Fig. 7.3A), with the highest values associated to the Roccamonfina potassic volcanic area in north-western Campania, and the lowest values associated to the eastern sedimentary carbonate rocks. Figure 7.5B has been compiled in the same way as Fig. 7.5A for Pb. Uranium values have been obtained by the decomposition of the U geochemical map (Fig. 7.1B), using the fractal filter selected on the S–A plot (Fig. 7.6B).

5. Pb BACKGROUND VALUES FOR THE VOLCANIC SOILS OF THE METROPOLITAN AND PROVINCIAL AREAS OF NAPOLI

Here the application of the C–A and S–A methods (Cheng, 2003) is reported, at local scale, to evaluate Pb background and anthropogenic anomaly in the top soils of the metropolitan and provincial areas of Napoli. Generally, at a local scale, geology and other environmental factors (e.g., climate) show more limited variations compared with a regional scale data distribution. In metropolitan areas, which are deeply influenced by human impact, it is very important (but also more difficult) to fix the borderline between concentrations of potentially harmful elements that naturally occurs in a soil (background) as opposed to concentrations of the same analytes as an anthropogenic source. The case of the Pb values in the Napoli metropolitan and provincial volcanic soils will be discussed.

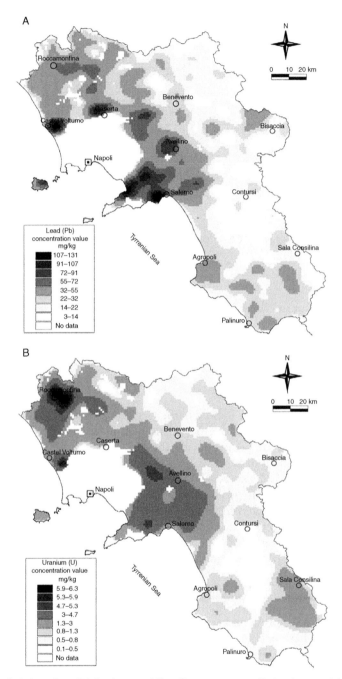

Figure 7.5 Pb (A) and U (B) background/baseline maps compiled using spatial and spectral analysis (S–A method). The values have been obtained by decomposition of the Pb and U geochemical maps (Fig. 7.1A and B, respectively) into separate components using the fractal filters selected on the S–A plot. Pb background/baseline map is generated only with signals with power spectrum (E) >15,713 (selected cutoff in Fig. 7.6A). U background/baseline map is generated only with signals with power spectrum (E)>706 (selected cutoff in Fig. 7.6B). For both maps pixel values have been reclassified by fractal C–A plot (see text).

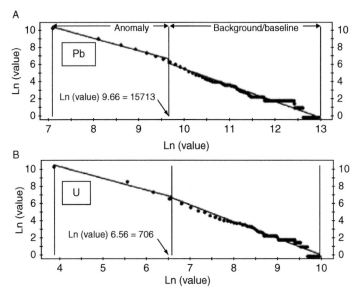

Figure 7.6 S–A plots for Pb and U. The vertical axis represents log area A (>E) and the horizontal axis the log-transformed power spectrum value itself (E). In this case, two straight-line segments have been fitted by means of least square; the two subsets of frequencies indicate high- and low-frequency power-spectrum components. The vertical line on each plot shows the cutoff applied to generate the corresponding filter used for background and anomaly separation.

The Neapolitan province is located in the Campania region, along the Tyrrhenian coastline (Fig. 7.3B), in one of the most densely populated active volcanic areas on Earth. More than 3 million people inhabit the 1171 km² of the Napoli metropolitan area. The communications network has expanded considerably in the last few decades, with the construction of motorways and an increase in traffic which has now practically reached saturation level. The geology of the area (Fig. 7.3B) is predominantly characterized by volcanics erupted from the Upper Pleistocene to Recent, mainly from two historically active volcanic complexes (Mt. Somma–Vesuvius on the east and the Campi Flegrei on the west) and by eruptive fissure events located in the Campania Plain, which gave rise to various ignimbrite eruptions (De Vivo, 2006; De Vivo et al., 2001, and reference therein). In the Neapolitan province, 982 soil samples were collected over an area of ~1171 km² with a grid of 0.5 km × 0.5 km in the urbanized areas and a grid of 1.5 km × 1.5 km in the suburban areas.

As described in the previous paragraphs, background/baseline and anomaly maps (Fig. 7.7A and B) have been obtained by GeoDAS software (Cheng, 2003) using the S–A method to the Pb geochemical MIDW interpolated map (not shown here). The S–A plot (Fig. 7.8A), showing the relationship between the cumulative area and the power spectrum values of the Fourier transformed map, has been modeled by fitting straight lines using least squares. Distinct classes are shown: the lower-, intermediate-, and high-power spectrum values. After the irregular filter has been applied on these distinct patterns to remove the anomalies and noise (related to the

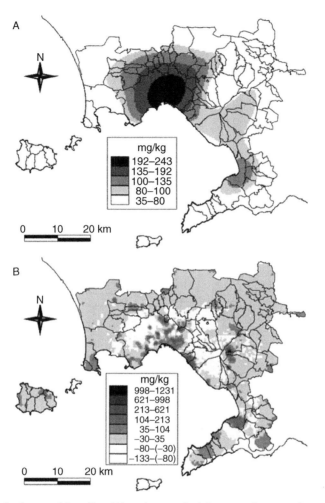

Figure 7.7 Pb background/baseline (A) and anomaly (B) maps of metropolitan and provincial areas of Napoli, compiled using the S–A plot (Fig. 7.8A).

intermediate- and lower-power-spectrum values), the image, converted back to a spatial domain (by inverse Fourier transformation), represents Pb background/ baseline geochemical patterns (plotted in Fig. 7.7A). In the same way, the Pb anomaly map (Fig. 7.7B) has been obtained applying a band-type filter (selecting the first and the third break as the cutoff on the S–A plot of Fig. 7.8A) that removes noise and background/baseline related to the lower- and high-power-spectrum values. The map shown in Fig. 7.7A is named background/baseline map because it represents both background and baseline values, being the high Pb concentrations, located mostly in the metropolitan area, anthropogenic. The real background Pb concentration values, reflecting geogenic sources are <80 mg/kg and are located outside the Napoli suburban area.

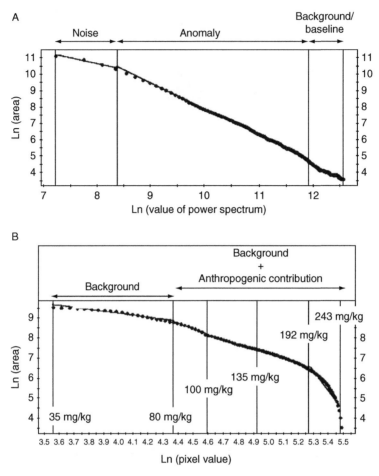

Figure 7.8 Volcanic soils of metropolitan and provincial areas of Napoli: (A) Pb S–A plot, straight lines are fitted by means of least squares. The three segments (representing noise, anomaly, and background) have slopes of –0.69, –1.59 and –1.54, respectively. (B) Fractal C–A plot for Pb background/baseline values. The cutoff corresponding to 80 mg/kg has been utilized to extrapolate background from baseline values. Figure 7.7A shows the areas affected by increasing anthropogenic source contributions.

In this case, the upper limit of background values (80 mg/kg) has been evaluated on the base of C–A plot (Fig. 7.8B). As already pointed out, this method allows images to be subdivided into components for symbolizing distinct image zones, representing specific features on the ground. Figure 7.8B shows C–A plot utilized to reclassify the pixel concentration values of Pb background/baseline map (Fig. 7.7A). As indicated on the C–A plot (Fig. 7.8B), the cutoff corresponding to 80 mg/kg has been utilized to extrapolate background from baseline values and on Pb background/baseline map (Fig. 7.7A) it is possible to see the areas affected by increasing anthropogenic source contributions, which correspond to the highly urbanized part of Napoli.

The evaluation of the background upper limit value (80 mg/kg) is consistent with the known concentration of Pb in Neapolitan volcanic rocks (average 42.5 mg/kg; Paone *et al.*, 2001) and also with the Pb concentration values obtained in areas characterized by similar lithologies and with very limited human impact. Figure 7.9 (A and B) shows Pb histogram and probability plot for the Neapolitan province 982 soil samples. On the probability plot are also reported: the background value, the percent cumulative value corresponding to the residential/recreational

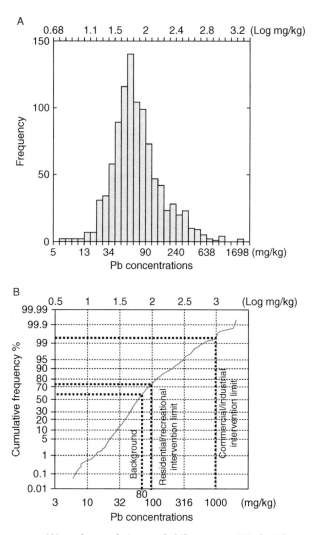

Figure 7.9 Histogram (A) and cumulative probability curve (B) for Pb in volcanic soils of metropolitan and provincial areas of Napoli. The probability curve also shows: the evaluated range of background values, the percent cumulative value corresponding to the residential/ recreational land use intervention limit, and the percent cumulative value corresponding to the commercial/industrial land use intervention limit given by the environmental Italian Law (471/99).

intervention limit, and the percent cumulative value corresponding to the commercial/ industrial intervention limit fixed by Italian Law (D.M. 471/99). In the investigated Neapolitan province, about 58% of soil samples have Pb content within the evaluated background, about 27% have Pb concentration above the residential/recreational intervention limit and about 0.6% Pb concentration higher than commercial/industrial intervention limit fixed by Italian Law (D.M. 471/99).

6. Conclusions

It is emphasized that by means of the statistical frequency analysis it is possible to separate dataset in populations, even if can be difficult to select one or two general threshold to utilize as reference to identify background values. In addition, the value distributions are not sufficient to divide the dataset into different populations, because they show an overlap, and give no idea of the spatial distribution and geometry of the geochemical anomalies. This is true both at local and regional scale.

Geochemical maps prepared, using MIDW for environmental purposes, show data distribution that reflects in detail both the lithologic and other natural processes. By applying the C–A method, the user can interact with the software and select cutoffs that separate geochemical values into different components, and identify areas with similar features. S–A filtering method together with the C–A method is useful for discriminating background values from anthropogenic anomalies.

In theory decomposing the MIDW interpolated map, background and anomaly maps should be obtained. In the majority of the applications for environmental purposes, it happens that the background map (obtained by fractal filtering) shows areas where concentration values are still affected by contaminations and they have hence to be interpreted as baseline values. For this reason, it is more appropriate to define such maps as background/baseline maps. The latters are very useful, along with other information on the investigated area, to evaluate background values, both at regional and local scale.

At a regional scale, as shown in the case of Pb and U distribution in the Campania region, these fractal techniques (Cheng, 2003) offer a good method, very efficient, for defining geochemical background and anomalies. High Pb values (>107 mg/kg) occurring in small areas around the big cities and in general, Pb values >91 mg/kg, in the north-western volcanic sector of the region, do not reflect geogenic sources but anthropogenic pollution. For U, a close relationship has been observed between background values and the different lithologies present in the region and likely the only significant U anthropogenic anomaly occurs in one area (SE of the Castel Volturno) affected probably by illegal waste disposal.

At a local scale, in a densely populated area, where lithologies are predominantly characterized by volcanics and the human impact is very strong, by means of the C–A method background have been extrapolated from baseline values. The results obtained indicate that Pb shows baseline values with an evident anthropogenic control mostly in the soils of the metropolitan areas of the city of Napoli or in areas characterized by industrial or agricultural activities. These areas affected by different anthropogenic source contributions are well delimitated on the Pb background/ baseline map (Fig. 7.7A). The provincial suburban areas show Pb concentration pattern where the geogenic control is more evident; in addition in selected areas, known to have a very limited human impact and characterized by the same volcanic lithology, Pb concentrations are always lower than background upper limit of 80 mg/ kg. This background upper limit is consistent with the known concentration of Pb in Neapolitan volcanic rocks.

ACKNOWLEDGMENTS

The author is grateful to C. C. Johnson (BGS, UK) and S. Pirc (University of Ljubljana, Slovenia) for the very useful comments, which helped to improve the final manuscript.

REFERENCES

Agterberg, F. P. (2001). Multifractal simulation of geochemical map patterns. *In* "Geologic Modeling and Simulation: Computer Applications in the Earth Sciences" (D. F. Merriam and J. C. Davis, eds.), pp. 31–39. Plenum Press, New York.

Albanese, S., Cicchella, D., De Vivo, B., and Lima, A. (2006). Geochemical background and baseline values of toxic elements in stream sediments of Campania region (Italy). *J. Geochem. Explor.* **93,** 21–34. doi:10.1016/j.gexplo.2006.07.006.

Albanese, S., Lima, A., De Vivo, B., Cicchella, D., and Fedele, L. (2007). Atlante geochimico-ambientale dei suoli dell'area urbana di Avellino. Aracne Editrice, Roma, pp.188. ISBN 978–88–548–1305–2.

Cheng, Q. (1999). Spatial and scaling modelling for geochemical anomaly separation. *J. Geochem. Explor.* **65,** 175–194.

Cheng, Q. (2001). Decomposition of geochemical map patterns using scaling properties to separate anomalies from background. *In* "Proceedings of the 53rd Session of the international Statistics Institute," Seoul, Korea, 22/8–29/8, 2000.

Cheng, Q. (2003). "GeoData Analysis System (GeoDAS) for Mineral Exploration and Environmental Assessment, User's Guide (GeoDAS Phase III)." York University, Toronto, Ontario, Canada.

Cheng, Q., Agterberg, F. P., and Ballantyne, S. B. (1994a). The separation of geochemical anomalies from background by fractal methods. *J. Geochem. Explor.* **51,** 109–130.

Cheng, Q., Bonham-Carter, G. F., Agteberg, F. P., and Wright, D. F. (1994b). Fractal modelling in the goscience and implementation with gis. *In* "Proceedings 6th Canadian Conference on geographic Information Systems," Ottawa, Vol. 1, pp. 565–577.

Cheng, Q., Agterberg, F. P., and Bonham-Carter, G. F. (1996). A spatial analysis method for geochemical anomaly separation. *J. Geochem. Explor.* **56,** 183–195.

Cheng, Q., Xu, Y., and Grunsky, E. (2000). Integrated spatial and spectrum method for geochemical anomaly separation. *Nat. Resour. Res.* **9,** 43–56.

Cicchella, D., De Vivo, B., and Lima, A. (2003). Palladium and platinum accumulation in Napoli metropolitan soils from catalytic exausts. *Total Sci. Environ.* **308,** 119–129.

Cicchella, D., De Vivo, B., and Lima, A. (2005). Background and baseline concentration values of elements harmful to human health in the volcanic soils of the metropolitan and provincial area of Napoli (Italy). *Geochem. Explor. Environ. Anal.* **5,** 29–40.

De Vivo, B. (Ed.) (2006). "Volcanism in the Campania Plain: Vesuvius, Campi Flegrei and Ignimbrites." Developments in Volcanology, Vol. 9. pp. 324. Elsevier, Amsterdam.

De Vivo, B., Rolandi, G., Gans, P. B., Calvert, A., Bohrson, W. A., Spera, F. J., and Belkin, H. E. (2001). New constraints on the pyroclastic eruptive history of the Campanian volcanic Plain (Italy). *Miner. Petrol.* **73,** 47–65.

De Vivo, B., Lima, A., Albanese, S., and Cicchella, D. (2006a). Atlante geochimico-ambientale della Regione Campania. Aracne Editrice, Roma. ISBN 88–548–0819–9.

De Vivo, B., Cicchella, D., Lima, A., and Albanese, S. (2006b). Atlante geochimico-ambientale dei suoli dell'area urbana e della provincia di Napoli. Aracne Editrice, Roma. ISBN 88–548–0563–7. pp. 315

Fedele, L., De Vivo, B., Lima, A., Cicchella, D., and Albanese, S. (in press). Atlante geochimico-ambientale dei suoli dell'area comunale di Salerno. Aracne Editrice.

Frattini, P., De Vivo, B., Lima, A., and Cicchella, D. (2006a). Background and baseline concentration values of human health harmful elements and gamma ray survey in the volcanic soils of Ischia island (Italy). Geochem. Explor. Environ. Anal. 6, 325–339.

Frattini, P., Lima, A., De Vivo, B., Cicchella, D., and Albanese, S. (2006b). Atlante geochimico-ambientale dei suoli dell'Isola d'Ischia Aracne Editrice, Roma. ISBN 88–548–0818–0.

Garrett, R. G. (1989). A cry from the earth. Explore 66, 18–20.

Govett, G. J. S., Goodfellow, W. D., Chapman, A., and Chork, C. Y. (1975). Exploration geochemistry distribution of elements and recognition of anomalies. Math. Geol. 7, 415–446.

Lima, A., De Vivo, B., Cicchella, D., Cortini, M., and Albanese, S. (2003). Multifractal IDW interpolation and fractal filtering method in environmental studies: An application on regional stream sediments of Campania Region (Italy). Appl. Geochem. 18, 1853–1865.

Lima, A., Albanese, S., and Cicchella, D. (2005). Geochemical baselines for the radioelements K, U, and Th in the Campania region, Italy: A comparison of stream-sediment geochemistry and gamma-ray surveys. Appl. Geochem. 20, 611–625.

Lima, A., De Vivo, B., Grezzi, G., Albanese, S., Cicchella, D., and Fedele, L. (2007). Atlante geochimico-ambientale dei suoli dell'area urbana di Caserta. Aracne Editrice, Roma. ISBN 978–88–548–1051–8.

Lima, A., Plant, J. A., De Vivo, B., Tarvainen, T., Albanese, S., and Cicchella, D. (2008). Interpolation methods for geochemical maps: A comparative study using arsenic data from European stream waters. In "Special Issue on Environmental Geochemistry" (B. De Vivo, J. A. Plant, and A. Lima, eds.), Vol. 8, pp. 1–9. Geochemistry: Exploration, Environment, Analysis.

Paone, A., Ayuso, R. A., and De Vivo, B. (2001). A metallogenic survey of alkalic rocks of Mt. Somma-Vesuvius volcano. Mineral. Petrol. 73, 201–233.

Matschullat, J., Ottenstein, R., and Reimann, C. (2000). Geochemical background—Can we calculate it? Environ. Geol. 39, 990–1000.

Miesch, A. T. (1981). Estimation of the geochemical threshold and its statistical significance. J. Geochem. Explor. 16, 49–76.

Salminen, R., and Gregorauskiene, V. (2000). Considerations regarding the definition of a geochemical baseline of elements in the surficial materials in areas differing in basic geology. Appl. Geochem. 15, 647–653.

Sinclair, A. J. (1974). Estimation of the geochemical threshold and its statistical significance. J. Geochem. Explor. 3, 129–149.

Sinclair, A. J. (1976). Application of probability graphs in mineral exploration. Assoc. Explor. Geochem. 4, 1–95.

Sinclair, A. J. (1991). A fundamental approach to threshold estimation in exploration geochemistry. Probability plots revisited. J. Geochem. Explor. 41, 1–22.

Stanley, C. R. (1988). Comparison of data classification procedures in applied geochemistry using Monte Carlo simulation Unpublished Ph.D. Thesis. University of British Columbia, Vancouver.

Stanley, C. R., and Sinclair, A. J. (1987). Anomaly recognition for multi-element geochemical data—A background characterization approach. J. Geochem. Explor. 29, 333–351.

Stanley, C. R., and Sinclair, A. J. (1989). Comparison of probability plots and gap statistics in the selection of threshold for exploration geochemistry data. J. Geochem. Explor. 32, 355–357.

Turcotte, D. L. (1997). "Fractals in Geology and Geophysics," 2nd edn. Cambridge University Press, New York.

CHAPTER EIGHT

URBAN GEOCHEMICAL MAPPING

Stefano Albanese,* Domenico Cicchella,† Annamaria Lima,*
and Benedetto De Vivo*

Contents

Abstract

Urban soil geochemistry is strongly influenced by the effect of human activities on the environment. Building and industrial activities, motor vehicle emissions and many other factors are responsible for releasing large amounts of organic and inorganic pollutants to the environment. At the urban scale, trace metals are the most diffuse contaminants, and soils can be considered the main receptacle of these elements. Many different approaches to urban soil mapping are reported in the literature; thus, the aim of this chapter is to synthesise the main considerations necessary to undertake urban mapping activities in terms of planning, sampling, chemical analyses and data presentation.

In this context, modern geographical information systems (GIS) represent an indispensable tool for better understanding the distribution, dispersion and interaction processes of some toxic and potentially toxic elements. Discussion on the use of GIS in the urban environment is, therefore, also provided.

* Dipartimento di Scienze della Terra, Univerità degli Studi di Napoli 'Federico II', Via Mezzocannone 8, 80134 Napoli, Italy
† Dipartimento di Studi Geologici e Ambientali, Università del Sannio, Via Port'Arsa 11, 82100 Benevento, Italy

Environmental Geochemistry
DOI: 10.1016/B978-0-444-53159-9.00008-5

© 2008 Elsevier B.V.
All rights reserved.

1. INTRODUCTION

The composition and geochemistry of urban soils are generally highly influenced by anthropogenic activities. Urban soils are mainly characterised by strong lateral spatial variability in composition and structure due to mixing with dusts and rubble during building activities, compaction and contamination phenomena related to the presence of industrial settlements, human mobility (e.g., motor vehicle emissions) and time-limited events (waste disposal, watercourse malfunctioning, etc.). Essentially, urban soils do not have the typical vertical stratification associated with naturally developed soils. Over the short history of an urban area, paedogenetic processes do not play a fundamental role in urban soil development, and soil section is much more dependent on the local accumulation of building materials, excavation remnants or the presence of underground infrastructures.

As a result of the above factors, the planning of soil sampling activities in urban areas is challenging since there is the need for high-density sampling, to define a suitable framework of the environmental status, set against the difficulty of identifying physical spaces where sampling is possible or allowed.

Urban soils contain several contaminants: fertilizers (or pesticides from cultivated land brought from the countryside to refill flower beds or used in public parks and gardens); heavy metals deposited on the soil surface by motor vehicle emissions, industrial processes, organic wastes, etc. Depending on their chemical nature, pesticides can be evaporated, immobilized by colloidal processes and, in some cases, they can migrate to water-bearing strata or the rhizosphere affecting plant growth and health.

Heavy metals (mainly As, Pb, Zn, Ni, Hg, Cu, Cd and Cr) can be introduced into the urban environment by a number of different agents, for example, domestic activities, residential heating, incinerators, electrical power plants, industrial boilers, gasoline and diesel vehicles, factories, spreading of fertilizers and anti-cryptogams on agricultural soils surrounding residential areas. Although they are usually adsorbed by colloids or organic matter, heavy metals can enter the biological cycle through plant uptake processes, the effect of which is dependant on plant genres, soil nature and boundary conditions (the presence of water, temperature, pH and Eh).

Heavy metals concentration generally decreases in urban soils away from the main road network and with increasing depth of sampling. This can be explained by the strong dependence of these contaminants on the use of motor vehicles—leaded fuels for Pb, tire wear for Zn and Cd, brake pads for Sb, converters and exhaust systems for platinum group elements (PGEs).

It has been demonstrated that catalytic converters are responsible for releasing considerable amounts of Pt ant Pd to the urban environment (Cicchella *et al.*, 2003, 2008a; Schafer and Puchelt, 1998; Zereini *et al.*, 1994), which has caused an increase in the incidence of allergies and problems related to the respiratory system (Rosner and Merget, 2000; Von Hoff *et al.*, 1976).

Other organic contaminants, such as polychlorinated biphenyls (PCBs), polycyclic aromatic hydrocarbons (PAHs) and dioxins (PCDDs), released from domestic wastes, plastics ignition, malfunctioning domestic heating systems, industrial processes (Breivik *et al.*, 2002; Cretney *et al.*, 1985; Vizard *et al.*, 2006) contribute to the degradation of the urban environment. PCBs are also highly concentrated in building materials (plaster and paints) (Andersson *et al.*, 2004; Kuusisto *et al.*, 2007); PAHs are released to the environment by coal and wood burning processes, diesel engines and coke plants (Khalili *et al.*, 1995; Yang *et al.*, 1991).

2. Definition of Geochemical Background and Baseline at an Urban Scale

It is very difficult to determine the background (geogenic) value of an element in an urban soil since the diffuse nature of the pollution, mostly dependent on both the fallout processes and the 'mobility' of the sources (motor vehicles) along the road network, makes it virtually impossible to find places that may be considered completely uncontaminated. In general terms, the only available option for defining background reference values for the soils of an urban area is to characterise soil samples collected at presumed uncontaminated locations at a distance from the urban area that are of the same paedological and geological nature as the urban soils. Alternatively, especially if the original soil stratification has not been altered by human activities, geochemical background can be defined by characterising soil samples collected at a certain depth from the surface, since contamination of urban soils is mostly superficial. The C horizon of a stratified soil, if available, is the most suitable for this purpose, and the ratio between the elemental concentrations determined in the topmost layer of the soil (topsoil) and in the C (or base) horizon (subsoil) could be used to quantify the impact of human activities on the original environment (Fordyce *et al.*, 2005). In the urban environment, the concept of baseline, which indicates the actual content of an element in the superficial environment at a given point (Salminen and Gregorauskiene, 2000), and background are not coincident. Generally, if we consider an urban area as part of a regional context, we can define two baselines and, as consequence, two anomaly thresholds (Fig. 8.1):

- the regional anomaly threshold, which is, basically, coincident with the upper limit of the natural background concentration interval;
- the local anomaly threshold, which is referred to the baseline of the area influenced by diffuse urban pollution.

3. Planning Urban Geochemical Mapping

Planning of a sampling activity has to be anticipated by a careful study of the human and environmental variables likely to affect urban soils characteristics.

To realise an environmental study of an urban area, it is essential to have knowledge of city socio-economical development, to be aware of the location of

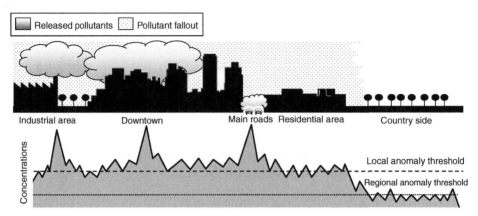

Figure 8.1 Regional and local anomaly threshold applied to an urban environment. The concentration graph shows the influence of anthropogenic sources (industrial area, downtown and main roads) on pollutant distribution in soils.

main industrial activities and to identify all latest changes in production activities. Moreover, even before detailed planning, it is useful to find out if the urban area has been influenced by extraordinary events, such as devastating blazes or explosions, which might have caused significant contamination of the surrounding soils.

Before sampling, it is important to be aware of the nature and origin of the soil (indigenous or imported). Gathering this information will lead to a more scientifically robust interpretation of data, being able to differentiate and characterise contaminated areas as distinct from baseline values.

Once all the above has been considered, it is possible for the urban area to be defined via a sampling grid, generally composed of square or triangular cells. The dimensions of the sampling cells should be capable of identifying elemental variance. Usually, at least, two orders of regular cells are used: sampling cells with larger dimensions for areas characterised by low anthropic pressure; and cells with smaller dimensions, which are generally obtained by dividing the larger cells into four identical sub-cells, to allow for higher sample density in potentially contaminated zones, for example, in downtown areas and where there is extensive industrial development.

Sampling of the main urban areas of Campania region (Albanese *et al.*, 2007; Cicchella *et al.*, in press; De Vivo *et al.*, 2006; Fedele *et al.*, in press; Lima *et al.*, 2007) has been based on grids with square cells of 1 km^2 and 500 m^2, respectively, for low anthropic pressure zones and for potentially polluted areas (downtown, industrial areas, etc.). Šajn *et al.* (1998) applied to the city of Ljubliana (Slovenia) a sampling grid with three cell orders: 1 km^2 for the surroundings, 500 m^2 for the town and 250 m^2 for clearly polluted zones.

Once the sampling grid has been defined, each cell should be labelled with an alphanumeric code later used to identify correspondent samples collected in the field.

4. SAMPLING PROTOCOLS AND FIELD ACTIVITIES

Before starting field activities, it is essential to define a sampling procedure designed to minimise human errors. This protocol should also respect international scientific community sampling guidance in order to encourage comparison with environmental research carried out in other urban areas.

In the framework of the activities carried out by the IUGS/IAGC Working Group on Global Geochemical Baselines with the aim of implementing the recommendations given by Darnley *et al.* (1995), internationally agreed methods for geochemical sampling were defined by the European scientific community under the aegis of the FOREGS Geochemistry Task Force and published in a field manual (Salminen *et al.*, 1998), which provides detailed instructions for sampling environmental media in Europe and other glaciated terrains. It should be noted, however, that this manual concentrates on regional rather than urban mapping.

Although Salminen *et al.* (1998) suggest the 0–25 cm depth range as suitable for topsoil sampling, in urban areas it is important to collect soil samples localised over a very shallow and narrow depth range in order to prevent exclusion of contaminants deposited by atmospheric fallout. There is no common accepted agreement on the sampling depth range for urban topsoils. For instance, Ottesen and Langedal (2001) used the 0–2 cm depth range for the city of Trondheim (Norway), Šajn *et al.* (1998) used the 0–5 cm depth range for the city of Ljubljana (Slovenia), Li *et al.* (2001) sampled the top 10 cm layer of the soil profile from the city of Hong Kong, De Vivo *et al.* (2006), Albanese *et al.* (2007), Lima *et al.* (2007), Cicchella *et al.* (in press) and Fedele *et al.* (in press) collected the top 15 cm from the main urban areas of Campania region (Italy), as did Kelly *et al.* (1996) for industrial and non-industrial areas of Britain. Fordyce *et al.* (2005) collected soil samples from both the 5–20 and 20–50 cm ranges to define, respectively, the chemistry of 'near-surface soil', that is, that likely to be contaminated, as distinct from the 'substrate materials', that is, basically geogenic material.

Under ideal conditions, soil samples should be collected from near to the centre of each sampling cell. This aspiration may not always be feasible because of the presence of buildings or other exclusion zones. In such cases, every attempt should be made to carry out the sampling respecting, at least, the cell boundary. Soil sampling adjacent to roads should be avoided since atmospheric agents (wind, rain, etc.) can produce an over-accumulation of contaminants.

Sampling tools (scoops and samples containers) should be manufactured from appropriate quality plastic materials or stainless steel. Operators should not wear or carry any jewellery or metallic apparel.

Before sampling, living surface vegetation, fresh litter, big roots and rock fragments (stones) should be removed from the sampling area. To minimise localised contaminant inhomogeneity effects, each sample should be a field composite sample based on 3 to 5 sub-samples with a minimum distance between any two sub-samples of not <5 m (Fig. 8.2). Each sub-sample should have an approximate weight of 0.5 kg to obtain a composite sample of about 2 kg. Depending on the quantity of soil

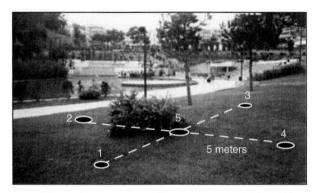

Figure 8.2   An example of five sub-samples collected at an urban sampling site.

required for analysis, a composite sample could be prepared by quartering —the process of reducing a representative soil sample to a convenient size, by homogenizing and splitting the composite sample into sub-aliquots (Schumacher *et al.*, 1991). The final soil sample should be stored in inert plastic bags and labelled with a water resistant pen.

It is important to retrieve geographic coordinates of the sampling point from both a topographic map and a GPS receiver (to compare their precision) and to assign an unequivocal location to each collected sample. Any necessary coordinate transformation and projection should be avoided in the field.

5. Sample Preparation and Analyses

According to the FOREGS recommendations (Salminen et al., 1998), soil samples should be dried at 40 °C or less to avoid Hg vaporisation. Samples can be air-dried or dried using IR lamps controlled by a thermal probe to keep the temperature of samples below 40 °C.

After drying, soil samples should be disaggregated and homogenised in a ceramic mortar, then sieved to retain the <2 mm fraction. After a further splitting process, one portion of the sample should be archived for future studies while the remaining aliquot should be pulverised in an agate shutter cone mill to a grain size <0.63 mm before chemical elemental analyses.

One of the most commonly used analytical methods for determining rare earth (REE) and major element concentrations in soils is inductively coupled plasma mass spectrometry (ICP-MS). This analytical method is highly sensitive and capable of the determination of a range of metals and several non-metals (elements with atomic mass ranging from 7 to 250, including both Li and U) at concentrations below one part in 10^{12}. It is based on coupling an ICP as a method of producing ions (ionization) with a mass spectrometer as a method of separating and detecting the ions.

Plasma source mass spectrometry has also been used in the identification and differentiation of trace metals between anthropogenic and natural sources using an isotopic technique. Although, in the past, isotopic studies were mainly concentrated on Pb (^{206}Pb, ^{207}Pb and ^{208}Pb) (Ault *et al.*, 1970; Callender and Rice, 2000; Hansmann and Koppel, 2000; Semlali *et al.*, 2001, 2004), modern techniques now also enable the characterisation of Cu and Zn isotopic concentrations in soils (Archer and Vance, 2004).

Cicchella *et al.* (2005, 2008b) demonstrated how the Pb concentrations in the volcanic soils of the city of Napoli, Italy, were derived from both the natural content of Pb from underlying parent materials and by the use of fuels containing Pb additives (Fig. 8.3).

Inductively coupled plasma atomic emission spectroscopy (ICP-AES) and x-ray fluorescence spectrometry (XRFS) are also used for elemental determination in environmental studies, although they are generally less sensitive than ICP-MS techniques.

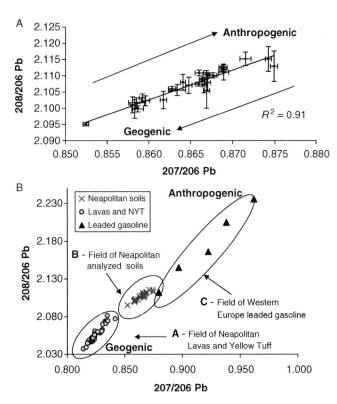

Figure 8.3 Pb isotopic concentrations in the urban area of Napoli, Italy (Cicchella *et al.*, 2008b). (A) Plot of ^{207}Pb/^{206}Pb vs. ^{208}Pb/^{206}Pb. In the diagram are also reported the 2σ errors and the regression coefficient (R^2). (B) ^{207}Pb/^{206}Pb vs. ^{208}Pb/^{206}Pb isotopic composition of samples collected as part of this study and other data reported in the literature.

With soil being a solid medium, it has to be solubilised through a leaching process before analysis. Leaching is the process of extracting a substance from a solid by dissolving it in a liquid. Depending on the solvent used, the leach extraction can result in either total or partial dissolution of the soil matrix. Typically in environmental studies, extraction methods with a range of different dissolution strengths are used.

Aqua regia extraction is a strong partial extraction method that dissolves carbonates, most sulphide minerals, some silicates like olivine and trioctahedral micas, clay minerals and primary and secondary salts and hydroxides (Salminen, 1995). It can be considered a 'quasi-total' extraction method, since actual total concentrations can be higher. On the other hand, this leaching method overestimates the bioavailable amount of toxic elements in a soil since metals trapped in the silicate lattice are released very slowly in the environment and are not easily involved in plant nutrition processes.

The definition of bioavailability of toxic metals in urban soils is considered an important topic since the high-demographic density of urban areas increases the likelihood of human beings (especially children) ingesting soils directly by hand-to-mouth activity (Ljung *et al.*, 2006) or indirectly through food produce (Siegel, 2002) grown in private gardens.

Extraction methods based on solutions of chelating agents, such as EDTA and DTPA, or salts of weak acids, such as ammonium acetate (CH_3COONH_4) (Lakanen and Erviö, 1971; Lindsay and Norwell, 1969), can be used to assess plant-available trace element contents of soils and to evaluate more robustly human exposure to environmental risks in an urban context. A study carried out in the main cities of Campania region in southern Italy (Albanese, 2008) demonstrated how the bioavailable concentrations of some trace elements such as Zn, Pb and Cu in soils, determined using ammonium acetate-EDTA extraction, are much lower than the elemental concentrations determined by an aqua regia extraction on the same samples (Table 8.1).

Apart from the extraction and analytical methods, it is fundamental that the quality of analytical data is demonstrated by running quality controls (QC). De Vivo *et al.* (2004) recommend to validate analytical data by evaluating precision (which expresses the extent of reproducibility of analytical determinations) and accuracy (which expresses the degree of correctness of the data compared with 'true' values) of results through the use of standard samples analysed together with the samples collected during the mapping activities. The accuracy and precision so determined should not exceed ±15% for the analytical results to be considered acceptable.

 ## 6. GEOCHEMICAL DATA PRESENTATION

Geochemical data from an urban environment can be elaborated and presented in different ways, depending on the aims of the research project.

Basically an urban geochemical mapping project should aim to:

- draw a picture of the distribution of elemental concentrations in the environment, that is, the geochemical baseline;

Table 8.1 Maximum, minimum, mean and standard deviation (expressed as milligram per kilogram) values of ammonium acetate-EDTA (AA-EDTA) and aqua regia elemental concentrations in 29 soil samples from the city of napoli (Albanese, 2008)

Elements	N	AA-EDTA				Aqua regia			
		Max.	Min.	Mean	St. Dev.	Max.	Min.	Mean	St. Dev.
As	29	1.5	0.3	0.66	0.29	79.7	7.1	14.3	12.8
Cd	29	1.6	0.06	0.55	0.32	2.03	0.18	0.8	0.39
Co	29	7.2	0.1	0.77	1.26	36.6	4.1	9.16	5.75
Cu	29	266	24.6	90.1	60.2	343	43.6	150	83.6
Ni	29	4.90	<0.1	1.26	0.95	36.3	4.1	15.6	7.25
Pb	29	389	31.9	158	88.5	734	103	335	140
Sb	29	0.57	0.03	0.23	0.14	9.53	1.18	4.87	2.44
Tl	29	0.16	0.03	0.09	0.03	2.15	0.57	1.07	0.34
V	29	7	<2	2.17	1.2	106	33	62.3	16.9
Zn	29	327	20	156	73.1	612	79.4	323	126

Note: It is clearly evident that bioavailable concentrations (evaluated through the use of ammonium acetate–EDTA extraction) are significantly lower than aqua regia determined concentrations for the elements considered. N = number of samples analysed, St. Dev. = one sigma.

- establish the natural geochemical background to determine the burden of the anthropogenic activities on the urban environment;
- delineate areas of the urban territory exceeding trigger and action levels (established by local governments) where remediation should be planned to restore the environmental 'status quo';
- individuate principal and 'hidden' sources of pollution to eliminate.

In many cases not all the above listed points are taken into account by researchers, depending on their own scientific approaches and specific needs.

Tijhuis (2003) exclusively chooses soil and sand from kindergartens and playgrounds of Oslo, Norway, as sample materials, aiming much more to evaluate the health risk to children rather than to generate geochemical maps.

Ottesen and Langedal (2001), basically in accordance with the example given by Birke and Rauch (1994, 2000), use geochemical data from surface soils of the city of Trondheim, Norway, to produce elemental distribution maps and to evaluate enrichment arising from anthropogenic activities by comparing them with the natural background represented by the deepest part of the overbank sediment profiles sampled in the surrounding region. The impact of potential pollution sources on soils is also evaluated by means of factor analysis and direct observations.

Cicchella *et al.* (2005) and De Vivo *et al.* (2006) approach the study of the urban area of Napoli, Italy, from a 'regional' point of view, using soil samples from the city territory and from its provincial area. While elemental distribution maps are produced separately for both the urban area and the provincial territory, the geochemical data from all 982 soil samples are used to generate single provincial baseline maps for each considered element. Baseline values, generated by a geostatistic process implemented by a geochemistry dedicated GIS software called Geo-DAS (Cheng, 2003), after elimination of statistically anomalous concentration values, have been used to present both 'natural background' data for the areas, outside the urban context, and as sum of background and diffuse contamination (baseline) in the urban areas.

6.1. GIS aided techniques for urban geochemical data presentation

The use of geographical information systems (GIS) has greatly enhanced the ability to manage and display geochemical data arising from urban soil mapping activities.

Apart from the geostatistical analysis capabilities offered by modern GIS software to improve geochemical modelling, a large array of geographically localised variables, such as residential buildings, infrastructures, road networks, industries and time-limited pollution sources, that may influence the distribution and mobility of elements in the urban surface can be represented and superimposed on geochemical data to improve the interpretation process (Thums and Farago, 2001; Wong *et al.*, 2006).

6.2. Dot maps

The distribution of geochemical data may be presented, in their most basic form, through the use of dot maps in which dots, or other symbols selected by the operator, of a size proportional to concentration are plotted at geographical coordinates for each sampled site. Dot sizes may be classified by means of concentration intervals based on percentiles or other statistical methods. Dot maps are often the most suitable method to represent isolated points of a discrete set of geochemical data.

In accordance with Siegel (2002), if geochemical data are characterised by a normal (Gaussian) statistical distribution, the intervals defined by the arithmetic mean ± 1 standard deviation ($x \pm 1\sigma$) can be used to map local fluctuation around the mean (which is considered the threshold value for background), intervals between $x \pm 1\sigma$ and $x \pm 2\sigma$ represent a mixing of background and anomalous values, while concentrations $>x \pm 2\sigma$ could be used to identify soil samples whose concentrations are strictly influenced by contamination sources (anomalies). Furthermore, since geochemical data from an urban area are derived from elemental contributions from various sources, cumulative frequency probability plots can be used as a method for defining data intervals by analysing curve inflection points (Fig. 8.4) (Albanese *et al.*, 2007; Cicchella *et al.*, in press; De Vivo *et al.*, 2006; Fedele *et al.*, in press; Lima *et al.*, 2007).

6.3. Interpolation

Spatial interpolation of geochemical data has greatly improved the quality of geochemical data presentation and subsequent interpretation over the past 20 years or so.

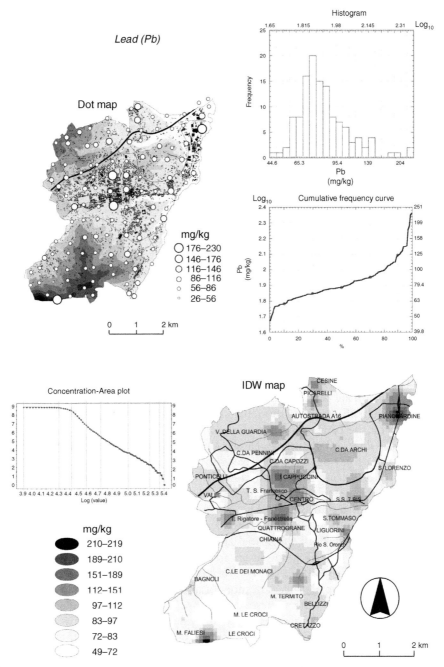

Figure 8.4 Lead dot and interpolated data distribution maps of the municipal area of Avellino, Italy (Albanese *et al.*, 2007). Histogram, frequency curve and C–A plot, respectively, used to reclassify dot intervals and MIDW interpolated grids are also shown in the map.

Interpolation, which is a process to generate a continuous surface from a discrete set of geochemical data by applying specific mathematical algorithms, does not provide deterministic information for non-sampled areas but does define, in a probabilistic way, the expected values. As a consequence, because of the considerable geochemical data variability in urban areas, any interpolation process should be based on a large amount of observed concentrations and be capable of retaining local singularities (expressed as dot anomalies).

Interpolation methods such as Trend Surface Analysis, Interpolated Distance Weighted (IDW) and Kriging, widely used in regional geochemical mapping, tend to smooth local singularities and to over- or under-estimate anomalies, since interpolated surfaces do not necessarily pass through the original sample concentrations (Cheng *et al.*, 1994b; Lima *et al.*, 2003, 2008).

For the geochemical mapping of the main urban areas of the Campania region of Italy (Albanese *et al.*, 2007; Cicchella *et al.*, in press; De Vivo *et al.*, 2006; Fedele *et al.*, in press; Lima *et al.*, 2007), a multifractal interpolation method, called Multifractal Inverse Distance Weighted (MIDW) (Cheng, 1999a, b), has been applied. The concepts of both fractality and multi-fractality have been widely used to describe complexity and self-similarity of natural phenomena. The multifractal interpolation process, implemented by the software GeoDAS (Cheng, 2003), is capable of retaining high-density information (anomalies), usually lost by 'conventional' interpolation methods, taking into account both the spatial association and the local singularity of data (Xu and Cheng, 2001).

Surfaces generated by the MIDW algorithm are reclassified by means of the concentration–area fractal method (C–A) developed by Cheng *et al.* (1994a) for geochemical anomaly separation.

C–A is a classification method that can be used to separate anomalies from background on the basis of geochemical concentration values, spatial variability of geochemical values and geometrical and scaling properties. It involves a plot of concentration values (C) against the area with concentration value above a cut-off (Fig. 8.4). From a 'multifractal' point of view, the extreme values, which are anomalous from a geochemical point of view, may follow a fractal rather than a normal or lognormal distribution. The C–A plot can be used to distinguish anomalies from background on the basis of concentration values (i.e., a frequency distribution of values) as well as the spatial and geometrical properties of geochemical patterns. A C–A plot on a log–log scale can be used to establish power–law relationships between the areas A (A $\geq$ C) with the concentration values greater than C and the concentration value C itself. A number of straight-line segments can be manually or automatically fitted to the values of A (A $\geq$ C) versus C for various values of C plotted on the logarithmic scale, each representing a power–law relationship between the area A and the cut-off concentration value C. The intersections of these straight-line segments provide a set of cut-off values for subdividing the concentration scale into discrete classes. On the map, these classes are zones that often highlight the effects of underlying geochemical processes such as contamination sources in an urban environment as well as patterns that are due to geological factors. Further details can be found in Cheng *et al.* (1994a, b, 1996, 2000, 2001), Lima *et al.* (2003) and Cicchella *et al.* (2005).

6.4. Background and baseline maps

Statistical methods based on histograms, cumulative frequency probability curves (see above), univariate and multivariate data analysis (Miesh, 1981; Sinclair, 1974, 1976, 1991; Stanley, 1987) are widely used to separate geochemical baseline (natural and/or anthropogenic) values from anomalies.

Cheng et al. (2000) propose a fractal filtering technique to separate geochemical anomalies from baseline values based on a Fourier spectral analysis. The interpolated maps generated from geochemical data by means of MIDW (see above) are transformed into the frequency domain in which a spatial concentration-area fractal (S–A) method is applied to distinguish the patterns on the basis of the power-spectrum distribution. A log–log plot shows the relationship between the area and the power-spectrum values on the Fourier-transformed map of the power spectrum. The values on the log–log plot are modelled by fitting straight lines using least squares. Distinct classes can be generated, such as lower-, intermediate- and high-power-spectrum values, which correspond approximately to baseline values, anomalies and noise of geochemical values in the spatial domain (Fig. 8.5). An irregular filter is applied on these distinct patterns to remove the anomalies and noise related to intermediate- and high-power-spectrum values. The image, converted back to a spatial domain with the filter applied, shows patterns that, after the removal of anomalies and noise, indicate distinct areas, representing baseline geochemical patterns for a considered geochemical element (Cheng et al., 1994a,b, 1996, 2000; Cheng, 1999a, b) (Fig. 8.6). In the same way, an anomaly map can be obtained by applying a band type filter (selecting the second and third break as the cut-off on the S–A plot) to remove noise and baseline related to lower- and high-power-spectrum values. The pixel values of baseline and anomaly geochemical maps can be sorted and reclassified using the C–A method.

This S–A method has already been applied in environmental studies on a regional basis by Lima et al. (2003) and Lima (2008), while at urban scale it has been applied by Cicchella et al. (2005) and De Vivo et al. (2006), as specified above. In Fig. 8.6, the baseline geochemical soil maps of Pb, Hg, Se, Tl, V and Zn, representing the

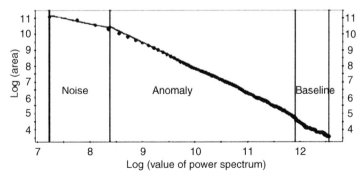

Figure 8.5 Log–log plot of the power-spectrum values for Pb in the urban and provincial area of Napoli, Italy (Cicchella et al., 2005). Straight lines are fitted by means of least-squares regression. The three segments, representing noise, anomaly and background, have slopes of −0.69, −1.59 and −1.54, respectively.

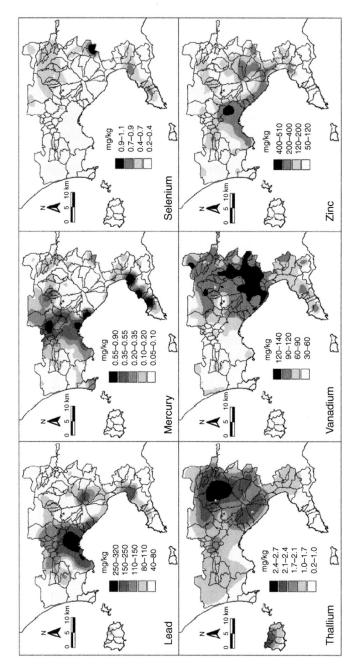

Figure 8.6 Pb, Hg, Se, Tl, V and Zn baseline geochemical soil maps of the metropolitan and provincial areas of Napoli, Italy (Cicchella *et al.*, 2005; De Vivo *et al.*, 2006).

metropolitan and provincial areas of Napoli (Cicchella *et al.*, 2005; De Vivo *et al.*, 2006), are shown.

At an urban scale, in some cases (Albanese *et al.*, 2007; Cicchella *et al.*, in press; Lima *et al.*, 2007), the use of this technique has been unsuitable due to both the small dimensions of the contaminated zones, characterised by a weak environmental impact on the surroundings, and the uniformity of the underlying geology along the whole urban area. Application of the C–A method to the MIDW interpolated grids and a careful observation of concentration intervals applied to the interpolated surfaces have led to a satisfactory definition of both background/baseline and anomalous values in most cases.

6.5. Multivariate analysis and scores mapping

Multivariate statistical analysis is considered a useful tool for evaluating the significance of geochemical anomalies in relation to both any individual variable and the mutual influence of variables on each other. In basic terms, when applied to geochemistry, multivariate analysis aims to identify spatial correlations between groups of elements—lithological characteristics, enrichment phenomena, anthropogenic pollution, etc.—in a complex system and reduce a multidimensional data set to more basic components.

Principal component analysis (PCA), factor analysis (FA) and cluster analysis (CA) are some of the most widely used multivariate analysis techniques applied to geochemistry.

The result of a multivariate analysis is an array of data in which elements are grouped as associations by means of their correlation coefficients or other measures of association.

Geochemists may interpret the correlations and relate each elemental association to specific phenomena (geology, pollution sources, geochemical processes, etc.).

The results of PCA and FA are usually discussed in terms of scores and loadings. Scores represent the incidence of a selected association of elements, expressed as a dimensionless value, at each sampled site, and can be mapped using dot or interpolated maps.

Šajn *et al.* (1998), De Vivo *et al.* (2006), Albanese *et al.* (2007), Lima *et al.* (2007), Cicchella *et al.* (in press) and Fedele *et al.* (in press) use the R-mode FA (Miesch Programs, 1990) to identify elemental associations and to explain data variability in urban soils of the main city of Campania region. Their results highlight how elements such as Cd, Cr, Cu, Pb, Sb and Zn are often associated and higher factor scores spatially relate to the presence of anthropogenic sources in an urban context (Fig. 8.7).

Lee *et al.* (2006) use both PCA and CA to determine metal associations arising from anthropogenic input to the soils of Hong Kong. The analysis identifies Cu, Pb and Zn as the main elements responsible for pollution in the area. The author uses the data to calculate and map the distribution of a soil pollution index (SPI), which is indicative of the traffic impact on the urban territory (Fig. 8.8).

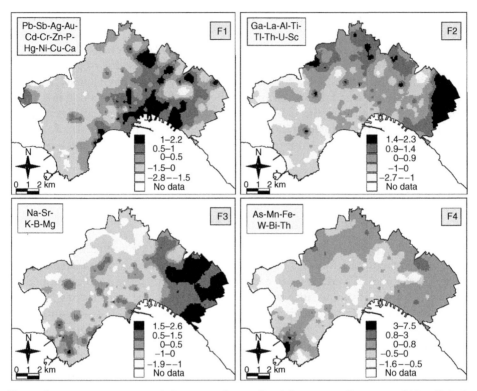

Figure 8.7 Factor scores association maps from soils of the metropolitan area of Napoli, Italy (Cicchella *et al.*, 2008b).

6.6. Risk mapping

In urban soils, concentration values of harmful elements often exceed trigger and action levels (TAL) established by local governments for both residential and industrial land use.

Risk maps do not need any peculiar mapping technique to be produced, but are based on the reclassification of geochemical data, represented as dots or interpolated, by means of TAL established by law for each element of concern. The main aim of a risk map is to highlight zones of urban areas that are in excess of TAL values and hence require potential remediation. It should be noted, however, that background values (unaffected by anthropogenic input) may exceed TAL values in some circumstances.

Thallium in volcanic soils of Napoli (Italy) is characterised by a range of background values between 0.2 and 1.5 mg/kg (Cicchella *et al.*, 2005; De Vivo *et al.*, 2006); the TAL for residential use established by the Italian law 152/06 (Ministero dell'Ambiente, 2006) is 1 mg/kg. As a consequence, the majority of the city territory should technically undergo remediation, even though there is strong evidence to suggest that the Tl is predominantly of geogenic origin. Albanese (2008) suggests that risk should be mapped on the basis of bioavailable

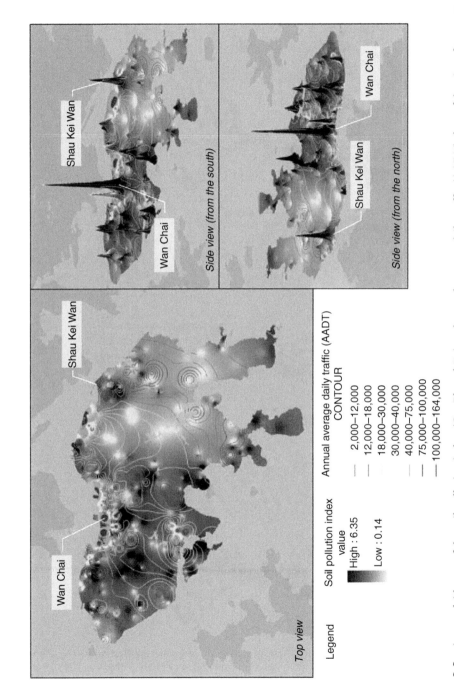

Figure 8.8 An overlaid map of the soil pollution index (Cu, Pb and Zn) and annual average daily traffic (AADT) data of the city of Hong Kong proposed by Lee et al. (2006).

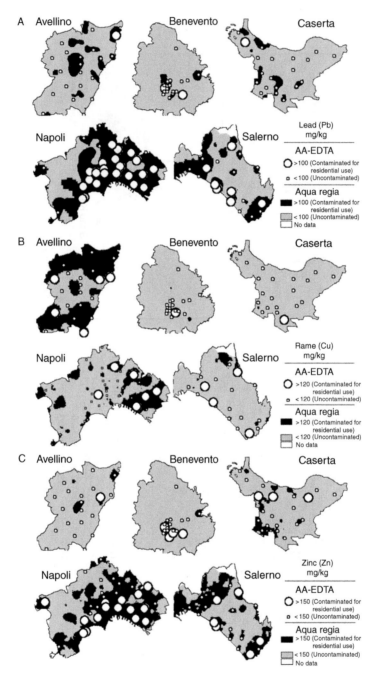

Figure 8.9 Interpolated distribution of 'quasi–total' elemental concentrations of Pb (A), Zn (B) and Cu (C) in the soils of the main cities of the Campania region of Italy (Albanese, 2008), based on aqua regia concentrations, compared with the 'bioavailable (ammonium acetate-EDTA)' concentrations (dots). The data were reclassified by means of the intervention levels established by Italian law D.M. 152/99 (Ministero dell'Ambiente, 2006).

concentrations of elements (using the EDTA method, see above) instead of their aqua regia concentrations. In the main cities of Campania region (including Napoli), bioavailable concentrations of geogenic elements such as Tl and As are always significantly below the corresponding TAL values. Contaminants shown to be derived principally from anthropogenic sources cover a significantly smaller area of land that would require subsequent remediation (Fig. 8.9). Furthermore, it is suggested that percent bioavailability of elements should be used as a marker of their geogenic or anthropogenic nature, thus giving to the environmental experts a more suitable tool for defining remediation plans mainly based on the evaluation of human exposure to environmental risks.

ACKNOWLEDGMENTS

The authors are grateful to S. Reeder (BGS, UK) and S. Pirc (University of Ljubljana, Slovenia) for the very useful comments, which helped to improve the final manuscript.

REFERENCES

Albanese, S. (2008). Evaluation of the bioavailability of potentially harmful elements in urban soils through ammonium acetate-EDTA extraction: A case study in southern Italy. *Geochem.: Explor. Environ. Anal.* **8,** 49–57.

Albanese, S., Lima, A., De Vivo, B., and Cicchella, D. (2007). "Atlante geochimico-ambientale dei suoli del territorio comunale di Avellino/Geochemical Environmental Atlas of the soils of Avellino" p. 192. Aracne Editrice, Roma. ISBN 978-88-548-1305-2.

Andersson, M., Ottesen, R. T., and Volden, T. (2004). Building materials as a source of PCB pollution in Bergen, Norway. *Sci. Total Environ.* **325**(1–3), 139–144.

Archer, C., and Vance, D. (2004). Mass discrimination correction in multiple collector plasma source mass spectrometry: An example using Cu and Zn isotopes. *J. Anal. At. Spectrom.* **19,** 656–665.

Ault, W. U., Senechal, R. G., and Erlebach, W. E. (1970). Isotopic composition as a natural tracer of lead in the environment. *Environ. Sci. Technol.* **4,** 305–313.

Birke, M., and Rauch, U. (1994). "Geochemical Investigation of the Urban Area of Berlin," p. 13. Federal Institute of Geosciences and Natural Resources, Branch Office Berlin, Germany.

Birke, M., and Rauch, U. (2000). Urban geochemistry: Investigations in the Berlin metropolitan area. *Environ. Geochem. Health* **22**(3), 233–248.

Breivik, K., Sweetman, A., Pacyna, J., and Jones, K. (2002). Towards a global historical emission inventory for selected PCB congeners—A mass balance approach 1. Global production and consumption. *Sci. Total Environ.* **290,** 181–198.

Callender, E., and Rice, K. C. (2000). The urban environmental gradient: Anthropogenic influences on the spatial and temporal distributions of lead and zinc in sediments. *Environ. Sci. Technol.* **34,** 232–238.

Cheng, Q. (1999a). Multifractality and spatial statistics. *Comput. Geosci.* **25**(10), 946–961.

Cheng, Q. (1999b). Spatial and scaling modeling for geochemical anomaly separation. *J. Geochem. Explor.* **65**(3), 175–194.

Cheng, Q. (2003). "GeoData Analysis System (GeoDAS) for Mineral Exploration and Environmental Assessment, User's Guide (GeoDAS Phase III)." York University, Toronto, Canada.

Cheng, Q., Agterberg, F. P., and Ballantyne, S. B. (1994a). The separation of geochemical anomalies from background by fractal methods. *J. Geochem. Explor.* **51,** 109–130.

Cheng, Q., Bonham-Carter, G. F., Agteberg, F. P., and Wright, D. F. (1994b). Fractal modelling in the geosciences and implementation with GIS. *In* "Proceedings 6th Canadian Conference on geographic Information Systems," Vol. 1, pp. 565–577. Ottawa.

Cheng, Q., Agterberg, F. P., and Bonham-Carter, G. F. (1996). A spatial analysis method for geochemical anomaly separation. *J. Geochem. Explor.* **56,** 183–195.

Cheng, Q., Xu, Y., and Grunsky, E. (2000). Integrated spatial and spectrum method for geochemical anomaly separation. *Nat. Res. Res.* **9,** 43–56.

Cheng, Q., Bonham-Carter, G. F., and Raines, G. L. (2001). GeoDAS-A new GIS system for spatial analysis of geochemical data sets for mineral exploration and environmental assessment. *In* "The 20th International Geochemical Exploration Symposium (IGES)," May 6th to 10th. pp. 42–43. Santiago de Chile.

Cicchella, D., De Vivo, B., and Lima, A. (2003). Palladium and platinum concentration in soils from the Napoli metropolitan area, Italy: Possible effects of catalytic exhausts. *Sci. Total Environ.* **308** (1–3), 121–131.

Cicchella, D., De Vivo, B., and Lima, A. (2005). Background and baseline concentration values of elements harmful to human health in the volcanic soils of the metropolitan and provincial area of Napoli (Italy). *Geochem.: Explor. Environ. Anal.* **5,** 1–12.

Cicchella, D., Albanese, S., De Vivo, B., Lima, A., Grezzi, G., and Zuppetta, A. (in press). "Atlante geochimico-ambientale dei suoli del territorio comunale di Benevento/Geochemical Environmental Atlas of the soils of Benevento." Aracne Editrice, Roma.

Cicchella, D., Fedele, L., De Vivo, B., Albanese, S., and Lima, A. (2008a). Platinum group element distribution in the soils from urban areas of Campania region (Italy). *Geochem.: Explor. Environ. Anal.* **8,** 31–40.

Cicchella, D., De Vivo, B., Lima, A., Albanese, S., Mc Gill, R. A. R., and Parrish, R. R. (2008b). Heavy metal pollution and Pb isotopes in urban soils of Napoli, Italy. *Geochem.: Explor. Environ. Anal.* **8,** 103–112.

Cretney, J. R., Lee, H. K., Wright, G. J., Swallow, W. H., and Taylor, M. C. (1985). Analysis of polycyclic aromatic hydrocarbons in air particulate matter from a lightly industrialised urban area. *Environ. Sci. Technol.* **19,** 397–404.

Darnley, A. G., Björklund, A., Bølviken, B., Gustavsson, N., Koval, P. V., Plant, Jane A., Steenfelt, A., Tauchid, M., and Xuejing, Xie (1995). "A Global Geochemical Database for Environmental and Resource Management. Recommendations for International Geochemical Mapping." Earth Science Report 19. Paris: UNESCO Publishing, Paris.

De Vivo, B., Lima, A., and Siegel, R. F. (2004). "Geochimica Ambientale. Metalli Potenzialmente Tossici," p. 449. Liguori editore, Napoli.

De Vivo, B., Cicchella, D., Lima, A., and Albanese, S. (2006). "Atlante geochimico-ambientale dei suoli dell'arean urbana e della Provincia di Napoli/Geochemical Environmental Atlas of the Urban and Provincial Soils of Napoli," p. 324. Aracne Editrice, Roma, ISBN 88–548–0563–7.

Fedele, L., De Vivo, B., Lima, A., Albanese, S., Cicchella, D., and Grezzi, G. (in press). "Atlante geochimico-ambientale dei suoli del territorio comunale di Salerno/Geochemical Environmental Atlas of the Soils of Salerno." Aracne Editrice, Roma.

Fordyce, F. M., Brown, S. E., Ander, E. L., Rawlins, B. G., O'Donnell, K. E., Lister, T. R., Breward, N., and Johnson, C. C. (2005). GSUE: Urban geochemical mapping in Great Britain. *Geochem.: Explor. Environ. Anal.* **5,** 325–336.

Hansmann, W., and Koppel, V. (2000). Lead isotopes as tracers of pollutants in soils. *Chem. Geol.* **171,** 123–144.

Kelly, J., Thornton, I., and Simpson, P. R. (1996). Urban geochemistry: A study of the influence of anthropogenic activity on the heavy metal content of soils in traditionally industrial and non-industrial areas of Britain. *Appl. Geochem.* **1,** 363–370.

Khalili, N., Sche, P., and Holsen, T. (1995). PAH source fingerprints for coke ovens, diesel and gasoline engines, highway tunnels, and wood combustion emissions. *Atmos. Environ.* **29,** 533–542.

Kuusisto, S., Lindroos, O., Rantio, T., Priha, E., and Tuhkanen, T. (2007). PCB contaminated dust on indoor surfaces—Health risks and acceptable surface concentrations in residential and occupational settings. *Chemosphere* **67**(6), 1194–1201.

Lakanen, E., and Erviö, R. (1971). A comparison of eight extractants for the determination of plant available micronutrients in soils. *Acta Agr. Fenn.* **123,** 223–232.

Lee, C. S., Li, X., Shi, W., Cheung, S. C., and Thornton, I. (2006). Metal contamination in urban, suburban, and country park soils of Hong Kong: A study based on GIS and multivariate statistics. *Sci. Total Environ.* **356,** 45–61.

Li, X., Poon, C., and Liu, P. S. (2001). Heavy metal contamination of urban soils and street dusts in Hong Kong. *Appl. Geochem.* **16,** 1361–1368.

Lima, A. (2008). Evaluation of geochemical background at regional and local scales by fractal filtering technique: Case studies in selected Italian areas. *In* "Environmental geochemistry: Site characterization, Data analysis, Case histories" (B. De Vivo, H. E. Belkin, and A. Lima, eds.). Elsevier (this volume).

Lima, A., Plant, J. A., De Vivo, B., Tarvainen, T., Albanese, S., and Cicchella, D. (2008). Interpolation methods for geochemical maps: A comparative study using arsenic data from European stream waters. *Geochem.: Explor. Environ. Anal.* **8,** 41–48.

Lima, A., De Vivo, B., Cicchella, D., Cortini, M., and Albanese, S. (2003). Multifractal IDW interpolation and fractal filtering method in environmental studies: An application on regional stream sediments of Campania region (Italy). *Appl. Geochem.* **18,** 1853–1865.

Lima, A., De Vivo, B., Grezzi, G., Albanese, S., and Cicchella, D. (2007). "Atlante geochimico-ambientale dei suoli del territorio comunale di Caserta/Geochemical Environmental Atlas of the Soils of Caserta," p. 205. Aracne Editrice, Roma, ISBN 88–548–1051–8.

Lindsay, W. L., and Norwell, W. A. (1969). Development of a DTPA micronutrient soil test. *Soil Sci. Soc. Am. Proc.* **35,** 600–602.

Ljung, K., Selinus, O., Otabbong, E., and Berglund, M. (2006). Metal and arsenic distribution in soil particle sizes relevant to soil ingestion by children. *Appl. Geochem.* **21,** 1613–1624.

Miesh, A. T. (1981). Estimation of the geochemical threshold and its statistical significance. *J. Geochem. Explor.* **16,** 49–76.

Miesh Programs (1990). "G-RFAC." Grand Junction, CO, USA.

Ministero dell'Ambiente (2006). Decreto Ministeriale n°152, 02/05/2006. Gazzetta Ufficiale (Suppl. Ordin.), 108 (del 11/05/2006).

Ottesen, R. T., and Langedal, M. (2001). Urban geochemistry in Trondheim, Norway. *Norges geologiske undersøkelse Bulletin* **438,** 63–69.

Rosner, G., and Merget, R. (2000). Evaluation of the health risk of platinum emissions from automotive emissions control catalysts. *In* "Anthropogenic Platinum-Group Element Emissions. Their Impact on Man and Environment" (F. Zereini and F. Alt, eds.), pp. 267–281. Springer-Verlag, Berlin.

Šajn, R., Bidovec, M., Andjelov, M., Pirc, S., and Gosar, M. (1998). "Geochemical Atlas of Ljubljana and Environs," p. 34. Inštitut za geologijo, geotehniko in geofiziko, Ljubljana.

Salminen, R. (ed.) (1995). *In* "Regional Geochemical Mapping in Finland in 1982–1994" (in Finnish, English Summary). Report of Investigation 1995, Geological Survey of Finland, p. 130.

Salminen, R., and Gregorauskiene, V. (2000). Considerations regarding the definition of a geochemical baseline of elements in the surficial materials in areas differing in basic geology. *Appl. Geochem.* **15,** 647–653.

Salminen, R., Tarvainen, T., Demetriades, A., Duris, M., Fordyce, F. M., Gregorauskiene, V., Kahelin, H., Kivisilla, J., Klaver, G., Klein, H., Larson, J. O., Lis, J., *et al.* (1998). FOREGS geochemical mapping. Field Manual. Guide 47, Geological Survey of Finland, p. 36.

Schafer, J., and Puchelt, H. (1998). Platinum-group-metals (PGM) emitted from automobile catalytic converters and their distribution in roadside soils. *J. Geochem. Explor.* **64,** 307–314.

Schumacher, B. A., Shines, K. C., Burton, J. V., and Papp, M. L. (1991). A comparison of soil sample homogenization techniques. *In* "Hazardous Waste Measurements" (M. S. Simmons, ed.), pp. 53–68. Lewis, Boca Raton, FL, USA.

Semlali, R. M., Van Oort, F., Denaix, L., and Loubet, M. (2001). Estimating distributions of endogenous and exogenous Pb in soils by using Pb isotopic ratios. *Environ. Sci. Technol.* **35,** 4180–4188.

Semlali, R. M., Dessogne, J. B., Monna, F., Bolte, J., Azimi, S., Navarro, N., Denaix, L., Loubet, M., Chateau, C., and Van Oort, F. (2004). Modelling lead input and output in soils using lead isotopic geochemistry. *Environ. Sci. Technol.* **38,** 1513–1521.

Siegel, F. R. (2002). "Environmental Geochemistry of Potentially Toxic Elements," p. 218. Springer-Verlag, Berlin.

Sinclair, A. J. (1974). Selection of threshold values in geochemical data using probability graphs. *J. Geochem. Explor.* **3,** 129–149.

Sinclair, A. J. (1976). "Applications of Probability Graphs in Mineral Exploration," Spec. Vol. 4, p. 95. Association of Exploration Geochemists, Rexdale, Ontario, Canada.

Sinclair, A. J. (1991). A fundamental approach to threshold estimation in exploration geochemistry: Probability plots revisited. *J. Geochem. Explor.* **41,** 1–22.

Stanley, C. R. (1987). Probplot. An interactive computer program to fit mixtures of normal (or lognormal) distributions with maximum likelihood of optimization procedures. *Assoc. Explor. Geochem.* Spec. Vol. 14, 39.

Thums, C., and Farago, M. E. (2001). Investigating urban geochemistry using geographical information systems. *Sci. Prog.* **84,** 183–204.

Tijhuis, L. (2003). "The Geochemistry of Topsoil in Oslo, Norway," p. 228. Norvegian University of Science and Technology, Oslo.

Vizard, C. G., Rimmer, D. L., Pless-Mulloli, T., Singleton, I., and Vivienne, S. (2006). Air identifying contemporary and historic sources of soil polychlorinated dibenzo-*p*-dioxins and polychlorinated dibenzofurans in an industrial urban setting. *Sci. Total Environ.* **370**(1), 61–69.

Von Hoff, D. D., Slavik, M., and Moggia, F. M. (1976). Allergic reactions to *cis*-platinum. *Lancet* **1,** 90–99.

Wong, C. S. C., Li, X., and Thornton, I. (2006). Urban environmental geochemistry of trace metals. *Environ. Pollut.* **142,** 1–16.

Xu, Y., and Cheng, Q. (2001). A fractal filtering technique for processing regional geochemical maps for mineral exploration. *Geochem.: Explor. Environ. Anal.* **1,** 147–156.

Yang, S. Y. N., Connell, D. W., Hawker, D. W., and Kayal, S. I. (1991). Polycyclic aromatic hydrocarbons in air, soil and vegetation in the vicinity of an urban roadway. *Sci. Total Environ.* **102,** 229–240.

Zereini, F., Skerstupp, B., Alt, F., Helmers, E., and Urban, H. (1994). Geochemical behaviour of platinum group elements (PGE) in particulate emissions by automobile exhaust catalysts: Experimental results and environmental investigations. *Sci. Total Environ.* **206,** 137–146.

CHAPTER NINE

CHEMICAL SPECIATION TO ASSESS POTENTIALLY TOXIC METALS' (PTMS') BIOAVAILABILITY AND GEOCHEMICAL FORMS IN POLLUTED SOILS

Paola Adamo* *and* Mariavittoria Zampella*

Contents

Abstract

Potentially toxic metals (PTMs) are persistent contaminants in the soil environment. The simple determination of their total or 'pseudototal' content in soil might minimise the risks for biota and human health, assuming that pollutants transferring to water resources or biota are simply correlated with contamination level. In contrast, relevant paradigms in environmental monitoring, risk assessment and remediation feasibility are the PTMs' mobility and bioavailability to microorganisms, plants, animals and humans. For a correct assessment of risk/toxicity (according to PTMs' content and availability) of a polluted soil and to predict its reduction after application of remediation techniques, it is crucial to establish the speciation, mobility and biogeochemistry of the contaminants.

* Dipartimento di Scienze del Suolo, della Pianta, dell'Ambiente e delle Produzioni Animali, Università degli Studi di Napoli Federico II, Via Università 100, 80055 Portici, Italy

Environmental Geochemistry
DOI: 10.1016/B978-0-444-53159-9.00009-7

© 2008 Elsevier B.V.
All rights reserved.

In this sense, a requirement exists for analytical methods and strategies that provide information on the dynamics and behaviour of PTMs in soil.

Speciation science seeks to characterise the various forms in which PTMs occur or, at least, the main metal pools present in soil. This chapter provides a review of the single and sequential chemical extraction procedures that have been more widely applied to determine the plant and the human bioavailability of PTMs from contaminated soil and their presumed geochemical forms. Examples of complementary use of chemical and instrumental techniques and applications of PTMs' speciation for risk and remediation assessment are illustrated.

1. INTRODUCTION

Contamination of soil with potentially toxic metals (PTMs) poses serious risks for biota and human health (Förstner, 1995). At low concentrations, some PTMs (e.g. copper, chromium, molybdenum, nickel, selenium and zinc) are essential to healthy functioning and reproduction of microorganisms, plants and animals (including man) (Alloway, 1995). However, at high concentrations, these same essential elements may cause direct toxicity or reproductive effects. Some elements are also non-essential (e.g. arsenic, lead and mercury) and even low concentrations of these elements in the environment can cause toxicity to both plants and animals (Alloway, 1995). Adverse effects are not necessarily manifested in the environment only when PTMs have an anthropogenic origin. Naturally high concentrations of some elements also cause toxicity (e.g. serpentine soils) and lead to natural adaptation of the biota to these high concentrations.

Despite new regulations and technological improvements, the soil pollution problem persists (Wong *et al.*, 2006). Worldwide, not only numerous urban-industrial sites but also agricultural sites and soils need to be investigated to define environmental hazards and to propose new occupation plans and eventually remediation treatments. Risk management based only on the total or 'pseudototal' content of PTMs in soil might minimise the risks, assuming that pollutants transferring to water resources or biota increase with contamination level. This analysis gives no idea of the extent to which elements are really transferable or bioavailable. For a correct assessment of risk/toxicity (according to PTMs' content and availability) of a polluted soil and to predict its decrease after application of remediation techniques, it is crucial to establish the speciation, mobility and biogeochemistry of the contaminants. In this sense, a requirement exists for analytical methods and strategies that provide information on the dynamics and behaviour of PTMs in soil.

2. PTMs' FORMS IN SOIL AND BIOAVAILABILITY

It is generally accepted that the distribution, mobility, biological availability and toxicity of PTMs in soil depend not simply on their total concentrations but, critically, on their forms (Morgan and Stumm, 1995). These may be soluble, readily exchangeable, complexed with organic matter, or hydrous oxides, substituted in stoichiometric compounds or occluded in mineral structures (Fig. 9.1) (Ritchie and Sposito, 2002). The soil is a dynamic system and any changes in environmental conditions, whether natural or anthropogenic, can alter the forms of PTMs, thereby affecting their behaviour in soil. The main controlling factors include pH, redox potential, ionic strength of the soil solution, the solid and solution components and their relative concentrations and affinities for an element and the time (Alloway, 1995).

In most countries, the soil pollution standards are based on the pseudototal metal fraction, which, in contrast to the theoretical total amount of PTMs in soil, is the amount that can be extracted with the help of strong acid solution. It represents an excellent criterion to define the extent of metal contamination in soil, but it is of little value for the prediction of ecological impact (Gupta *et al.*, 1996). An increasing need is felt not only to analyse metal concentrations in soils but also to assess their influence on the terrestrial ecosystem itself, such as the toxicity of metals to soil microorganisms, and on other boundary ecosystems such as the ground water, air, plants, animals and humans (Gupta, 1991a, b; Gupta and Aten, 1993).

Metal bioavailability is the fraction of the total metal occurring in the soil matrix, which can be taken up by an organism and can react with its metabolic system (Campbell, 1995). Metals can be plant-bioavailable, if they come in contact with plants (physical accessibility) and have a form which can be uptaken by plant roots (chemical accessibility). Soil metals become accessible for humans by ingestion, inhalation and dermal contact. Available forms of PTMs are not necessarily associated with one particular chemical species or a specific soil component. Main soil PTMs' pools of different mobility, target organisms and routes of transfer are sketched in Fig. 9.2. The most labile fraction, corresponding to the soluble metal pool, occurs as either free ions or soluble complexed ions and is considered the

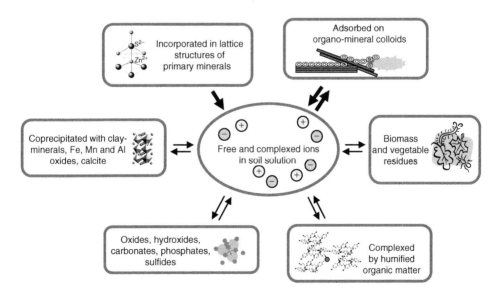

Figure 9.1 The major forms of potentially toxic metals (PTMs) in soil.

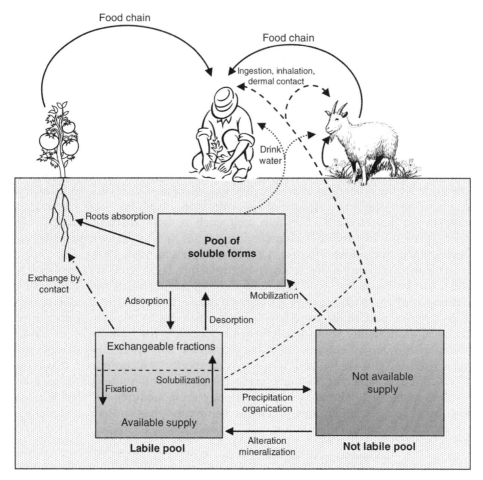

Figure 9.2 Potentially toxic metals' (PTMs') pools of different mobility, target organisms and routes of transfer in and from the soil.

readily bioavailable fraction to receptor organisms. Although plant roots absorb nutrients from the soil solution, availability is rarely equated with solubility. This is because metal solubility in soil is a function of interacting biological, chemical and physical factors. Ions that are weakly adsorbed on exchange surface in soils are also fairly labile and can easily become bioavailable. Along with exchange and desorption, release from the solid phase may result also from processes such as decomposition and dissolution. Rate of release as well as the capacity to release nutrients, the buffer power, has a great effect on the bioavailability of the ion and may strongly differ among soils, reflecting differences in bonding strengths. Estimation of metal bioavailability in soil relies on understanding the rate at which bioavailable fractions are replenished from the soil matrix.

3. The Need of Speciation and Speciation Methods

According to International Union of Pure and Applied Chemistry (IUPAC), the terms 'speciation' and 'chemical species' should be reserved for the forms of an element defined as to isotopic composition, electronic or oxidation state and/or complex or molecular structure (Templeton *et al.*, 2000). This 'classical' definition, appropriate to speciation in solution samples, would exclude most speciation studies on solid materials, such as soils and sediments, more properly defined as fractionation studies. The terminology used in this chapter is based on the broader definition of 'speciation' given by Ure and Davidson (2002), which encompass the IUPAC's narrow definition and includes the selective extraction and fractionation techniques of solid samples.

In this context, speciation science seeks to characterise the various forms in which PTMs occur in soil or, at least, the main metal pools present in soil. Understanding speciation is important for assessing the potential of soil to supply micronutrients for plant growth or to contain toxic quantities of PTMs, and for determining amelioration procedures for soils at risk of causing the PTMs' contamination of waterways. The residence time of an element in a soil depends on the mobility of its predominant forms. Speciation science is relevant to scientists with many different backgrounds and should be taken into consideration by legislators in the field of environmental protection.

PTMs' speciation in soil can be achieved using either direct or indirect analytical methods. Direct methods include X-ray diffraction (XRD) and microanalysis [energy/wavelength dispersive X-ray (EDS/WDS)], infrared absorption [Fourier transform infrared spectroscopy (FTIR)] and Mössbauer spectrometry, scanning/transmission electron microscopy (SEM/TEM), magnetic resonance (nuclear magnetic resonance, electron paramagnetic resonance) and photoelectron [extended X-ray absorption fine structure (EXAFS), X-ray absorption near edge structure] spectroscopy (Goodman and Glidewell, 2002). With such techniques, the combinational forms of major elements in soil components, such as clay minerals, iron, manganese and aluminium oxyhydroxides and humic materials, and the chemical structures of these soil components can be precisely elucidated. Nevertheless, these direct, mainly non-destructive, methods for speciation are usually qualitative and often not sensitive enough to detect forms present in small amounts. Furthermore, their correct application requires sophisticated instrumentations and specialised operators.

More widely applied to determine the potential, plant and human bioavailability are the methods of PTMs' speciation which involve selective chemical extraction techniques. Estimation of the plant- or human-available element content of soil using single chemical extractants is an example of *functionally defined speciation*, in which the 'function' is plant or human availability. In *operationally defined speciation*, single extractants are classified according to their ability to release elements from specific soil phases. Selective sequential extraction procedures are examples of operational speciation (Ure and Davidson, 2002).

A number of other approaches to PTMs' speciation exists. Isotopic dilution methods, developed primarily for the study of P availability in agricultural soils, have also been used to study PTMs, in particular Cd. These methods enable a direct examination of the available pool of metals in soil, estimating exchangeable and labile metal pools (Hammer *et al.*, 2006). Modelling of the solid–liquid phase partitioning is also done to describe soil solution concentrations. Its application requires, however, a great deal of data and analysis. A comprehensive analysis of many chemical models developed for PTMs' speciation has been done by Waite (1989) and Lumsdon and Evans (2002).

This chapter provides a review of the single and sequential chemical extraction procedures that have been more widely applied to estimate the plant and the human bioavailability of PTMs from contaminated soil and their presumed geochemical forms.

The review is not intended to provide a discussion of chemical methodologies design nor an exhaustive list of all the studies that have been conducted to date, but it aims to examine methodologies to obtain data that may be used to estimate PTMs' risks, mainly in term of bioavailability to plants and humans. It does not include any consideration on the impact of soil PTMs' contamination on groundwaters and soil microbes.

4. Plant Bioavailability

Soil–plant transfer of PTMs is a part of chemical element cycling in the nature. It is a very complex process governed by several factors of geochemical, climatic and biological origin. Soil conditions play a crucial role in PTMs' behaviour. It can be generalised that in well-aerated (oxidising) acid soils, several metals (especially Cd and Zn) are easily mobile and available to plants, while in poorly aerated (reducing) neutral or alkaline soils, metals are substantially less available (Kabata-Pendias, 2004). In polluted areas, the transfer of PTMs from soils to plants is of great concern. Legal directives define the permissible content of these elements in plants for consumption (Chojnacka *et al.*, 2005). Legislations start to consider a reduction of bioavailable metal concentrations as an important step in soil remediation programs, providing valuable indications of the phytoextraction techniques efficiency (Keller and Hammer, 2004).

The plant-bioavailable fraction of PTMs can be defined as the fraction of a metal total content in the soil that can be absorbed by plants via roots uptake (Kabata-Pendias and Pendias, 2001). Usually, this fraction is only a small proportion of the total element content of soils and shows much higher spatio-temporal variability than the total concentration. A plant uptakes mobile ion from the soil solution and the soil element fraction which is in solution is that which is considered immediately available. Nevertheless, the soil solid phases, inorganic as well as organic, take part in the supply and buffering of elements and allow their retention under wet conditions, which would otherwise leach all soluble elements from the soil. Therefore, the solid-bound elements take part to the available pool and this is why element concentrations in the soil solution are one to three orders of magnitude lower than those in plants (Bargagli, 1998).

4.1. Metals in soil solution

The chemistry of the soil solution provides useful information for the prediction of bioavailability and toxic effects of PTMs on plants and biological activities in soils. A good estimation of 'availability' can be achieved by measuring the elements concentration in soil pore water. The chapter of Di Bonito et al. (2008, this book) describes in detail properties of soil pore water and some of the current methodologies used to extract soil pore water. A large number of analytical techniques are available for the characterisation of the soil liquid phase and quantification of elements therein: electro-analytical techniques (e.g. selective ion electrode, polarography, differential pulse anodic and cathodic stripping voltametry), cation/anion exchange resins and chemical adsorbents (e.g. Al and Mn oxides) to fractionate ionic and non-ionic (i.e. complexed) forms, ultrafiltration, dialysis and gel permeation techniques for molecular size fractionation, spectroscopic techniques (e.g. nuclear magnetic resonance, electron spin resonance, FTIR) for measuring elements oxidation state, X-ray techniques (e.g. electron microprobe, SEM) to measure element distribution, chromatographic techniques (high performance liquid chromatography, gas chromatography–mass spectrometry) to measure phase distribution of PTMs, synchrotron-based spectroscopic techniques to examine both geochemical transformation and speciation (Adriano et al., 2004). The ranges of some elements measured in the solution obtained by various techniques from uncontaminated and contaminated soils are reported in Table 9.1. However, the soil solution is constantly and rapidly changing in amount and in chemical composition due to the contact with the highly diverse soil solid phases and by the uptake of ions and water by plant roots. In order to predict PTMs' availability, either we have to develop methods that determine specifically the forms involved or we have to establish an empirical relationship between an accepted diagnostic metal measure and plant growth. In reality, a combination of these two approaches might produce the most accurate predictions.

4.2. Single chemical methods to assess phytoavailable metals in soil

The plant-bioavailable PTMs' content can be simulated by chemical extractions or can be measured directly, analysing plant tissues. Several methods for the assessment of the plant-available PTMs' content in soils have been developed (Table 9.2) (Houba et al., 1996; Kabata-Pendias, 2004; Ure and Davidson, 2002). They are mainly based on extractions by various solutions: dilute mineral or organic acids (e.g. acetic acid, HCl, HCl + HNO_3), chelating agents [e.g. Na_2EDTA, EDTA, DTPA (+TEA), citric acid, NH_4-citrate], buffered salts (e.g. CH_3COONH_4) and neutral salts (e.g. $CaCl_2$, $MgCl_2$, NH_4Cl, $NaNO_3$, NH_4NO_3). These speciation methodologies use a single extractant to dissolve soil phases, chemical forms or binding types whose element content correlates with the availability of the element to plants.

Neutral salts of strong acids are likely to release soluble and non-specifically adsorbed metals. These solutions have the advantage that the pH of the soil suspension is determined by the soil mainly, and not by the extractant. Displacing

Table 9.1 Amounts of some potentially toxic metals (PTMs) measured in the soil matrix (mg·kg⁻¹) and soil solution ($\mu g\cdot L^{-1}$) obtained by various techniques from uncontaminated and contaminated soils

Soil	Soil sol. sampling technique	Cd Soil	Cd Soil sol.	Cu Soil	Cu Soil sol.	Ni Soil	Ni Soil sol.	Pb Soil	Pb Soil sol.	Zn Soil	Zn Soil sol.	Ref.
Arable	SMS	0.4	0.2	11.9	18	8.7	<5	n.d.	n.d.	56.1	24	1
Arable + ss[a]	SMS	1.1	1.8	36.9	79	15.6	54	n.d.	n.d.	216.3	859	1
Ex-wood	SMS	0.3	1.9	9.8	11	8.6	6	n.d.	n.d.	43.0	168	1
Ex-wood + ss[a]	SMS	0.9	21.1	32.3	108	12.3	313	n.d.	n.d.	184.7	7437	1
S soil + ss[b]	SMS	2.4	41.3	31.3	138	7.7	67	n.d.	n.d.	329	1289	2
	centr	1.7	4.6	n.d.	n.d.	n.d.	n.d.	n.d.	n.d.	320	2450	3
SL soil + ss[c]	SMS	4.4	38	52.2	106	25.5	88	n.d.	n.d.	210	287	3
	centr	0.77–10.3	1.4–16	n.d.	n.d.	n.d.	n.d.	n.d.	n.d.	81–397	140–2370	3
L soil + ss[d]	SMS	2.2	34.7	92.6	86	36.3	59	n.d.	n.d.	383	36	2
	centr	0.85	0.6	n.d.	n.d.	n.d.	n.d.	n.d.	n.d.	340	180	3
Zn smelter	SMS	2.9–23.8	36.3–46.7	44.8–74.6	46–69	12.6–46.0	50–53	n.d.	n.d.	400–2166	39–448	2
	centr	1.2–16.5	0.6–18	n.d.	n.d.	n.d.	n.d.	n.d.	n.d.	317–370	210–1470	3
Mine spoil	SMS	58.1	87.3	41.8	34	61.8	55	n.d.	n.d.	3259	1087	2
Zn/Pb smelter	filtration	15.5	32–147	n.d.	n.d.	n.d.	n.d.	383	17–68	1819	5.6–23	4
	ultracentr	15.5	32–146	n.d.	n.d.	n.d.	n.d.	383	14–39	1819	5.6–23	4
Forest soil next to Ni smelter	centr	n.d.	n.d.	845–1739	890–1200	1979–4589	190–1100	n.d.	n.d.	n.d.	n.d.	5

[a] 300 m³ year⁻¹ of sewage sludge for 10 years. [b] 10 t ha⁻¹ year⁻¹ of sewage sludge for 20 years. [c] 19 t ha⁻¹ year⁻¹ of sewage sludge for 19 years. [d] 4.9 t ha⁻¹ year⁻¹ of sewage sludge for 40 years. Soil sol., soil solution; ss, sewage sludge; Ex-wood, ex-woodland; S, sandy; L, loam; SMS, rhizon soil moisture sampler; centr, centrifugation; ultracentr, ultracentrifugation; Ref, reference; n.d. = not determined.

1. McGrath et al. (1999), 2. Knight et al. (1998), 3. Holm et al. (1998), 4. Denaix et al. (2001), 5. Paton et al. (2006).

Table 9.2 Examples of extractants and presumed extracted pools of plant-available potentially toxic metals (PTMs) in soil

Extractant/presumed PTMs pools	Elements	References
Soluble + readily exchangeable		
0.1/0.01/0.05 M $CaCl_2$	Cd, Pb, Zn	1–14
1 M NH_4Cl	Cd	15–17
0.1 M $NaNO_3$	Cd, Cu, Pb, Zn	9, 12–14, 18–20
1 M NH_4NO_3	Cd, Ni	14, 17, 21–27
1 M ammonium acetate, pH 7	Mo, Ni, Pb, Zn	28–33
2% ammonium citrate	As, Cd, Cr, Hg, Pb	34
Exchangeable + carbonate		
Acetic acid 0.43 M	Cd, Co Cr, Cu, Pb, Ni, Zn	11, 24, 32, 35–43
Specifically adsorbed and organically complexed		
1 M ammonium acetate + 0.01 M EDTA, pH 7	Cu, Mn, Zn	33
0.5 M ammonium acetate + 0.02/0.05 M EDTA	Cd, Cu, Fe, Mn, Mo, Pb, Zn	19, 44–51
0.05 M EDTA, pH 7	Cd, Cu, Cr, Mo, Ni, Pb, Zn	14, 36, 42, 12–13, 52–54
0.005/0.5 M DTPA + 0.01 M $CaCl_2$ + 0.1 M TEA, pH 7.3	Cd, Cu, Fe, Mn, Ni, Zn	4, 12–14, 19, 32, 39, 43, 52–53, 55–65

1. Häni and Gupta (1985), 2. Sauerbeck and Styperek (1985), 3. Sanders et al. (1986), 4. Whitten and Ritchie (1991), 5. Novozamsky et al. (1993), 6. Andrews et al. (1996), 7. Houba et al. (1996), 8. Singh et al. (1996), 9. Houba et al. (1997), 10. Houba et al. (1999), 11. Sastre et al. (2004), 12. Feng et al. (2005a), 13. Feng et al. (2005b), 14. Gupta and Sinha (2006), 15. Krishnamurti et al. (1995b), 16. Gray et al. (1999), 17. Krishnamurti et al. (2000), 18. Häni and Gupta (1983), 19. Hammer and Keller (2002), 20. Keller and Hammer (2004), 21. Symeonides and McRae (1977), 22. Hornburg et al. (1995), 23. DIN (1995), 24. Merkel (1996), 25. Davies et al. (1989), 26. Düring et al. (2003), 27. Bhogal et al. (2003), 28. John (1972), 29. John et al (1972), 30. Sedberry and Reddy (1976), 31. Haq et al. (1980), 32. Soon and Bates (1982), 33. Sterckeman et al. (1996), 34. Chojnacka et al. (2005), 35. Mitchell et al. (1957), 36. Clayton and Tiller (1979), 37. Carlton-Smith and Davis (1983), 38. Ellis and Alloway (1983), 39. Tills and Alloway (1983), 40. Burridge and Berrow (1984), 41. Rauret et al. (2001), 42. Cappuyns and Swennen (2006), 43. Obrador et al. (2007), 44. Lakanen and Ervio (1971), 45. Sillanpaa (1982), 46. Sauerbeck and Styperek (1984), 47. Sillanpaa and Jansson (1992), 48. Ervio and Sippola (1993), 49. Tarvainen and Kallio (2002), 50. Díez Lázaro et al. (2006), 51. Kidd et al. (2007), 52. Davis (1979), 53. Sahuquillo et al. (1999), 54. Sastre et al. (2004), 55. Williams and Thornton (1973), 56. Street et al. (1977), 57. Latterell et al. (1978), 58. Linsday and Norvell (1978), 59. Soltanpour and Schwab (1979), 60. Sillanpaa (1982), 61. Houba et al. (1990), 62. Fuentes et al. (2004), 63. Fuentes et al. (2006), 64. Remon et al. (2005), 65. Mendoza et al. (2006).

activity of cations (Ca^{2+}, Mg^{2+}, Na^+, NH_4^+) generally tends to increase with concentration. The complexing ability of chloride increases the proportion of some metal removal (Shuman, 1985).

Calcium chloride has been frequently used to study PTMs (Zn, Cd, Cu, Ni, Mn and Pb) which accumulate in soils as a result of human activities and for risk assessment (Häni and Gupta, 1985; Houba et al., 1996, 1999; Novozamsky et al., 1993;

Sanders *et al.*, 1986; Sauerbeck and Styperek, 1985). The 0.01 M $CaCl_2$ single extraction is simple, cheap and environmental friendly. It matches the soil solution with respect to pH, concentration and composition (Houba *et al.*, 1996). Its extraction efficiency has been compared with that of other soil extractants and in some cases, analytical results have been judged against crop response (Houba *et al.*, 1996; Singh *et al.*, 1996). Correlations between soil $CaCl_2$ extractable PTMs and plant contents have been found weak or demonstrated for cadmium and lead limitedly to some vegetables or soil conditions (Andrews *et al.*, 1996; Sauerbeck and Styperek, 1985; Singh *et al.*, 1996; Whitten and Ritchie, 1991).

The 0.1 M sodium nitrate extraction, proposed in the early 1980s, is now the reference for the available pool in the Swiss ordinance on impacts on the soil (OIS, 1998). It is a quick extraction, with an uncertain direct link to plant uptake, which more likely gives the soluble content of metal in the soil (Keller and Hammer, 2004). $NaNO_3$ releases less metals compared with $CaCl_2$ (Häni and Gupta, 1983), nowadays easily accessible by ICP-OES and ICP-MS.

Ammonium chloride (1 M) was tested along with other six extractants (0.01 and 0.05 M $CaCl_2$, Na_2EDTA, DTPA-TEA, 1 M NH_4NO_3 and AAAC-EDTA) for assessing the plant-available cadmium by Krishnamurti *et al.* (2000). The amounts of Cd extracted were related to the Cd concentration in the stem and leaves of *Triticum aestivum* and 1 M NH_4Cl gave the best overall correlation ($r = 0.928$, $p < 0.001$) compared with the other extractants.

The 1 M ammonium nitrate removes soil solution and readily exchangeable metal forms (Davies *et al.*, 1989). Bhogal *et al.* (2003) used 1 M NH_4NO_3 extraction to predict metal availability in cultivated soils in order to evaluate the effects of past sewage sludge addition on crop yields. They found significant negative correlations between NH_4NO_3 extractable metals and crop yields. Gupta and Sinha (2006) found significant correlations between NH_4NO_3-extractable and metal accumulation in *Sesamum indicum* (L.) for Zn ($r = 0.981$, $p < 0.01$), Mn ($r = 0.817$, $p < 0.05$), Cr ($r = 0.889$, $p < 0.05$) and Cd ($r = 0.822$, $p < 0.05$).

Buffered (pH 7) 0.1 M ammonium acetate and 0.43 M acetic acid have also been adopted to extract soil plant-available PTMs (Burridge and Berrow, 1984; Ellis and Alloway, 1983; Haq *et al.*, 1980; John *et al*, 1972; Merkel, 1996; Obrador *et al.*, 2007; Sastre *et al.*, 2004; Soon and Bates, 1982; Sterckeman *et al.*, 1996). For both extractants, acetate complexation ability may contribute to increase the amounts of PTMs released (Adamo *et al.*, 1996; McLaren and Crawford, 1973). Only for acetic acid, in addition to readily exchangeable, also specifically adsorbed and carbonate-bound metals are likely displaced by hydrogen ions (Berrow and Mitchell, 1980). Acetic acid could partially extract metals bound to organic matter (Payà-Pérez *et al.*, 1993) and associated with minerals (kaolinite, K-feldspars and ferrihdrite) (Whalley and Grant, 1994).

Recently, Chojnacka *et al.* (2005) have reported an excellent linear relationship between available PTMs' (As, Cd, Cr, Cu, Hg, Mn, Pb and Zn) content in polluted industrial soils, as extracted by 2% (w/v) ammonium citrate solution, and wheat plants content. For some elements of environmental concern, such as Pb, As, Cd, Hg and Cr, also the results obtained by 0.1 M $NaNO_3$, 0.5 M KNO_3 and 0.05 M $CaCl_2$

were good, while for the other tested extractants [1 M $MgCl_2$, 0.1 M $K_2P_2O_7$, 2% (w/v) CH_3COOH, 2% (w/v) HCOOH, 2% (w/v) citric acid, 0.05 M Na_2EDTA, 0.1 M EDTA, 0.1 M HCl + HNO_3], the correlations were weak.

Soluble strong complexing compounds, such as ethylenediamine tetraacetic acid (EDTA) and diethylenetriamine pentaacetic acid (DTPA), remove from soil PTMs specifically adsorbed by the mineral fraction and fixed in organic and organo-metallic complexes. The EDTA is a strong chelating agent, can form stable chelates with many metal ions (Feng *et al.*, 2005b) and can remove organically bound metals, along with metals occluded in oxides and secondary clay minerals in part (Payà-Pérez *et al.*, 1993). The EDTA is a non-specific extractant and can remove both labile and non-labile fractions (Bermond *et al.*, 1998). The 0.01/0.02/0.05 M EDTA has been used with ammonium acetate, in acidic and neutral conditions, by many authors to extract bioavailable metals from soil (Berrow and Mitchell, 1980; Clayton and Tiller, 1979; Davis, 1979; Díez Lázaro *et al.*, 2006; Ervio and Sippola, 1993; Feng *et al.*, 2005a,b; Gupta and Sinha, 2006; Hammer and Keller, 2002; Kidd *et al.*, 2007; Lakanen and Ervio, 1971; Sastre *et al.*, 2004; Sillanpaa, 1982; Sillanpaa and Jansson, 1992; Sterckeman *et al.*, 1996; Tarvainen and Kallio, 2002; Tlustos *et al.*, 1994). The extracting power of this mixture is based on the combined action of ammonium acetate, which exchanges the ammonium ion with PTMs, and the EDTA, which acts as a chelating agent forming stable chelates with many metal ions and preventing secondary precipitation of phosphate compounds during extraction (Cottenie *et al.*, 1982). The DTPA procedure was developed by Linsday and Norvell (1978) to evaluate available Zn, Fe, Mn and Cu in carbonate-rich soils and has been used by many authors also to assess the bioavailability of other metals (Davis, 1979; Feng *et al.*, 2005a,b; Fuentes *et al.*, 2004, 2006; Gupta and Sinha, 2006; Hammer and Keller, 2002; Latterell *et al.*, 1978; Mendoza *et al.*, 2006; Obrador *et al.*, 2007; Remon *et al.*, 2005; Sillanpaa, 1982; Soltanpour and Schwab, 1979; Soon and Bates, 1982; Street *et al.*, 1977; Tills and Alloway, 1983; Williams and Thornton, 1973). The DTPA extraction is most suitable for near-neutral and calcareous soils where it prevents dissolution of carbonate-bound forms and release of occluded metals (Feng *et al.*, 2005b; Linsday and Norvell, 1978). In acidic soils, the DTPA procedure can potentially extract metals from iron and manganese oxides (O'Connor, 1988). Moreover, a higher extracting power in acidic soil could derive from the presence in the DTPA extractant of high $CaCl_2$ concentration, causing a rapid Ca exchange with bivalent cations, and the occurrence of triethanolamine (TEA), which is protonated at pH 7.3 and could exchange H with cations from the exchange sites (Hammer and Keller, 2002; Linsday and Norvell, 1978).

In order to improve comparability between results of extraction protocols for soil plant-available PTMs obtained by different laboratories, the Measurement and Testing Programme (formerly BCR) of the European Commission has recently proposed two harmonised procedures for extractable Cd, Cu, Cr, Ni, Pb and Zn in soils on the basis of 0.05 M EDTA (pH 7.0) and 0.43 M acetic acid solutions (Rauret *et al.*, 2001). Moreover, a suitable reference material (BCR-700 organic-rich soil) for these extractions was prepared, which enable the quality of the measurements to be controlled.

4.3. Contrasting aspects in the use of single chemical methods to define plant-available PTMs

Despite the large number of studies on plant-availability, the majority of chemical extraction systems used to measure the PTMs' plant-available soil content have been devised for agricultural and nutritional purposes. Moreover, although much work has been done to find universal extractants, many of them remain, to varying degrees, specific to one element, restricted in use to particular soil types and relevant only to specific crops. In fact, the plant-available PTMs' content depends on the intrinsic plant characteristics, which vary between species and ecotypes (Zha *et al.*, 2004), and on the chemical form and distribution of a metal in the soil. Plants can change metal availability directly (uptake) and indirectly by different mechanisms (e.g. exudation of complexing agents, respiration of roots which accounts for pH changes, high root density and high root to shoot ratio and so on) (Keller *et al.*, 2003; Schwartz *et al.*, 1999). Metal-resistant mycorrhizal fungi and rhizosphere bacteria can enhance or depress metal transfer from the soil into the higher plants (Christie *et al.*, 2004; Vivas *et al.*, 2003; Whiting *et al.*, 2001b). The respective effects of these factors, roughly outlined in Fig. 9.3, are difficult to quantify because of procedural difficulties (Hammer and Keller, 2002). However, when a statistically significant linear correlation exists between metal ion content in soil plant-available extracts and in plants, then it is possible to evaluate the soil–plant transfer factor (TF) for a given metal (Chojnacka *et al.*, 2005; Tome *et al.*, 2003). Recently, a new method for evaluation of metal plant-availability, based on the rhizosphere-plant interaction, has been developed. The rhizosphere-based method tries to simulate the real field conditions as closely as possible, and also to simulate the combined effects of rhizosphere–roots interactions as a whole. The two critical parameters of this new method are the use of fresh moist rhizosphere soil and the application of a combination of low–molecular weight organic acids usually exudates by plant roots and produced by fungi and bacteria activity (acetic, lactic, citric, malic and formic acids) (Feng *et al.*, 2005a,b).

Less well-established is the validity of the application of plant-available speciation methods in heavily polluted soils, for example, in cases of urban or industrial pollution. These soils are highly complex and heterogeneous mixtures of many contributing materials and associated contaminants, whose modality of interaction with chemical extractants may not be comparable with natural soil components. In fact, when these extractants are applied in polluted soils, they will not necessarily measure plant-available contents, but rather the labile or mobile species contents or the more specific exchangeable species contents (Ure and Davidson, 2002). These extracts often indicate potentially plant-available or leachable contents rather than actual contents. In sewage sludge-amended soils, despite increased total and EDTA-extractable PTMs' concentrations, wild plants and *Zea mays* uptake mechanisms were found to restrict metal transport to aerial parts, suggesting a greater risk of metal leaching losses than a risk of metal transfer into plant tissue and/or the food chain in the case of crops (Kidd *et al.*, 2007; Zheljazkov and Warman, 2004). Similarly, in a pluri-contaminated soil from Aznalcóllar (Seville, Spain), although the bioavailability

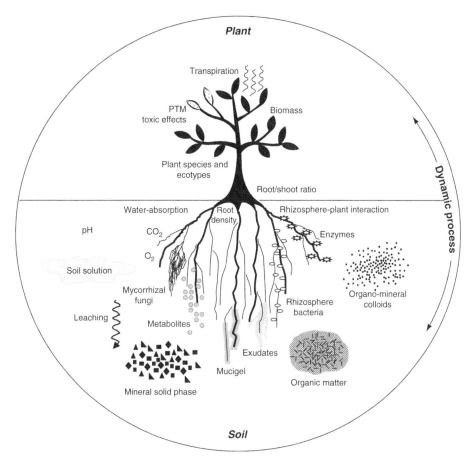

Figure 9.3 Major plant and soil factors affecting plant-available potentially toxic metals' (PTMs') content.

(DTPA-extractability) of PTMs was high, *Brassica juncea* feasibility for metal phytoextraction was too low for efficient soil remediation (Clemente *et al.*, 2005).

Taking into account these difficulties and all discussions and evaluations of recent research results, many authors propose the weak salt solutions of calcium chloride, sodium nitrate and ammonium nitrate as extraction reagents to predict the risk/toxicity associated to soil PTMs' pollution or residing after remediation actions (Kabata-Pendias, 2004; Keller and Hammer, 2004; Ure and Davidson, 2002).

In any case, in spite of intensive investigations, it is not easy to obtain good estimates of prediction of the bioavailability of PTMs. In fact, one-step soil-extracting procedures predict the phytoavailability of PTMs assuming a static equilibrium of the soil system. Nevertheless, reactions in the soil environment are rarely at equilibrium, but instead are in a state of continuous change because of the dynamic processes occurring. Keller and Hammer (2004), for example, in a study performed to assess the efficiency of repeated croppings of *Thlaspi caerulescens* to extract Cd and Zn from two contaminated soils, found that the amount of the two elements removed by *T. caerulescens* was larger than the bioavailable (0.1 M NaNO$_3$-extractable) amount measured at the beginning of the experiment. They explained their results by a large buffering capacity of the soils that replenished the bioavailable pool and therefore supplied PTMs at a rate that was nearly as fast as the PTMs' uptake by plants. Such a replenishment was observed also by Hammer and Keller (2002) and Whiting *et al.* (2001a). In general and especially for researches on phytoremediation by hyperaccumulator plants, useful information about metal distribution among the soil components, giving ideas on mechanisms responsible for PTMs' uptake by plants, might come from the use of sequential chemical extractions (as discussed in Section 6). This approach could be very interesting in the context of phytoextraction, in order to assess the respective effects of hyperaccumulators and crop plants on the soil chemically defined pools, and more specifically the potential of these plants to mobilise or deplete metals from these pools. This allows assessment of the potential risks or benefits of using plants to remediate PTMs-contaminated soils as well as the long-term efficiency of the technique.

5. HUMAN BIOAVAILABILITY

PTMs-contaminated soils pose a human health risk on the basis of the potential of the contaminant to leave the soil and enter the human bloodstream. In order to assess human health risk, several pathways of transfer of metals from soil to humans have to be taken into account. The most important metal intake takes place via the food chain in which plants or meat of animal play a key role. The direct ingestion of soil can be a major route of exposure for humans to many low mobile soil contaminants, particularly for small children through putting hands into the mouth (Gupta *et al.*, 1996). The contribution from the inhalation of particles smaller than 10 μm and from dermal contact with soil have little meaning compared with oral ingestion and are found to be less than 1% and 0.1% of the total intake, respectively (Paustenbach, 2000).

Oral bioavailability of soil-borne contaminants is defined as the contaminant fraction that reaches systemic circulation and derives from the combined result of soil ingestion, bioaccessibility, absorption and the first-pass effect (Fig. 9.4) (Oomen *et al.*, 2003; Wragg and Cave, 2003). The bioaccessible fraction is defined as the contaminant fraction that is mobilised from soil into the digestive juice chyme and represents the maximum amount of contaminant available for intestine absorption (Ruby *et al.*, 1996, 1999).

Bioaccessibility, and therefore oral bioavailability of soil contaminants, depends on soil type and contaminant (Davis *et al.*, 1997; Grøn and Anderson, 2003; Hamel *et al.*, 1998; Ruby *et al.*, 1999). PTMs occur in soil as a complex mixture of solid-phase chemical compounds of varying particle size and morphology, characterised by variable metal bioavailability. Mineral phases that form under acidic conditions (e.g. lead sulphate, iron-lead sulphate) will tend to be more stable in the acidic conditions of the stomach and hence less bioaccessible. By contrary, mineral phases

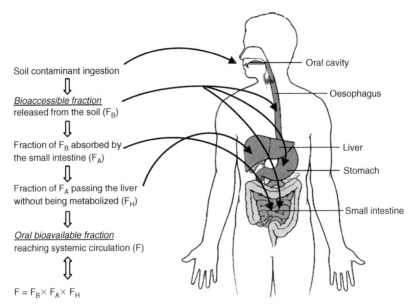

Soil contaminant ingestion

⇩

Bioaccessible fraction
released from the soil (F_B)

⇩

Fraction of F_B absorbed by
the small intestine (F_A)

⇩

Fraction of F_A passing the liver
without being metabolized (F_H)

⇩

Oral bioavailable fraction
reaching systemic circulation (F)

⇧⇩

$F = F_B \times F_A \times F_H$

Oral cavity

Oesophagus

Liver

Stomach

Small intestine

Figure 9.4 Oral bioavailability of soil contaminants as the combined result of soil ingestion, bioaccessibility, absorption and first-pass effect (modified from Oomen *et al.*, 2003).

that form under alkaline conditions (e.g. lead carbonate, lead oxide) will be less stable in the acidic conditions of the stomach and more bioaccessible (Ruby *et al.*, 1999). Other minerals, in particular phosphates, have highly variable compositions (Ruby *et al.*, 1996), which result in a wide range of bioaccessibility values (Ruby *et al.*, 1999). Small (<10 μm) particles are likely to be more easily inhaled, ingested and dissolved (more bioaccessible) in the gastrointestinal tract than larger ones, unless they are encapsulated (non-bioaccessible) within refractory phases (Davis *et al.*, 1997). The variable bioaccessibility of PTMs in different mineral phases and fractions involve that the use of total concentration in soils to estimate the exposure to immobile contaminants for humans (i.e. assuming that all the metal present in the soil can enter the bloodstream) provides a conservative approach, which can overestimate the exposure to soil contaminants. The estimation of the bioaccessible fraction, usually greater than the bioavailable fraction, provides a still conservative measure, which better relates with bioavailability (Paustenbach, 2000).

In the last decade, several *in vitro* digestion models based on human gastrointestinal tract physiology have been developed as simple, cheap and reproducible tools to investigate bioaccessibility of soil contaminants (Table 9.3) (Wragg and Cave, 2003). This type of systems was originally used in nutrition studies for assessing the iron bioavailability in food (Miller and Schricker 1982; Miller *et al.*, 1981) or the PTMs' bioaccessibility from children's toys (European Committee for Standardization, 1994). In these systems, PTMs-contaminated soils are treated with acidic (gastric) and/or neutral (intestinal) solutions, containing enzymes and organic acids, for a variable period intended to mimic residence time in the stomach and in the small intestine. Extractions are carried out at the normal temperature of the human body, 37 °C. Models can

Table 9.3 Schematic overview of different *in vitro* digestion models for assessing bioaccessible potentially toxic metals' (PTMs') content of soils

Model	Simulated compartments	Extractants	Tested elements	References
SBET	Stomach	0.4 M glycine, pH 1.5	As, Cd, Pb	1–5
PBET	Stomach	0.4 M glycine, pH 1.5	As, Pb	2–3, 5–9
	Small intestine	0.4 M glycine, pH 7		
DIN	Oral cavity	Saliva, pH 6.4	As, Cd,	5, 10–12
	Stomach	Gastric juice (pepsin, mucin), pH 2	Pb, Cr,	
	Small intestine	Intestinal juice (porcine bile, trypsin, pancreatine), pH 7.5	Hg	
RIVM	Oral cavity	Saliva, pH 6.5	As, Cd,	5, 10
	Stomach	Gastric juice (pepsin, mucin, BSA), pH 1.1	Pb	
	Small intestine	Intestinal juice (bovine bile, pancreatine, lipase, BSA), pH 8.0		
SHIME	Stomach	SHIME medium (pectin, mucin, cellobiose, proteose peptone, starch), pH 5.2	As, Cd, Pb	5, 13
	Small intestine	Pancreatic fluid (bovine bile, pancreatine), pH 6.5		
TIM	Oral cavity	Saliva, pH 5	As, Cd,	5, 14
	Stomach	Gastric juice (lipase, pepsin), pH 2	Pb	
	Small intestine	Intestinal juice (porcine bile, pancreatine), pH 7.2		

SBET, simple bioaccessibility extraction test; PBET, physiologically based extraction test; DIN, E DIN 19738 German method; RIVM, *in vitro* digestion model, National Institute of Public Health and the Environment, The Netherlands; SHIME, simulator of human intestinal microbial ecosystems of infants; TMO, TNO gastrointestinal model.
1. Ruby *et al.* (1993), 2. Ruby *et al.* (1996), 3. Ruby *et al.* (1999), 4. Davis *et al.* (1997), 5. Oomen *et al.* (2002), 6. Oomen *et al.* (2003), 7. Wragg and Cave (2003), 8. Yang *et al.* (2003), 9. Ruby (2004), 10. Rotard *et al.* (1995), 11. Hack and Selenka 1996), 12. Hack *et al.* (1998), 13. Molly *et al.* (1993), 14. Minekus *et al.* (1995).

simulate the transit through the human gastrointestinal tract in static [simple bioaccessibility extraction test (SBET), DIN, *in vitro* digestion model (RIVM), simulator of human intestinal microbial ecosystem (SHIME)] and dynamic (SHIME, TIM) systems, and can be extended to more compartments (i.e. oral cavity, stomach, small intestine and colon) (Molly *et al.*, 1993). They can involve simple one-stage extraction scheme or more complex multi-stage sequential extraction methods. Milk powder (DIN) or cream (SHIME) can be added to digestion systems to simulate the influence of food on mobilisation of contaminants from soil. Experimental design of *in vitro* digestion models can have a strong influence on PTMs' bioaccessibility (Oomen *et al.*, 2002). The low gastric pH, and the absence of an intestinal compartment with more neutral pH levels, is likely to be the main cause of the higher liberation of contaminants out of the soil matrix of the SBET compared with the other methods. Differences in soil pre-treatments (e.g. size cut-off value of soil sieving), solid mass (0.5 in RIVM to 10 g in SHIME and TIM), solid to liquid ratio (1:4 in SHIME to 1:100 in SBET), residence time of solid matrix in gastrointestinal juices (1 h in SBET to 8 h in SHIME), association of metals with particles or their complexation by food constituents, may lead to a wide range of bioaccessibility values (Table 9.4) (Davis *et al.*, 1997; Hamel *et al.*, 1998; Oomen *et al.*, 2002). Because it is unknown to which extent bioaccessibility values are accurate for the human *in vivo* situation, any general conclusion on the best model can be made. So far, validation of *in vitro* digestion models by *in vivo* studies has been done only occasionally and exclusively for lead and arsenic bioavailability (Maddaloni *et al.*, 1998; Medlin, 1997; Minekus *et al.*, 1995; Rodriguez *et al.*, 1999; Ruby *et al.*, 1996, 1999; Smith *et al.*, 1997) and at present there is no standard method of estimating bioaccessibility. In spite of these uncertainties, *in vitro* digestion models are likely to become increasingly important in risk assessment of polluted soils, especially for determining the urgency of remediation and for developing and evaluating remedial technologies, which may have an impact on metals bioavailability. According to the influence that mineralogy, particle-size distribution and soil chemical properties (i.e. pH, Eh and CEC) have on PTMs' bioavailability, for health risk assessment, site-specific results of oral bioavailability are required (Ruby *et al.*, 1999).

6. PTMs' PARTITIONING BETWEEN SOIL GEOCHEMICAL PHASES

When PTMs' concentration is well in excess of normal soil content, extraction method validation, in terms of direct correlation between soil extractable contents and plant contents, is less easy to achieve. In these cases, it may be adequate to develop an operational estimate of the mobile and potentially mobile metal species rather than plant-available species. It is necessary to analyse metal partitioning between such fractions as exchangeable sites, organic matter and minerals of varying solubility.

6.1. Sequential chemical extractions

Single and sequential chemical extractions have been commonly used in environmental studies for fractionation of PTMs among the various geochemical forms in which they might exist in soil. This operational approach is based on the use of a series of reagents selected for their ability to react with different major soil components and release associated metals. The metal-binding types or soil phases and single extractants used for their attack have been individually considered and comprehensively discussed by Rauret (1998) and Ure and Davidson (2002). The selectivity of the various reagents

Table 9.4 Examples of potentially toxic metals' (PTMs') bioaccessibility values (% of soil total content in mg·kg^{-1}) in contaminated soils determined with different *in vitro* digestion models

Matrix analysed	Method	As Total	As Bioacc.	Cd Total	Cd Bioacc.	Pb Total	Pb Bioacc.	Ref.
Soil	SBET	72–213	11–59	15–35	92–99	634–6380	56–91	1, 4
	DIN	77–235	18–50	17–38	62–90	730–5742	23–68	1
	DIN–WM[a]	77–235	11–41	17–38	38–62	730–5742	16–46	1
	RIVM	55–206	19–95	14–39	51–78	612–5454	11–66	1
	SHIME	82–227	1–10	18–41	5–7	725–6230	1–4	1
	TIM	74–236	15–52	14–38	50–58	616–6063	4–17	1
Soil with pottery flakes	RIVM	n.d.	n.d.	n.d.	n.d.	50–2400	28–73	2
Pottery flakes	RIVM	n.d.	n.d.	n.d.	n.d.	11000	0.3	2
Residential soil	PBET[b]	410–3900	34	n.d.	n.d.	1388–2090	22	3
Mine waste	PBET[b]	n.d.	n.d.	n.d.	n.d.	3908–3940	3.8–13	3
Copperton tailing	PBET[b]	n.d.	n.d.	n.d.	n.d.	6890–7220	6–8	3
Stream sediment	PBET[b]	n.d.	n.d.	n.d.	n.d.	10230	24	3
House dust	PBET[b]	170	31–32	n.d.	n.d.	n.d.	n.d.	3
Industrial slag	SBET	n.d.	n.d.	n.d.	n.d.	422–6490	3–27	4
Slag/soil mixture	SBET	n.d.	n.d.	n.d.	n.d.	2080–4620	30–51	4
Street dust	SBET	n.d.	n.d.	n.d.	n.d.	2630	77	4

[a] DIN–WM = DIN digestion without milk powder addition.

[b] Data only at pH 2.5 and for Pb only stomach data are reported.

Bioacc., bioaccessible; Ref., reference; n.d. = not determined; SBET, simple bioaccessibility extraction test; RIVM, *in vitro* digestion model; PBET, physiologically based extraction test; SHIME, simulator of human intestinal microbial ecosystem; TIM, TNO gastrointestinal model.

1. Oomen *et al.* (2002), 2. Oomen *et al.* (2003), 3. Ruby *et al.* (1996), 4. Davis *et al.* (1997).

towards specific geochemical phases is improved by their sequential application in schemes that generally follow an order of increasing extractants strength and decreasing phases solubility (Table 9.5). The extractants are generally applied according with the following order: unbuffered salts, weak acids, reducing agents, oxidising agents and strong acids.

A detailed description of the numerous extraction schemes reported in the literature with their applications to soil, sediments and various solid materials of environmental concern is reported in Das *et al.* (1995), Kersten and Förstner (1995), Sheppard and Stevenson (1997), Filgueiras *et al.* (2002), Kersten (2002) and Ure and Davidson (2002). Many of the sequential extraction schemes employed are based on the five-stage procedure of Tessier *et al.* (1979), which delineates the metal species sequentially as exchangeable, carbonate-bound, Fe and Mn oxide-bound, organic matter-bound and residual, mainly hold within the crystal structure of primary and secondary silicate minerals. Subsequent main modifications of Tessier scheme consist in the use of additional reagents to differentiate the Fe and Mn oxide-bound species into three distinct species: easily reducible metal oxide-bound, amorphous metal oxide-bound and crystalline metal oxide-bound species (Elliott *et al.*, 1990; Gibson and Farmer, 1986; Kersten and Förstner, 1986; Shuman, 1985; Sposito *et al.*, 1982). The differentiation of the metal–organic-complex-bound metal species, as distinct from the organically bound species, using 0.1 M sodium pyrophosphate (pH 10) is the innovative proposal of the scheme suggested by Krishnamurti *et al.* (1995a). Further subfractionation of the metal–organic complexes as metal-fulvate and metal-humate complexes was introduced by Krishnamurti and Naidu (2000) in their novel but highly sophisticated nine-step sequential extraction scheme.

6.2. The BCR sequential extractions

So far, fractionation schemes have not been standardised and each researcher produces his own scheme or uses a modification of the one developed earlier. The number of extraction steps has been generally defined on the basis of the study purpose and of the soil contamination status (Davidson *et al.*, 2004). The reagents and the extraction conditions (i.e. reagent strength, volume and extraction time) utilised at each stage have been chosen taking into account (1) the metals involved, (2) the nature of inorganic and organic soil components and (3) the physical and chemical properties of the soil. A general lack of uniformity, therefore, exists among sequential extraction methods, which makes extremely difficult the comparison of results obtained by different laboratories. Moreover, no effective quality control system can be implemented in the absence of reference materials. In order to harmonise sequential extraction procedures between European Union member states, the Measurements and Testing Programme—MAT (formerly BCR)—of the European Commission, has proposed two simple, three-stage sequential extraction protocols: the original one (Ure *et al.*, 1993a,b) and a modified one (Rauret *et al.*, 1999). Reference materials, certified for metals extractable by the procedures, have also been prepared (sediments CRM 601, for the original, and

Table 9.5 Extractants and forms of potentially toxic metals (PTMs) separated by various sequential extraction schemes (the number indicates the order of successive steps)

Supposed forms of metals	Extractant	Tessier et al. (1979)	Förstner et al. (1981)	Shuman (1985)	Krishnamurti et al. (1995a)	Ma and Uren (1998)	Hammer and Keller (2002)	Burt et al. (2003)	Silveira et al. (2006)
Water soluble (Soluble)-Exchangeable	DDI water								
	1 M MgCl$_2$/NH$_4$OAc/Mg(NO$_3$)$_2$, pH 7	1	1	1	1	1		1	1
	0.1 M NaNO$_3$/CaCl$_2$				1				1
Specifically adsorbed	1% Na Ca HEDTA in 1 M NH$_4$OAc, pH 8.3					2	1	2	
Carbonates-bound	0.5/1 M NaOAc/NH$_4$OAc, pH 5	2	2		2	4	2	3	2
Mn and Fe oxides-bound	0.04/1 M NH$_2$OH·HCl in 25% HOAc	3					5	4	2
(Easily reducible) Mn–oxides-bound	0.05/0.1/0.25 M NH$_2$OH·HCl, pH 2		3	3	4	3	4		4
	0.2% Quinol in 1 M NH$_4$OAc, pH 7								
Amorphous (Fe)–oxides-bound	0.2 M NH$_4$Oxa–HOxa, pH 3 in dark		4	4	6	6		6	5
	0.175 M NH$_4$Oxa–HOxa, pH 3.25								
Crystalline (Fe)–oxides-bound	0.2 M NH$_4$Oxa–HOxa, pH 3 in 0.1 M ascorbic acid			5	7				
	6 M HCl								6

Fraction (reagents)								
Organic/sulphide bound								
0.1 M $Na_4P_2O_7$								
0.7 M NaOCl, pH 8.5	4	5	2	5		3	5	3
30% H_2O_2 pH 2, then 3.2 M NH_4OAc/1 M $Mg(NO_3)_2$ in 20% HNO_3					5			
30% H_2O_2 pH 4.74, then 0.5 M NaOAc pH 4.74					5			
Metal–organic complex-bound								
0.1 M $Na_4P_2O_7$, pH 10				3				
Residual (silicate minerals)								
HF–HNO_3–HCl/$HClO_4$	5	6				6		
HF–$HClO_4$				8			5	
HNO_3–HCl								
HNO_3		6					6	7
Total—sum of extractable		6				7		7

DDI water, double deionised water.

CRM 701, for the modified method) and made available (Pueyo *et al.*, 2001; Queauviller *et al.*, 1997). The BCR original procedure is based on extractions with

- 0.11 M acetic acid (step 1)
- 0.1 M hydroxylamine hydrochloride, pH 2 (step 2)
- 8.8 M hydrogen peroxide, then 1 M ammonium acetate, pH 2 (step 3)

For quality control, the residue from step 3 has to be digested in *aqua regia* ('step 4') and the cumulative amount of metal extracted [i.e. Σ(step 1 + step 2 + step 3 + 'step 4')] is compared with the pseudototal amount of metal obtained by *aqua regia* digestion of a separate 1-g soil sample. In the modified BCR procedure, a 0.5 M hydroxylamine hydrochloride solution at pH 1.5 is applied in step 2.

The metal phases which are presumed to be sequentially extracted are: step 1, soluble, exchangeable and carbonate-bound metals; step 2, metals occluded in easily reducible manganese and iron oxides; step 3, organically bound metals and sulphides; 'step 4', metals in non-silicate minerals lattice structure. It is important to emphasise that these metal phases are nominal target only, operationally defined by the extraction used. Consequently, is highly desirable and recommended to refer to the sequentially extracted metal fractions as 'easily extractable', 'reducible', 'oxidisable' and 'residual', respectively.

The BCR procedures have been applied to a variety of matrices, including soils, sediments and sewage sludges (Ure and Davidson, 2002). Uncertainties in reproducibility occur when the original procedure is performed on field-moist soil samples and by different analysts (Davidson *et al.*, 1999). To address the lack of soil reference materials, recently Žemberyová *et al.* (2006) have established valuable BCR-extractable concentrations of Cd, Cr, Cu, Ni, Pb and Zn in Slovak soil reference materials. The possibility to replace the 0.5 M hydroxylamine hydrochloride (pH 1.5) solution in step 2 with 0.2 M ammonium oxalate (pH 3) has been proved to be convenient in presence of iron-containing soil components, unless calcium and lead concentrations are to be measured (Davidson *et al.*, 2004). The introduction of a preliminary step with buffered NH_4NO_3 in the protocol of Ure *et al.* (1993) improved the determination of the most labile pools in calcareous soil (Renella *et al.*, 2004). A problem of the BCR scheme, as well as of many other schemes, is the inability to distinguish between metal associations with organic matter and sulphides, a distinction which would appear to be highly desirable from an environmental point of view (Adamo *et al.*, 1996). An attempt, using linear multiple regression, to describe the dependence of Cd, Pb, Ni, Cu and Zn uptake by various vegetables on the concentrations of their mobile forms in soil defined by BCR speciation procedure was done by Moćko and Wacławek (2004). In a recent study, the selectivity of each extraction step was tested by observing the effect of reversing the extraction order in the procedure (Kim and McBride, 2006).

6.3. Problems and options of sequential extractions

Two experimental problems have been recognised in the application of sequential extractions for PTMs' fractionation in soil and sediments. One is the PTMs' readsorption and redistribution among phases during extraction, meaning that PTMs released by one extractant could associate with other undissolved solid components or freshly exposed surfaces within the timescale of the extraction step (Gómez–Ariza et al., 1999). The readsorption and redistribution processes have been mainly considered and discussed in studies on natural sediments (Ajayi and Vanloon, 1989; Belzile et al., 1989; Raksasataya et al., 1996; Whalley and Grant, 1994), where they are influenced by the geochemical characteristics of the matrix and may lead to serious misinterpretation of extraction data, especially those related to metal mobility (i.e. the most labile fractions) (Gómez–Ariza et al., 1999). In a study using radioisotopes, Ho and Evans (2000) found, however, that in soil, the degree of readsorption was less than expected and that it did not invalidate the sequential extraction results. The results of a long-term study of the association of PTMs with soil components in an upland catchment in Scotland confirmed this finding (Bacon et al., 2005).

The second and major problem, strictly linked with *operational speciation*, is that the detailed nature of the material being extracted is not known. Idealised pure phases are often named in various extraction schemes on the basis of their response to the extraction reagents, but this approach may be misleading in that the behaviour of natural samples may be variable according to the detailed nature of the inorganic and organic constituents with which the metals are associated. In practice, therefore, the operationally defined metal forms may be much less specific than desired and may be made up of a variety of different or even unexpected element associations. On the contrary, more direct methods for solid-state speciation [i.e. instrumental techniques such as SEM/TEM, EDS/WDS analysis, XRD, proton-induced X-ray emission (PIXE) and Auger electron spectroscopy (AES)] may yield more precise information regarding the actual nature of the soil constituents involved in metal retention, so that the various metal forms present can be more precisely characterised. Direct methods are more sophisticated than chemical methods and need a high level of specialisation to be routinely included in PTMs' speciation studies. Nevertheless, a complementary use of the two approaches may provide a more realistic picture of the actual forms of PTMs in soils.

A first attempt in this direction was reported by Adamo et al. (1996). In this study, BCR sequential extraction and SEM/EDS observations were complementarily used to define Cu and Ni forms in Sudbury soils. Most Cu (on average 75%) was associated with 'non-residual' soil forms, whereas Ni was mainly (on average 60%) associated with inorganic 'residual' forms of a sulphide and oxide nature (Fig. 9.5). The SEM/EDS analysis confirmed the Cu association with organic matter and clay fraction surfaces, highlighted that both sulphides and Fe oxides were not completely decomposed by BCR steps 2 and 3 and, more essential, shown associations of Cu and Ni which would not have been predicted by the sequential extractions alone, such as carbonaceous material, silicate spheres and carbonate particles (Figs. 9.6 and 9.7). On the basis of whole speciation results, authors could conclude that a much greater risk of Cu mobilisation from the Sudbury soils exists. With a similar approach (sequential extractions plus SEM and XRD), Henderson et al. (1998) studied the PTMs' geochemistry of humus and till from the Flin Flon–Snow Lake area to assess smelter contamination in northern Manitoba and Saskatchewan environment.

Adamo et al. (2002a) also determined the Fe, Al, Cu, Co, Cr, Pb, Zn, Ni and Mn speciation in a representative industrial soil profile, from the dismantled ILVA

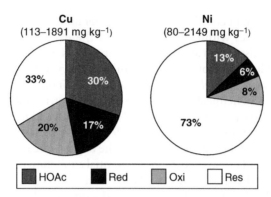

Figure 9.5 Copper and nickel distribution among BCR sequentially extracted fractions in Sudbury soils around the Copper Cliff smelter. (HOAc = acetic acid-extractable; red = reducible; oxi = oxidisable; res = residual) (from Adamo *et al.*, 1996).

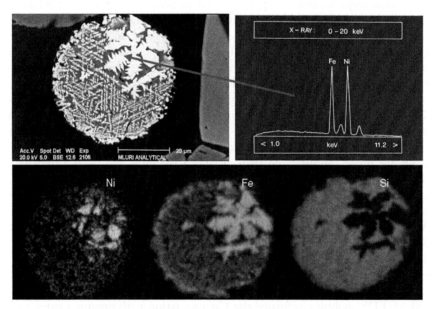

Figure 9.6 The scanning electron microscopy (SEM) in the backscattered mode, the energy dispersive X-ray (EDX) spectrum and X-ray distribution maps of a spherical particle from Sudbury soil showing Ni and Fe microstructures in a silicate matrix (from Adamo *et al.*, 1996).

iron–steel plant of Bagnoli (Naples). The large presence of elements trapped in the mineralogical structure of oxides and silicates and occluded in easily reducible manganese or iron oxides was identified by BCR sequential extraction. A constant amount of Cu was associated with organic compounds. A significant amount of Zn (20%) was extracted in diluted acetic acid solution, indicating that the element was present in a more readily and potentially available form (Fig. 9.8). In the clay fraction (<2 μm), PTMs were associated with both amorphous and crystalline iron forms. Despite the numerous chemical evidences of low metal mobility, the presence of

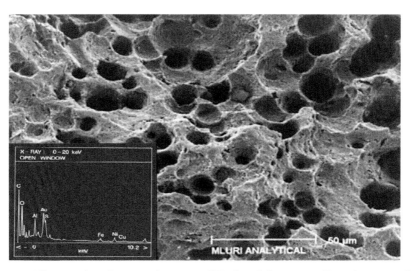

Figure 9.7 The scanning electron microscopy (SEM) and the energy dispersive X-ray (EDX) spectrum of a carbonaceous particle with a porous texture separated from Sudbury soil (from Adamo *et al.*, 1996).

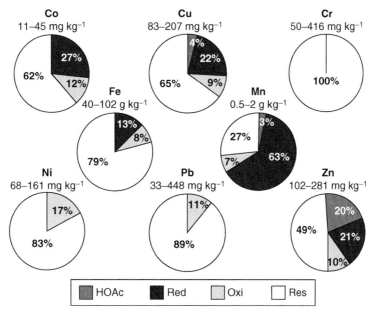

Figure 9.8 Potentially toxic metals' (PTMs') distribution in the BCR sequentially extracted fractions from the Bagnoli industrial ex-ILVA soil (0–72-cm depth). (HOAc = acetic acid-extractable; red = reducible; oxi = oxidisable; res = residual) (from Adamo *et al.*, 2002a).

Fe Cu, Co, Cr and Zn-rich clay coatings was evident in the illuvial pores of deeper horizons by microscopical (OM, SEM) and microanalytical (EDS, WDS) analysis of thin sections of undisturbed soil samples. Possible translocation of metals down through the soil profile mainly bound to fine particles of relatively inert forms of iron was hypothesised (Fig. 9.9). Magnetite, goethite, hematite, calcite and quartz mixed with K-feldspars, clynopyroxenes and mica occurred in the coarse sand fractions (2–0.2 mm) of the soil samples from all the surface horizons. Talcum and goethite together with clay minerals at 1.4 nm, kaolinite and illite were found in the clays (<2 μm). A similar risk of Cu- and Cr-rich sediment transfer along the soil pore network during water movement was also shown by combination of sequential extractions and quantitative WDS analysis in the overflowed agricultural volcanic soils of the Solofrana river valley (Adamo et al., 2003, 2006; Zampella, 2006).

The bioavailability of PTMs depends not only on their activity in the solution phase but also on the relation that exists between solution ions and solid-phase ions (Kabata-Pendias and Pendias, 2001). In order to understand the role of solid-phase fractions in determining the availability of PTMs to northern leopard frog (Rana pipiens) in marshes sediments from western Vermont, USA, Ryan et al. (2002) used quantitative mineralogical analysis [quantitative X-ray diffraction (QXRD)] in conjunction with BCR procedure. Differences in speciation between Pb on the one hand, and Co, Cr, Cu, Ni and Zn on the other, suggested only for Pb an anthropogenic origin. Trioctahedral clay minerals, including chlorite, high charge smectite and vermiculite, were found to be the most common reservoir of Co, Cr, Cu, Ni and Zn in the studied marshes sediments. Lead appeared to be contained in poorly crystalline or amorphous oxyhydroxides, which changing redox conditions might cause the release of the metal into pore water. The QXRD also indicated that expandable trioctahedral clay dissolves progressively throughout the BCR sequential extraction procedure, underscoring the importance of corroborating QXRD analyses in the interpretation of metal speciation by sequential extractions.

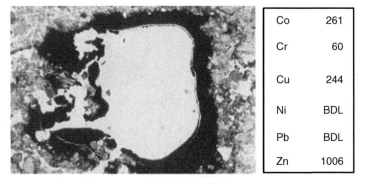

Co	261
Cr	60
Cu	244
Ni	BDL
Pb	BDL
Zn	1006

Figure 9.9 Thin-section micrograph (PPL) of a pore in the 72–80-cm depth horizon from Bagnoli soil completely covered by dark clay coatings and potentially toxic metals' (PTMs') quantification (mg·kg^{-1}) in the coatings as determined by the wavelength dispersive spectroscopy (WDS) (average of 40 points around the pore) (from Adamo et al., 2002a).

In assessing environmental risks, however, the lack of specificity can be not so crucial. It might be more interesting to get information on the possibility of mobilisation or on bioavailability of PTMs than to identify the exact chemical species of metals in soil. It could be not so important to know whether the metals come from sulphides or organic matter, but could be more useful to know that they can be 'released under reducing or oxidising conditions'.

7. APPLICATIONS OF PTMs' SPECIATION FOR RISK AND REMEDIATION ASSESSMENT

A multidisciplinary approach, including chemical, physical and mineralogical characterisation of contaminants and soils, has been shown to be fundamental to understand comprehensively the history of industrial soil contamination, to assess the potential and the actual risk associated with elevated levels of PTMs in soil (i.e. the relation of PTMs with soil matrix and their chemical accessibility and mobility) and to define the feasibility of potential remediation treatments (Adriano et al., 2004; Banat et al., 2005; Gasser and Dahlgren, 1994; Sahuquillo et al., 2003; Venditti et al., 2000a,b).

The synchrotron-based X-ray radiation microfluorescence (μ-SXRF) and the microfocused and powder EXAFS spectroscopy measurements were combined with chemical extractions and thermodynamic modelling to identify and quantify the solid-state forms of Zn in near-neutral (pH 6.5–7.0) soil contaminated by sewage long-term irrigation, and to evaluate the modification of metal speciation upon adding organic ligands (Kirpichtchikova et al., 2006). In the field, the mobility and the bioavailability of Zn was mainly controlled by iron oxyhydroxide ($\sim$30%), phosphate ($\sim$28%) and phyllosilicate ($\sim$24%). These three secondary phases were preferentially ($\sim$80%) solubilised with citrate, when the extracting solution was renewed, indicating that this cheap and easily biodegraded reagents may be used to remediate the soil.

Several experimental works on the basis of the relative mobility of PTMs and their distribution among the different soil phases have inferred that PTMs from anthropogenic sources are potentially more mobile/available than those that are ultimately inherited from the geological parent material (Adamo et al., 2002b; Chlopecka et al., 1996; Kabata-Pendias and Pendias, 2001; Lee, 2006; Ma and Rao, 1997; Renella et al., 2004). The soil type and management as well as the organic matter, clay or carbonate content, and CEC do not appear to influence the elements distribution in any consistent manner. Nevertheless, the influence of pH and the extent of contamination is clear. Where the PTMs are present at background levels, usually the residual form is dominant suggesting that PTMs are associated with primary minerals. In most of the works based on sequential extractions data, it is assumed that the mobility and the bioavailability of PTMs are related to their solubility and geochemical forms and, therefore, decrease in the order of extraction sequence.

The mobility and the bioavailability of PTMs and thereby the potential risks to human health and the environment of contaminated paddy soil near mine tailings dumps in Korea were found mainly related to the presence of solubility controlling solid phases for Cd and Pb, while Zn and Cu were controlled by adsorption/desorption processes (Lee, 2006). Either iron sulphide or iron oxide, forming under reducing flooded and oxidising dry conditions/cycles, were likely the solid phases precipitating/adsorbing PTMs and lowering their concentration in soil solution. In the context of paddy soils, the control of the redox condition and pH is, therefore, extremely important for monitoring and remediation purposes.

A detailed characterisation (particle-size analysis, PTMs' total content and speciation by sequential extractions, SEM/EDS) of the Cu and Pb contaminants in sandy soils from three hazardous industrial-waste sites from western New York provided information pertinent to soil washing and washing solutions feasibility assessment (Yarlagadda et al., 1995). Lead and Cu were not highly concentrated in the smaller soil size fractions, as it is often assumed by remedial approaches (e.g. soil washing) achieving contamination reduction through clay/silt separation. High-density PTMs containing particulates occurred in soil, suggesting remediation through density separation processes. Results from sequential extractions indicated PTMs' amounts likely removable by water washing (i.e. soluble and exchangeable metals) and helped to formulate possible PTMs' extraction solutions for removal of Cu and Pb bound to carbonate and Fe/Mn oxides.

Quantitative chromium solid-state speciation in chromite ore processing residue (COPR) has defined the mineral species and the processes controlling the retention and release of Cr(VI) from COPR-contaminated sites (Hillier et al., 2003). Information that, used within a process-based modelling framework, has helped to predict the impact of changes in physicochemical conditions on the COPR, to test the extent to which the system may be considered at equilibrium and that, therefore, need to be considered within the context of informed remediation (Geelhoed et al., 2001).

The efficiency of Cd and Zn phytoextraction from a calcareous and an acidic-contaminated soil with hyperaccumulating plants (*Thlaspi caerulescens* J.&C. Presl., les Avinières population) was assessed by Keller and Hammer (2004) by different approaches in order not to overlook any potential hazard. Precisely, the total extraction efficiency (2 M HNO$_3$-extractable metals) changes in metal bioavailability (0.1 M NaNO$_3$-extractable metals and lettuce uptake) and toxicity (lettuce biomass and BIOMET$^®$ biosensor), and metals in the soil solution were monitored over the whole experiment. Soil characteristics were found to determine phytoextraction efficiency. *T. caerulescens* was the most efficient in the acidic soil, where significantly decreased Cd and Zn contamination level, bioavailability, toxicity and soil solution concentrations. However, in the same soil re-equilibration processes and the possible release of metals from the decaying roots remaining in the soil could lead in a long term to a leaching of metals from soil.

The possibility of using *Lolium perenne* for re-vegetation of soil from a former ferrous metallurgical site contaminated by Cu, Pb and Zn was demonstrated in a greenhouse study by Arienzo et al. (2004). Plants were healthy with 100% survival in all cases, with no macroscopic symptoms of metal toxicity. The high pH of the soil (8.4) was the most important parameter responsible for low metal solubility/toxicity. The PTMs' distribution, and therefore the mobility/availability of metals, defined by sequential extractions, was only slightly affected by *L. perenne* growth with changes regarding only the organic-bound Cu and Zn pool (reduction up to 24%).

8. CONCLUDING REMARKS

The necessity to have guidelines for PTMs in soils which consider not only their total contents but also mobility, availability and toxicity of contaminants, in order to predict adverse effects and define appropriate strategies of remediation, is beyond any doubt. The effort made by the scientific community to search for harmonised analytical procedures and certified reference materials is from this point of view remarkably consequential, increasing the quality of produced data. Nevertheless, the bioavailability, as currently understood, is a concept that is more meaningful to legislators than to scientists. In many respects, the limitations stem largely from the lack of ability of most widely used chemical extraction methods to assess the appropriate chemical forms, the potential transport of PTMs through the soil and the transfer to plants, animals and humans. Correlations between extractable PTMs and response in one soil generally do not hold across different soils, as the latter may well show variations in soil properties and PTMs' sources. Moreover, the differences in the metal bioavailability for the range of plant species growing in the same polluted soil illustrates the problem of drafting a legal framework that can be applied consistently across the board. In many cases, the bioavailability of soil metals has a negligible impact on the quality of human life; vegetation can colonise and develop in metal-polluted soils, stabilising and covering exposed surfaces and significantly reducing direct exposure risk. However, agricultural soils pose a greater risk when crops grown on polluted soil form a significant part of the human diet. Identification and cultivation of crop plants that exclude metals might be a significant method of further reducing risk.

ACKNOWLEDGMENT

Authors gratefully acknowledge the work of two anonymous referees for their helpful review of the manuscript.

REFERENCES

Adamo, P., Dudka, S., Wilson, M. J., and McHardy, W. J. (1996). Chemical and mineralogical forms of Cu and Ni in contaminated soils from the Sudbury mining and smelting region, Canada. *Environ. Pollut.* **91**, 11–19.
Adamo, P., Arienzo, M., Bianco, M. R., and Violante, P. (2002a). Heavy metal contamination of the soils devoted to stocking raw materials in the former ILVA iron-steel industrial plant of Bagnoli (south Italy). *Sci. Total Environ.* **295**, 17–34.

Adamo, P., Dudka, S., Wilson, M. J., and McHardy, W. J. (2002b). Distribution of trace elements in soils from the Sudbury smelting area (Ontario, Canada). *Water Air Soil Pollut.* **135,** 95–116.

Adamo, P., Denaix, L., Terribile, F., and Zampella, M. (2003). Characterization of heavy metals in contaminated volcanic soils of the Solofrana river valley (southern Italy). *Geoderma* **117,** 347–366.

Adamo, P., Zampella, M., Gianfreda, L., Renella, G., Rutigliano, F. A., and Terribile, F. (2006). Impact of river overflowing on trace element contamination of volcanic soils in south Italy: Part I. Trace element speciation in relation to soil properties. *Environ. Pollut.* **144**(1), 308–316.

Adriano, D. C., Wenzel, W. W., Vangronsveld, J., and Bolan, N. S. (2004). Role of assisted natural remediation in environmental cleanup. *Geoderma* **122,** 121–142.

Ajayi, S. O., and Vanloon, G. W. (1989). Studies on redistribution during the analytical fractionation of metals in sediments. *Sci. Total Environ.* **87/88,** 171–187.

Alloway, B. J. (1995). Soil processes and the behaviour of metals. *In* "Heavy Metals in Soils" (B. J. Alloway, ed.). 2nd edn., pp.11–37. Blackie Academic and Professional, London.

Andrews, P., Town, R. M., Hedley, M. J., and Laganathan, P. (1996). Measurement of plant-available cadmium in New Zealand soils. *Aust. J. Soil Res.* **34,** 441–452.

Arienzo, M., Adamo, P., and Cozzolino, V. (2004). The potential of *Lolium perenne* for revegetation of contaminated soil from a metallurgical site. *Sci. Total Environ.* **319,** 13–25.

Bacon, J. R., Hewitt, I. J., and Cooper, P. (2005). Reproducibility of the BCR sequential extraction procedure in a long-term study of the association of heavy metals with soil components in an upland catchment in Scotland. *Sci. Total Environ.* **337,** 191–205.

Banat, K. M., Howari, F. M., and Al-Hamad, A. A. (2005). Heavy metals in urban soils of central Jordan: Should we worry about their environmental risks? *Environ. Res.* **97,** 258–273.

Bargagli, R. (1998). "Trace Elements in Terrestrial Plants. An Ecophysiological Approach to Biomonitoring and Biorecovery," pp. 324. Springer, Berlin.

Belzile, L., Lecomte, P., and Tessier, A. (1989). Testing readsorption of trace elements during partial chemical extractions of bottom sediments. *Environ. Sci. Technol.* **23,** 1015–1020.

Bermond, A., Yousfi, I., and Ghestem, J. P. (1998). Kinetic approach to the chemical speciation of trace metals in soils. *Analyst* **123,** 785–789.

Berrow, M. L., and Mitchell, R. L. (1980). Location of trace elements in soil profiles: Total and extractable contents of individual horizons. *Trans. R. Soc. Edinburgh: Earth Sci.* **71,** 103–121.

Bhogal, A., Nicholson, F. A., Chambers, B. J., and Shepherd, M. A. (2003). Effects of past sewage sludge additions on heavy metal availability in light textured soils: Implications for crop yields and metal uptakes. *Environ. Pollut.* **121,** 413–423.

Burridge, J. C., and Berrow, M. L. B. (1984). Long-term effects of metal-contaminated sewage sludges on soils and crops. *In* "Proceedings of the International Conference on Environmental Contamination," London, pp. 215–224. CEP Consultants, Edinburgh.

Burt, R., Wilson, M. A., Keck, T. J., Dougherty, B. D., Strom, D. E., and Lindahl, J. A. (2003). Trace element speciation in selected smelter-contaminated soils in Anaconda and Deer Lodge Valley, Montana, USA. *Adv. Environ. Res.* **8,** 51–67.

Campbell, P. G. C. (1995). Interactions between trace metals and aquatic organisms: A critique of the free-ion activity model. *In* "Metal Speciation and Bioavailability in Aquatic Systems" (A. Tessier and D. R. Turner, eds.), pp. 45–102. John Wiley, New York, NY.

Cappuyns, V., and Swennen, R. (2006). Comparison of metal release from recent and aged Fe-rich sediments. *Geoderma* **137,** 242–251.

Carlton-Smith, C. H., and Davis, R. D. (1983). An interlaboratory comparison of metal determinations in sludge-treated soil. *Water Pollut. Control* **82,** 544–556.

Chlopecka, A., Bacon, J. R., Wilson, M. J., and Kay, J. (1996). Forms of cadmium, lead, and zinc in contaminated soils from southwest Poland. *J. Environ. Qual.* **25,** 69–79.

Chojnacka, K., Chojnacki, A., Górecka, H., and Górecki, H. (2005). Bioavailability of heavy metals from polluted soils to plants. *Sci. Total Environ.* **337,** 175–182.

Christie, P., Li, X., and Chen, B. (2004). Arbuscular mycorrhiza can depress translocation of zinc to shoots of host plants in soils moderately polluted with zinc. *Plant Soil* **261,** 209–217.

Clayton, P. M., and Tiller, K. G. (1979). A chemical method for the determination of the heavy metal content of soils in environmental studies. *CSIRO Division of Soils Technical Papers* **41,** 17.

Clemente, R., Walker, D. J., and Bernal, M. P. (2005). Uptake of heavy metals and As by Brassica juncea grown in a contaminated soil in Aznacóllar (Spain): The effect of soil amendments. *Environ. Pollut.* **138,** 46–58.

Cottenie, A., Velghe, G., Verloo, M., and Kiekens, L. (1982). "Biological and Analytical Aspects of Soil Pollution." Lab. of Analytical and Agrochemistry, State University, Ghent, Belgium.

Das, A. K., Chakraborty, R., Cervera, M. L., and de la Guardia, M. (1995). Metal speciation in solid matrices. *Talanta* **42,** 1007–1030.

Davidson, C. M., Ferreira, P. C. S., and Ure, A. M. (1999). Some sources of variability in application of the three-stage sequential extraction procedure recommended by BCR to industrially-contaminated soil. *Fresenius J. Anal. Chem.* **363,** 446–451.

Davidson, C. M., Hursthouse, A. S., Tognarelli, D. M., Ure, A. M., and Urquhart, G. J. (2004). Should acid ammonium oxalate replace hydroxylammonium chloride in step 2 of the revised BCR sequential extraction protocol for soil and sediment? *Anal. Chim. Acta* **508,** 193–199.

Davies, B. E., Lear, J. M., and Lewis, N. J. (1989). Plant availability of heavy metals in soils. *In* "Pollutant Transport and Fate in Ecosystems" (P. J. Coughtrey, M. H. Martin, and M. H. Unsworth, eds.), pp. 267–275. Blackwell, Oxford.

Davis, R. D. (1979). Uptake of Cu, Ni and Zn by crops growing in contaminated soil. *J. Sci. Food Agric.* **30,** 937–947.

Davis, A., Ruby, M. V., Goad, P., Eberle, S., and Chryssoulis, S. (1997). Mass balance on surface-bound, mineralogic, and total lead concentrations as related to industrial aggregate bioaccessibility. *Environ. Sci. Technol.* **31,** 37–44.

Denaix, L., Semlali, R. M., and Douay, F. (2001). Dissolved and colloidal transport of Cd, Pb, and Zn in a silt loam soil affected by atmospheric industrial deposition. *Environ. Pollut.* **113,** 29–38.

Di Bonito, M., Breward, N., Crout, N., Smith, B., and Young, S. (2008). Soil pore water extraction methods for trace metals determination in contaminated soils. *In* "Environmental Geochemistry: Site Characterization, Data Analysis, and Case Histories" (B. De Vivo, H. E. Belkin, and A. Lima, eds.). Elsevier Science Publishers, Amsterdam, The Netherlands.

Díez Lázaro, J., Kidd, P. S., and Monterroso Martínez, C. (2006). A phytogeochemical study of the Tra's-os-Montes region (NE Portugal): Possible species for plant-based soil remediation technologies. *Sci. Total Environ.* **354,** 265–277.

DIN (Deutsches Instutut fur Normung) (1995). "Bodenbeschaffenhit. Extraktion von Spurenelemente mit Ammonium-nitratlosung." Vornorm DIN V19730. Boden-Chemische Bodenuntersuchungsverfahren, Berlin, Germany.

Düring, R. A., Hoß, T., and Gath, S. (2003). Sorption and bioavailability of heavy metals in long-term differently tilled soils amended with organic wastes. *Sci. Total Environ.* **313,** 227–234.

Elliott, H. A., Dempsey, B. A., and Maille, M. J. (1990). Content and fractionation of heavy metals in water treatment sludges. *J. Environ. Qual.* **19,** 330–334.

Ellis, R. H., and Alloway, B. J. (1983). Factors affecting the availability of cadmium, lead and nickel in soils amended with sewage sludge. *In* "Proceedings of the International Conference on Heavy Metals in the Environment," Heidelberg, pp. 358–361. CEP Consultants, Edinburgh.

Ervio, R., and Sippola, J. (1993). Micronutrient concentration of Italian ryegrass (*Lolium multiflorum* L.) grown on different soils in a pot experiment. *Agric. Sci. Finl.* **2,** 141–148.

European Committee for Standardization (1994). "Safety of Toys-Part 3: Migration of Certain Elements: 1994." European Standard EN 71–3, Brussels.

Feng, M. H., Shan, X. Q., Zhang, S., and Wen, B. (2005a). A comparison of the rhizosphere-based method with DTPA, EDTA, CaCl$_2$, and NaNO$_3$ extraction methods for prediction of bioavailability of metals in soil to barley. *Environ. Pollut.* **137,** 231–240.

Feng, M. H., Shan, X. Q., Zhang, S., and Wen, B. (2005b). Comparison of rhizosphere-based method with other one-step extraction methods for assessing the bioavailability of soil metals to wheat. *Chemosphere* **59,** 939–949.

Filgueiras, A. V., Lavilla, I., and Bendicho, C. (2002). Chemical sequential extraction for metal partitioning in environmental solid samples. *J. Environ. Monit.* **4,** 823–857.

Förstner, U. (1995). Land contamination by metals—Global scope and magnitude of problem. *In* "Metal Speciation and Contamination of Soil" (H. E. Allen, C. P. Huang, G. W. Bailey, and A. R. Bowers, eds.), pp. 1–33. Lewis Publishers, Boca Raton, FL.

Förstner, U., Calmano, K., Conrad, H., Jaksch, H., Schimkus, C., and Schoer, J. (1981). Chemical speciation of heavy metals in solid waste materials (sewage sludge, mining wastes, dredged materials, polluted sediments) by sequential extraction. *In* "Proceedings of the International Conference on Heavy Metals in the Environment," pp. 698–704. WHO/EED.

Fuentes, A., Lioréns, M., Sàez, J., Soler, A., Aguilar, I., Ortuño, J. F., and Meseguer, V. F. (2004). Simple and sequential extractions of heavy metals from different sewage sludges. *Chemosphere* **54,** 1039–1047.

Fuentes, A., Lioréns, M., Sàez, J., Aguilar, I., Belén, Pérez-Marín, A., Ortuño, J. F., and Meseguer, V. F. (2006). Ecotoxicity and extractability of heavy metals from different stabilised sewage sludge. *Environ. Pollut.* **143,** 355–360.

Gasser, U. G., and Dahlgren, R. A. (1994). Solid-phase speciation and surface association of metals in serpentinitic soils. *Soil Sci.* **158,** 409–420.

Geelhoed, J. S., Meeussen, J. C. L., Lumsden, D. G., Hillier, S. J., Roe, M. J., Thomas, R. P., Bewley, R. J. F., Farmer, J. G., and Paterson, E. (2001). Modelling of chromium behaviour and transport at sites contaminated with chromite ore processing residue: Implications for remediation methods. *Environ. Geochem. Health* **23,** 261–265.

Gibson, J. J., and Farmer, J. G. (1986). Multi-step chemical extraction of heavy metals from urban soils. *Environ. Pollut.* (Series B). **11,** 117–135.

Gómez-Ariza, J. L., Giráldez, I., Sánchez-Rodas, D., and Morales, E. (1999). Metal readsorption and redistribution during the analytical fractionation of trace elements in oxic estuarine sediments. *Anal. Chim. Acta* **399,** 295–307.

Goodman, B. A., and Glidewell, S. M. (2002). Direct methods of metal speciation. *In* "Chemical Speciation in the Environment" (A. Ure and C. M. Davidson, eds.), pp. 30–66. Blackwell, Oxford.

Gray, C. W., McLaren, R. G., Roberts, A. H. C., and Condron, L. M. (1999). Cadmium phytoavailability in some New Zealand soils. *Aust. J. Soil Res.* **37,** 461–477.

Grøn, C., and Anderson, L. (2003). Human bioaccessibility of heavy metals and PAH from soil. 840/2003, Danish Environmental Protection Agency.

Gupta, S. K. (1991a). Assessment of ecotoxicological risk of accumulated metals in soils with the help of chemical methods standardised through biological tests. *In* "Trace Metals in the Environment" (J. P. Vernet, ed.). Vol. 1, pp. 55–56. Elsevier Science Publishers, Amsterdam, The Netherlands.

Gupta, S. K. (1991b). Mobilisable metal in anthropogenically contaminated soils and its ecological significance. *In* "Trace Metals in the Environment" (J. P. Vernet, ed.). Vol. 2, pp. 229–310. Elsevier Science Publishers, Amsterdam, The Netherlands.

Gupta, S. K., and Aten, C. (1993). Comparison and evaluation of extraction media and their suitability in a simple model to predict the biological relevance of heavy metal concentrations in contaminated soils. *Int. J. Anal. Chem.* **51,** 26–46.

Gupta, A. K., and Sinha, S. (2006). Chemical fractionation and heavy metal accumulation in the plant of *Sesamum indicum* (L.) var. T55 grown on soil amended with tannery sludge: Selection of single extractants. *Chemosphere* **64,** 161–173.

Gupta, S. K., Vollmer, M. K., and Krebs, R. (1996). The importance of mobile, mobilisable and pseudo total heavy metal fractions in soil for three-level risk assessment and risk management. *Sci. Total Environ.* **178,** 11–20.

Hack, A., and Selenka, F. (1996). Mobilization of PAH and PCB from contaminated soil using a digestive tract model. *Toxicol. Lett.* **88,** 199–210.

Hack, A., Kraft, M., Mackrodt, F., Selenka, F., and Wilhem, M. (1998). Mobilisierung von Blei und Quecksilber aus real kontaminiertem Bodenmaterial durch synthetische Verdauungssäfte unter besonderer Berücksichtigung des Einflusses von Lebensmitteln. *Umweltmedizin in Forschung und Praxis* **3,** 297–305.

Hamel, S. C., Buckley, B., and Lioy, P. J. (1998). Bioaccessibility of metals in soils for different liquid to solid ratios in synthetic gastric fluid. *Environ. Sci. Technol.* **32,** 358–362.

Hammer, D., and Keller, C. (2002). Changes in the rhizosphere of metal-accumulating plants evidenced by chemical extractants. *J. Environ. Qual.* **31,** 1561–1569.

Hammer, D., Keller, C., McLaughlin, M. J., and Hamon, R. E. (2006). Fixation of metals in soil constituents and potential remobilization by hyperaccumulating and non-hyperaccumulating plants: Results from an isotopic dilution study. *Environ. Pollut.* **143,** 407–415.

Häni, H., and Gupta, S. (1983). Total and biorelevant heavy metal contents and their usefulness in establishing limiting values in soils. *In* "Environmental Effects of Organic and Inorganic Contaminants in Sewage Sludge" (R. D. Davis, G. Hucker, P. l'Hermite, and Dordrecht, eds.), pp. 121–129. "Proceedings of a Workshop held at Stevenage," UK, May 25–26, 1982, Holland.

Häni, H., and Gupta, S. (1985). Reason to use neutral salt solution to assess the metal impact on plants and soils. *In* "Chemical Methods for Assessing Bio-Available Metals in Sludges and Soils" (R. Leschber, R. D. Davis, and R. l'Hermite, eds.), pp. 42–48. Elsevier, New York, NY.

Haq, A. U., Bates, T. E., and Soon, Y. K. (1980). Comparison of extractants for plant-available zinc, cadmium, nickel, and copper in contaminated soils. *Soil Sci. Soc. Am. J.* **72**, 772–777.

Henderson, P. J., McMartin, I., Hall, G. E., Percival, J. B., and Walker, D. A. (1998). The chemical and physical characteristics of heavy metals in humus and till in the vicinity of the base metal smelter at Flin Flon, Manitoba, Canada. *Environ. Geol.* **34**, 39–58.

Hillier, S., Roe, M. J., Geelhoed, J. S., Fraser, A. R., Farmer, J. G., and Paterson, E. (2003). Role of quantitative mineralogical analysis in the investigation of sites contaminated by chromite ore processing residue. *Sci. Total Environ.* **308**, 195–210.

Ho, M. D., and Evans, G. J. (2000). Sequential extraction of metal contaminated soils with radiochemical assessment of readsorption effects. *Environ. Sci. Technol.* **34**, 1030–1035.

Holm, P. E., Christensen, T. H., Lorenz, S. E., Hamon, R. E., Domingues, H. C., Sequeira, E. M., and McGrath, S. P. (1998). Measured soil water concentrations of cadmium and zinc in plant pots and estimated leaching outflows from contaminated soils. *Water Air Soil Pollut.* **102**, 105–115.

Hornburg, V., Welp, G., and Brummer, G. W. (1995). Behaviour of heavy metals in soils. 2. Extraction of mobile heavy metals in soils with $CaCl_2$ and NH_4NO_3. *Zeitschrift für Pflanzenernährung und Bodenkunde* **158**, 137–145.

Houba, V. J. G., Novozamsky, I., Lexmond, T. M., and van der Lee, J. J. (1990). Applicability of 0.01 M $CaCl_2$ as a single extraction solution for the assessment of nutrient status of soils and other diagnostic purposes. *Commun. Soil Sci. Plant Anal.* **21**, 2281–2290.

Houba, V. J. G., Lexmond, T. M., Novozamsky, I., and van der Lee, J. J. (1996). State of the art and future developments in soil analysis for bioavailability assessment. *Sci. Total Environ.* **178**, 21–28.

Houba, V. J. G., Novozamsky, I., and Temminghof, E. J. M. (1997). "Soil and Plant Analysis, Part 5." Department of Soil Science and Plant Nutrition, Wageningen Agricutural University, The Netherlands.

Houba, V. J., Temminghof, E. J. M., and van Varrk, W. (1999). "Soil Analysis Procedure Extraction with 0.01 M $CaCl_2$." Univ. Wageningen, Wageningen.

John, M. K. (1972). Lead availability related to soil properties and extractable lead. *J. Environ. Qual.* **1**, 295–298.

John, M. K., Van Laerhover, J., and Chauh, H. H. (1972). Factors affecting plant uptake and phytotoxicity of cadmium added to soils. *Environ. Sci. Technol.* **6**, 1005–1009.

Kabata-Pendias, A. (2004). Soil-plant transfer of trace elements—An environmental issue. *Geoderma* **122**, 143–149.

Kabata-Pendias, A., and Pendias, H. (2001). "Trace Elements in Soils and Plants." CRC Press, London.

Keller, C., and Hammer, D. (2004). Metal availability and soil toxicity after repeated croppings of *Thlaspi caerulescens* in metal contaminated soils. *Environ. Pollut.* **131**, 243–254.

Keller, C., Hammer, D., Kayser, A., Richner, W., Brodbeck, M., and Ennhauser, M. (2003). Root development and heavy metal phytoextraction efficiency: Comparison of different plant species. *Plant Soil* **249**, 67–81.

Kersten, M. (2002). Speciation of trace metals in sediments. *In* "Chemical Speciation in the Environment" (A. Ure and C. M. Davidson, eds.), 2nd edn., pp. 301–321. Blackwell, Oxford.

Kersten, M., and Förstner, U. (1986). Chemical fractionating of heavy metals in anoxic estuarine and costal sediments. *Water Sci. Technol.* **18**, 121–130.

Kersten, M., and Förstner, U. (1995). Speciation of trace metals in sediments and combustion wastes. *In* "Chemical Speciation in the Environment" (A. Ure and C. M. Davidsond, eds.). 1st edn. Blackwell, Oxford.

Kidd, P. S., Domínguez-Rodríguez, M. J., Díez, J., and Monterroso, C. (2007). Bioavailability and plant accumulation of heavy metals and phosphorous in agricultural soils amended by long-term application sewage sludge. *Chemosphere* **66,** 1458–1567.

Kim, B., and McBride, M. B. (2006). A test of sequential extractions for determining metal speciation in sewage sludge-amended soils. *Environ. Pollut.* **144,** 475–482.

Kirpichtchikova, T. A., Manceau, A., Spadini, L., Panfili, F., Marcus, M. A., and Jacquet, T. (2006). Speciation and solubility of heavy metals in contaminated soil using X-ray microfluorescence, EXAFS spectroscopy, chemical extraction, and thermodynamic modeling. *Geochim. Cosmochim. Acta* **70,** 2163–2190.

Knight, B. P., Chaudri, A. M., McGrath, S. P., and Giller, K. E. (1998). Determination of chemical availability of cadmium and zinc in soils using inert soil moisture samplers. *Environ. Pollut.* **99,** 293–298.

Krishnamurti, G. S. R., and Naidu, R. (2000). Speciation and phytoavailability of cadmium in selected surface soils of South Australia. *Aust. J. Soil Res.* **38,** 991–1004.

Krishnamurti, G. S. R., Huang, P. M., Van Rees, K. C. J., Kozak, L. M., and Rostad, H. P. W. (1995a). Speciation of particulate-bound cadmium of soils and its bioavailability. *Analyst* **120,** 659–665.

Krishnamurti, G. S. R., Huang, P. M., Van Rees, K. C. J., Kozak, L. M., and Rostad, H. P. W. (1995b). A new soil test method for the determination of plant available Cd in soils. *Commun. Soil Sci. Plant Anal.* **26,** 2857–2867.

Krishnamurti, G. S. R., Smith, L. H., and Naidu, R. (2000). Method for assessing plant-available cadmium in soils. *Aust. J. Soil Res.* **38,** 823–836.

Lakanen, E., and Ervio, R. (1971). A comparison of eight extractants for the determination of plant available nutrients in soil. *Suomen Maataloustieteellisen Seuran Julkaisuja* **123,** 223–232.

Latterell, J. J., Dowdy, R. H., and Larson, W. E. (1978). Correlation of extractable metals and metal uptake of snap beans grown on soil amended with sewage sludge. *J. Environ. Qual.* **7,** 435–440.

Lee, S. (2006). Geochemistry and partitioning of trace metals in paddy soils affected by metal mine tailings in Korea. *Geoderma* **135,** 26–37.

Linsday, W. L., and Norvell, W. A. (1978). Development of a DTPA soil test for zinc, iron, manganese and copper. *Soil Sci. Soc. Am. J.* **42,** 421–428.

Lumsdon, D. G., and Evans, L. J. (2002). Predicting chemical speciation and computer simulation. *In* "Chemical Speciation in the Environment" (A. M. Ure and C. M. Davidson, eds.), pp. 89–131. Blackwell, Oxford.

Ma, L. Q., and Rao, G. N. (1997). Chemical fractionation of cadmium, copper, nickel, and zinc in contaminated soils. *J. Environ. Qual.* **26,** 259–264.

Ma, Y., and Uren, N. (1998). Transformation of heavy metals added to soil—Application of a new sequential extraction procedure. *Geoderma* **84,** 157–168.

Maddaloni, M., Lolacono, N., Manton, W., Blum, C., Drexler, J., and Graziano, J. (1998). Bioavailability of soilborne lead in adults, by stable isotope dilution. *Environ. Health Perspect.* 1589–1594.

McGrath, S. P., Knight, B., Killham, K., Preston, S., and Paton, G. I. (1999). Assessment of the toxicity of metals in soils amended with sewage sludge using a chemical speciation technique and a lux-based biosensor. *Environ. Toxicol. Chem.* **18,** 659–663.

McLaren, R. G., and Crawford, D. V. (1973). Studies on soil copper. I. The fractionation of copper in soils. *J. Soil Sci.* **24,** 172–181.

Medlin, E. A. (1997). An *in vitro* method for estimating the relative bioavailability of lead in humans. Master's Thesis, Department of Geological Sciences, University of Colorado at Boulder.

Mendoza, J., Garrido, T., Castillo, G., and San Martin, N. (2006). Metal availability and uptake by sorghum plants grown in soils amended with sludge from different treatments. *Chemosphere* **65,** 2304–2312.

Merkel, D. (1996). Cadmium, copper, nickel, lead and zinc contents of wheat grain and soils extracted with $CaCl_2/DTPA$ (CAD), $CaCl_2$ and NH_4NO_3. *Agribiological Research—Zeitschrift Fur Agrarbiologie Agrikulturchemie Okologie* **49,** 30–37.

Miller, D. D., and Schricker, B. R. (1982). Nutritional bioavailability of iron. *In* "ACS Symposium Series 23" (C. Kies, ed.). American Chemical Society, Washington, DC.

Miller, D. D., Schricker, B. R., Rasmussed, R. R., and Van Campen, D. (1981). An *in vitro* method for estimation of iron availability from meals. *Am. J. Clin. Nutr.* **34,** 2248–2256.

Minekus, M., Marteau, P., Havenaar, R., and Huis in't Veld, J. H. J. (1995). A multicompartmental dynamic computer-controlled model simulating the stomach and small intestine. *Altern. Lab. Anim.* **23**, 197–209.

Mitchell, R. L., Reith, J. W. S., and Johnson, I. M. (1957). Soil copper status and plant uptake. *J. Sci. Food Agric.* **8**, 52–60.

Moćko, A., and Wactawek, W. (2004). Three-step extraction procedure for determination of heavy metals availability to vegetables. *Anal. Bioanal. Chem.* **380**, 813–817.

Molly, K., Van de Woestyne, M., and Verstraete, W. (1993). Development of a 5-step multi-chamber reactor as a simulation of the human intestinal microbial ecosystem. *Appl. Microbiol. Biotechnol.* **39**, 254–258.

Morgan, J. J., and Stumm, W. (1995). Chemical processes in the environment, relevance of chemical speciation. *In* "Metals and Their Compounds in the environment" (E. Merian, ed.). pp. 67–103. VCH, Weinheim.

Novozamsky, I., Lexmond, T. M., and Houba, V. J. G. (1993). A single extraction procedure of soil for evaluation of uptake of some heavy metals by plants. *Int. J. Environ. Anal. Chem.* **51**, 47–58.

O'Connor, G. A. (1988). Use and measure of DTPA soil test. *J. Environ. Qual.* **17**, 715–718.

Obrador, A., Alvarez, J. M., Lopez-Valdivia, L. M., Gonzalez, D., Novillo, J., and Rico, M. I. (2007). Relationships of soil properties with Mn and Zn distribution in acidic soils and their uptake by a barley crop. *Geoderma* **137**, 432–443.

OIS (1998). Ordinance relating to impacts on the soil. *Collection of Swiss Federal Legislation* SR 814. 12.

Oomen, A. G., Hack, A., Minekus, M., Zeijdner, E., Cornelis, C., Schoeters, G., Verstraete, W., Van de Wiele, T., Wragg, J., Rompelberg, C. J. M., Sips, A. J. A. M., and Van Wijnen, J. H. (2002). Comparison of five *in vitro* digestion models to study the bioaccessibility of soil contaminants. *Environ. Sci. Technol.* **36**, 3326–3334.

Oomen, A. G., Rompelberg, C. J. M., Bruil, M. A., Dobbe, C. J. G., Pereboom, D. P. K. H., and Sips, A. J. A. M. (2003). Development of an *in vitro* digestion model for estimating the bioaccessibility of soil contaminants. *Arch. Environ. Contam. Toxicol.* **44**, 281–287.

Paton, G. I., Viventsova, E., Kumpene, J., Wilson, M. J., Weitz, H. J., and Dawson, J. C. (2006). An ecotoxicity assessment of contaminated forest soils from the Kola Peninsula. *Sci. Total Environ.* **355**, 106–117.

Paustenbach, D. J. (2000). The practice of exposure assessment: A state of the art review. *J. Toxicol. Environ. Health Part B* **3**, 179–291.

Payà-Pérez, A., Sala, J., and Mousty, F. (1993). Comparison of ICP-AES and ICP-MS for the analysis of trace elements in soil extracts. *Int. J. Environ. Anal. Chem.* **51**, 223–230.

Pueyo, M., Rauret, G., Luck, D., Yli-Halla, M., Muntau, H., Quevauville, P., and Lopez-Sanchez, J. F. (2001). Certification of the extractable contents of Cd, Cr, Cu, Ni, Pb and Zn in a fresh water sediment following a collaboratively tested and optimised three-step sequential extraction procedure. *J. Environ. Monit.* **3**, 243–250.

Queauviller, P., Rauret, G., Lopez-Sanchez, J. F., Rubio, R., Ure, A. M., Muntau, H., Fiedler, H. D., and Griepink, B. (1997). The certification of the extractable contents (mass fractions) of Cd, Cr, Ni, Pb and Zn in sediment following a three-step sequential extraction procedure. *CRM* 601. Report EUR, 17554 EN, ISBN 92-828-0127-6.

Raksasataya, M., Langdon, A. G., and Kim, N. D. (1996). Assessment of the extent of lead redistribution during sequential extraction by two different methods. *Anal. Chim. Acta* **332**, 1–14.

Rauret, G. (1998). Extraction procedure for the determination of heavy metals in contaminated soil land sediment. *Talanta* **46**, 449–455.

Rauret, G., López-Sánchez, J. F., Sahuquillo, A., Davidson, C., Ure, A., and Quevauviller, P. (1999). Improvement of the BCR 3-step sequential extraction procedure prior to the certification of new sediment and soil reference materials. *J. Environ. Monit.* **1**, 57–61.

Rauret, G., López-Sánchez, J. F., Bacon, J., Gomez, A., Muntau, H., and Queaviller, P. (2001). Certification of the contents (mass fraction) of Cd, Cr, Cu, Ni, Pb and Zn in an organic-rich soil following harmonised EDTA and acetic acid extraction procedures. BCR-700. *BCR information, reference materials*. Report EUR 19774 EN.

Remon, E., Bouchardon, J. L., Cornier, B., Guy, B., Leclerc, J. C., and Faure, O. (2005). Soil characteristics, heavy metal availability and vegetation recovery at a former metallurgical landfill: Implications in risk assessment and site restoration. *Environ. Pollut.* **137,** 316–323.

Renella, G., Adamo, P., Bianco, M. R., Landi, L., Violante, P., and Nannipieri, P. (2004). Availability and speciation of cadmium added to a calcareous soil under various managements. *Eur. J. Soil Sci.* **55,** 123–133.

Ritchie, G. S. P., and Sposito, G. (2002). Speciation in soils. *In* "Chemical Speciation in the Environment" (A. Ure and C. M. Davidson, eds.), pp. 237–264. Blackwell, Oxford.

Rodriguez, R. R., Basta, N. T., Casteel, S. W., and Pace, L. W. (1999). An *in vitro* gastrointestinal method to estimate bioavailable arsenic in contaminated soils and solid media. *Environ. Sci. Technol.* **33,** 642–649.

Rotard, W., Christmann, W., Knoth, W., and Mailahn, W. (1995). Bestimmung der resorptionsver-fügbaren PCDD/PCDF aus Kieselrot. *Umweltwissenschaften und Schadstoff-Forschung Zeitschrift für Umweltchemie und Ökotoxikologie* **7,** 3–9.

Ruby, M. V. (2004). Bioavailability of soil-borne chemicals: Abiotic assessment tools. *Hum. Ecol. Risk Assess.* **10,** 647–656.

Ruby, M. V., Davis, A., Link, T. E., Schoof, R., Chaney, R. L., Freeman, G. B., and Bergstrom, P. (1993). Development of an *in vitro* screening test to evaluate the *in vivo* bioaccessibility of ingested mine-waste lead. *Environ. Sci. Technol.* **27,** 2870–2877.

Ruby, M. V., Davis, A., Schoof, R., Eberle, S., *et al.* (1996). Estimation of lead and arsenic bioavailability using a physiologically based extraction test. *Environ. Sci. Technol.* **30,** 422–430.

Ruby, M. V., Schoof, R., Brattin, W., Goldade, M., Post, G., Harnois, M., Mosby, D. E., Casteel, S. W., Berti, W., Carpenter, M., Edwards, D., Cragin, D., *et al.* (1999). Advances in evaluating the oral bioavailability of inorganics in soil for use in human health risk assessment. *Environ. Sci. Technol.* **33,** 3697–3705.

Ryan, P. C., Wall, A. J., Hillier, S., and Clark, L. (2002). Insights into sequential chemical extraction procedures from quantitative XRD: A study of trace metal partitioning in sediments related to frog malformities. *Chem. Geol.* **184,** 337–357.

Sahuquillo, A., Lopez-Sanchez, J. F., Rubio, R., Rauret, G., Thomas, R. P., Davidson, C. M., and Ure, A. M. (1999). Use of a certified reference material for extractable trace metals to assess sources of uncertainty in the BCR three-stage sequential extraction procedure. *Anal. Chim. Acta* **382,** 317–327.

Sahuquillo, A., Rigol, A., and Rauret, G. (2003). Overview of the use of leaching/extraction tests for risk assessment of trace metals in contaminated soils and sediments. *Trends Analyt. Chem.* **22,** 152–159.

Sanders, J. R., McGrath, S. P., and Adams, T. M. (1986). Zinc, copper and nickel concentrations in ryegrass grown on sewage sludge contaminated soils. *J. Sci. Food Agric.* **37,** 961–968.

Sastre, J., Hernàndez, E., Rodriguez, R., Alcobé, X., Vidal, M., and Rauret, G. (2004). Use of sorption and extraction tests to predict the dynamics of the interaction of trace elements in agricultural soils contaminated by a mine tailing accident. *Sci. Total Environ.* **329,** 261–281.

Sauerbeck, D., and Styperek, P. (1984). Predicting the cadmium availability from different soils by CaCl$_2$ extraction. *In* "Processing and Use of Sewage Sludge" (P. L'Hermite, H. Ott, and D. Reidel, eds.), pp. 431–435. Publishing Company, Dordrecht, The Netherlands.

Sauerbeck, D. R., and Styperek, P. (1985). Evaluation of chemical methods for assessing the cadmium and zinc availability from different soils and sources. *In* "Chemical Methods for Assessing Bioavailable Metals in Sludges and Soils" (R. Leschber, R. D. Davis, and P. L'Hemite, eds.), p. 49. Elsevier, Amsterdam.

Schwartz, C., Morel, J. L., Saumier, S., Whiting, S. N., and Baker, A. J. M. (1999). Root architecture of the Zn-hyperaccumulator plant *Thlaspi caerulescens* as affected by metal origin, content and localisation in soil. *Plant Soil* **208,** 103–115.

Sedberry, J. E., and Reddy, C. N. (1976). The distribution of zinc in selected soils in Indiana. *Commun. Soil Sci. Plant Anal.* **7,** 787–795.

Sheppard, M. I., and Stevenson, M. (1997). Critical evaluation of selective extraction methods for soils and sediments. *In* "Contaminated Soils" (R. Prost, ed.). "Proceedings of 3rd International Conference on the Biogeochemistry of Trace Elements," May 15–19, 1995, pp. 69–97. INRA, Paris, France.

Shuman, L. M. (1985). Fractionation method for soil microelements. *Soil Sci.* **140,** 11–22.

Sillanpaa, M. (1982). Micronutrients and the nutrient status of soils: A global study. *FAO Soils Bull.* **48,** 444 FAO Rome, Italy.

Sillanpaa, M., and Jansson, H. (1992). Status of cadmium, lead, cobalt and selenium in soils and plants of thirty countries. *FAO Soils Bull.* **65,** Food and Agricultural Organization of the United Nations, Rome, Italy.

Silveira, M. L., Alleoni, L. R. F., O'Connor, G. A., and Chang, A. C. (2006). Heavy metal sequential extraction methods—A modification for tropical soils. *Chemosphere* **64,** 1929–1938.

Singh, S. P., Tack, F. M., and Verloo, M. G. (1996). Extractability and bioavailability of heavy metals in surface soils derived from dredged materials. *Chem. Spec. Bioavail.* **8,** 105–110.

Smith, L. S., Oliver, D. P., McLaughlin, M. J., Naidu, R., Maynard, T., and Calder, I. (1997). Solubility characteristics of Lead in household dusts derived from smelter-contaminated sources. *In* "Proceedings from the 4th International Conference on the Biogeochemeistry of Trace Elements," June 23–26, pp. 281–282. Clark Kerr Campus, Berkeley, CA.

Soltanpour, P. N., and Schwab, A. P. B. (1979). A new soil test for simultaneous extraction of macro- and micro-nutrients in alkaline soils. *Commun. Soil Sci. Plant Anal.* **8,** 195–207.

Soon, Y. K., and Bates, T. E. (1982). Chemical pools of Cd, Ni, and Zn in some polluted soils and some preliminary indications of their availability to plants. *J. Soil Sci.* **33,** 477–488.

Sposito, G., Lund, L. J., and Chang, A. C. (1982). Trace metal chemistry in arid-zone field soils amended with sewage sludge: I. Fractionation of Ni, Cu, Zn, Cd, and Pb in solid phases. *Soil Sci. Soc. Am. J.* **46,** 260–264.

Sterckeman, T., Gomez, A., and Cielsielski, H. (1996). Soil and waste analysis for environmental risk assessment in France. *Sci. Total Environ.* **178,** 63–69.

Street, J. J., Lindsay, W. L., and Sabey, B. R. (1977). Solubility and plant uptake of cadmium in soils amended with cadmium and sewage sludge. *J. Environ. Qual.* **6,** 72–77.

Symeonides, C., and McRae, S. G. (1977). The assessment of plant available cadmium in soils. *J. Environ. Qual.* **6,** 120–123.

Tarvainen, T., and Kallio, E. (2002). Baselines of certain bioavailable and total heavy metal concentrations in Finland. *Appl. Geochem.* **17,** 975–980.

Templeton, D. M., Ariese, F., Cornelis, R., Danielson, L. G., Munau, H., van Leeuwen, H. P., and Lobinski, R. (2000). Guidelines for terms related to chemical speciation and fractionations of elements. Definitions, structural aspects and methodological approaches. (IUPAC Recommendations 2000). *Pure Appl. Chem.* **72,** 1453–1470.

Tessier, A., Campbell, P. G. C., and Bisson, M. (1979). Sequential extraction procedure for the speciation of particular trace metals. *Anal. Chem.* **51,** 844–850.

Tills, A. R., and Alloway, B. J. (1983). An appraisal of currently used tests for available copper. *J. Sci. Food Agric.* **34,** 1190–1196.

Tlustos, P., van Dijk, D., Szakova, J., and Parlikova, D. (1994). Cd and Zn release through the use of selected extractants. *Rostl. Vyroba* **40,** 1107–1121.

Tome, F. V., Rodriguez, M. P. B., and Lozano, J. K. (2003). Soil-to-plant transfer factors for natural radionuclides and stable elements in a Mediterranean area. *J. Environ. Radioact.* **65,** 161–175.

Ure, A., and Davidson, C. M. (2002). Chemical speciation in soils and related materials by selective chemical extraction. *In* "Chemical Speciation in the Environment" (A. Ure and C. M. Davidson, eds.), pp. 265–300. Blackwell, Oxford.

Ure, A., Queauviller, P., Muntau, H., and Griepink, B. (1993). *EUR Report,* 14763.

Ure, A., Quevauviller, H., Muntau, H., and Griepink, B. (1993a). Improvements in the determination of extractable contents of trace metals in soil and sediment prior to certification. Chemical analysis. BCR information. Commission of the European Communities, Luxembourg.

Ure, A., Quevauviller, H., Muntau, H., and Griepink, B. (1993b). Speciation of heavy metals in soils and sediments. An account of the improvement and harmonization of extraction techniques undertaken under the auspices of the BCR of the CEC. *Int. J. Environ. Anal. Chem.* **51,** 135–151.

Venditti, D., Durécu, S., and Berthelin, J. (2000a). A multidisciplinary approach to assess history, environmental risks, and remediation feasibility of soils contaminated by metallurgical activities. Part A: Chemical and physical properties of metals and leaching ability. *Arch. Environ. Contam. Toxicol.* **38,** 411–420.

Venditti, D., Berthelin, J., and Durécu, S. (2000b). A multidisciplinary approach to assess history, environmental risks, and remediation feasibility of soils contaminated by metallurgical activities. Part B: Direct metal speciation in the solid phase. *Arch. Environ. Contam. Toxicol.* **38,** 421–427.

Vivas, A., Vörös, A., Biró, B., Barea, J. M., Ruiz-Lozano, J. M., and Azcón, R. (2003). Beneficial effects of indigenous Cd-tolerant and Cd-sensitive Glomus mossae associated with a Cd-adapted strain of Brevibacillus sp. In improving plant tolerance to Cd contamination. *Appl. Soil Ecol.* **24,** 177–186.

Waite, T. D. (1989). Mathematic modelling of trace metal speciation. *In* "Trace Metal Speciation: Analytical Methods and Problems" (J. E. Batley, ed.). pp. 117–184. CRC Press, Boca Raton, FL.

Whalley, C., and Grant, A. (1994). Assessment of the phase selectivity of the European community Bureau of Reference (BCR) sequential extraction procedure for metals in sediment. *Anal. Chim. Acta* **291,** 287–295.

Whiting, S. N., Leake, J. R., McGrath, S. P., and Baker, A. J. M. (2001a). Assessment of Zn mobilization in the rhizosphere of *Thlaspi caerulescens* by bioassay with non-accumulator plants and soil solution. *Plant Soil* **237,** 147–156.

Whiting, S. N., De Souza, M. P., and Terry, N. (2001b). Rhizosphere bacteria mobilize Zn for hyperaccumulation by *Thlaspi caerulescens*. *Environ. Sci. Technol.* **35,** 3144–3150.

Whitten, M. G., and Ritchie, G. S. P. (1991). Calcium chloride extractable cadmium as an estimate of cadmium uptake by subterranean clover. *Aust. J. Soil Res.* **29,** 215–221.

Williams, C., and Thornton, I. (1973). The use of soil extractants to estimate plant-available molybdenum and selenium in potentially toxic soils. *Plant Soil* **39,** 149–159.

Wong, C. S. C., Li, X., and Thornton, I. (2006). Urban environmental geochemistry of trace metals. *Environ. Pollut.* **142,** 1–16.

Wragg, J., and Cave, M. R. (2003). *In-vitro* methods for the measurement of the oral bioaccessibility of selected metals and metalloids in soils: A critical review. P5–062/TR/01, British Geological Survey.

Yang, J. K., Barnett, M. O., Jardine, P. M., and Brooks, S. C. (2003). Factors controlling the bioaccessibility of Arsenic (V), and Lead (II) in soil. *Soil Sediment. Contam.* 165–179.

Yarlagadda, P. S., Matsumoto, M. R., Van Benschoten, J. E., and Kathuria, A. (1995). Characteristics of heavy metals in contaminated soils. *J. Environ. Eng. ASCE* **121,** 276–286.

Zampella, M. (2006). Distribuzione e biodisponibilità di metalli in traccia in suoli e sedimenti della valle del torrente Solofrana (Italia meridionale). PhD Thesis in Valutazione e Mitigazione del Rischio Ambientale, Università degli Studi di Napoli Federico II, CIRAM, Napoli.

Žemberyová, M., Barteková, J., and Hagarová, I. (2006). The utilization of modified BCR three-step sequential extraction procedure for the fractionation of Cd, Cr, Cu, Ni, Pb and Zn in soil reference materials of different origins. *Talanta* **70,** 973–978.

Zha, H. G., Jiang, R. F., Zhao, F. J., Vooijs, R., Schat, H., Barkers, J. H. A., and McGrath, S. P. (2004). Co-segregation analysis of cadmium and zinc accumulation in *Thlaspi caerulescens* inter-ecotypic crosses. *New Phytol.* **163,** 299–312.

Zheljazkov, V. D., and Warman, P. R. (2004). Phytoavailability and fractionation of copper, manganese and zinc in soil following application of two composts to four crops. *Environ. Pollut.* **131,** 187–195.

CHAPTER TEN

Overview of Selected Soil Pore Water Extraction Methods for the Determination of Potentially Toxic Elements in Contaminated Soils: Operational and Technical Aspects

Marcello Di Bonito,* Neil Breward,[†] Neil Crout,[‡] Barry Smith,[†] *and* Scott Young[‡]

Contents

Abstract

Chemical elements that are either present naturally in the soil or introduced by pollution are more usefully estimated in terms of 'availability' of the element, because this property can be related to mobility and uptake by plants. A good estimation of 'availability' can be achieved by measuring the concentration of the element in soil pore water. Recent achievements in analytical techniques allowed to expand the range of interest to trace elements, which play a crucial role both in contaminated and uncontaminated soils and include those defined as potentially toxic elements (PTE) in environmental studies. A complete chemical analysis of soil pore water represents a powerful diagnostic tool for the interpretation of many soil chemical phenomena relating to soil fertility, mineralogy and environmental fate. This chapter describes some of the current methodologies

* Environment Agency, Trentside Offices, Scarrington Road, West Bridgford, Nottingham, NG2 5FA, UK
† British Geological Survey, Kingsley Dunham Centre, Keyworth, Nottingham, NG12 5GG, UK
‡ School of Biosciences, University of Nottingham, University Park, Nottingham, NG7 2RD, UK

Environmental Geochemistry
DOI: 10.1016/B978-0-444-53159-9.00010-3

© 2008 Elsevier B.V.
All rights reserved.

used to extract soil pore water. In particular, four laboratory-based methods, (1) high-speed centrifugation-filtration, (2) low (negative-)-pressure Rhizon™ samplers, (3) high-pressure soil squeezing and (4) equilibration of dilute soil suspensions, are described and discussed in detail. A number of operational factors are presented: pressure applicable (i.e. pore size involved), moisture prerequisites of the soil, pore water yielding, efficiency, duration of extraction, materials and possible contaminations for PTE studies. Some consideration is then taken to assess advantages and disadvantages of the methods, including costs and materials availability.

1. INTRODUCTION

1.1. Soil pore water and the concept of (bio)availability

Many studies have examined the concentration and retention of metals in soils and the effect of various parameters on their adsorption and solubility, including pH (Cavallaro and McBride, 1980; Green *et al.*, 2003; Harter, 1983; McBride and Blasiak, 1979; Robb and Young, 1999), redox conditions (Davranche and Bollinger, 2001; Davranche *et al.*, 2003; Qafoku *et al.*, 2003), amount of metals (Basta and Tabatabai, 1992; Garcia-Miragaya, 1984; Sauvé *et al.*, 2000), cation exchange capacity (Ziper *et al.*, 1988), organic matter (OM) content (Benedetti *et al.*, 1996a,b; Elliot *et al.*, 1986; Gerritse and Vandriel, 1984; Kashem and Singh, 2001; Kinniburgh *et al.*, 1999), soil mineralogy (Cavallaro and McBride, 1984; Jenne, 1968; Kinniburgh *et al.*, 1976; Kuo, 1986; Lindroos *et al.*, 2003; Tiller *et al.*, 1963), biological and microbial conditions (Dumestre *et al.*, 1999; Gerritse *et al.*, 1992; Warren and Haack, 2001) as well as developing assemblage models to mechanistically predict these processes (Celardin, 1999; Dzombak and Morel, 1987; Haworth, 1990; Impellitteri *et al.*, 2003; McBride *et al.*, 1997; Tye *et al.*, 2003, Weng *et al.*, 2002). From these studies, it has emerged that total soil metal content alone is not a good measure of short-term bioavailability and not a very useful tool to determine the potential risks from soil contamination (Sauvé *et al.*, 1998; Tack and Verloo, 1995). In fact, because plants take up most nutrients from the soil pore water, it is often assumed that the dissolved potentially toxic elements (PTE) are readily available to organisms (Barber, 1984; Vig *et al.*, 2003). The definition of bioavailability (or phytoavailability) as given by Sposito (1989) suggests, 'a chemical element is bioavailable if it is present as, or can be transformed readily to, the free-ion species, if it can move to plant roots on a time scale that is relevant to plant growth and development, and if, once absorbed by the root, it affects the life cycle of the plant'. It is clear that the concentration and speciation of metals in the pore water may provide more useful information on metal bioavailability and toxicity than total soil concentration (Cances *et al.*, 2003; Hani, 1996; Knight *et al.*, 1998; Percival, 2003; Prokop *et al.*, 2003; Shan *et al.*, 2003). Traditionally, however, the soil pore water has not been utilized as a means of assessing bioavailability. This has probably been due to analytical and technical difficulties related to sampling of the soil pore water. Instead, most assessments of metal availability have involved chemical extractants [e.g. ethylenediaminetetraacetic acid (EDTA), acetic acid] intended to remove the entire reservoir of reactive metal. This pool may involve a total amount of metal that is several orders of magnitude greater than that found in the soil pore water.

1.1.1. Soil pore water definition

The soil liquid phase has a composition and reactivity defined by the properties of the incoming water and fluxes of matter and energy originating from the local (neighbouring) soil solid phase, biological system and atmosphere (Fig. 10.1).

The current view is that in a porous medium, two liquid-phase regions can be identified on functional grounds (Yaron et al., 1996). The first is near the solid phase and is considered the most important surface reaction zone of the porous medium system. This near-surface water also controls the diffusion of the mobile fraction of the solute in contact with (sorbed on) the solid phase. The second region covers the 'free' water zone, which governs the water flow and solute transport in soils (Fig. 10.2).

Both phases represent what can be defined as 'soil pore water': this term is preferred to the more specific 'soil solution' and will be used throughout this chapter.

1.1.2. Bioavailability and soil pore water sampling

To assess the environmental bioavailability, mobility and geochemical cycling of trace elements in soil, analyses of soil pore water composition are frequently more instructive than those from whole soil or soil extracts. The validity of this concept

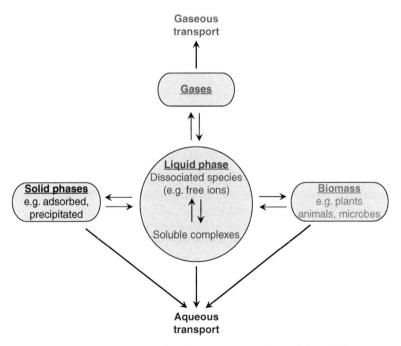

Figure 10.1 Biogeochemical cycling of soil contaminants: the soil liquid phase is acting as a regulator of contaminant fate (modified from Hesterberg, 1998, based on Lindsay, 1979; Mattigod et al., 1981).

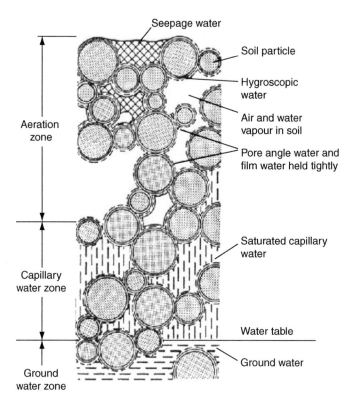

Figure 10.2   Schematic representation of soil water states and their definitions (from Shaw, 1993).

has led to the development of several models that attempt to predict solid–solution partitioning of elements and their solution speciation. These 'assemblage models' include an increasing number of variables as they develop greater mechanistic capability. Soil pore water analysis can be used to model the nature, direction, extent and rate of chemical reactions. In fact we can assume that

1. if soil pore water represents the natural medium for plant growth, then soil pore water analysis allows for prediction of plant response to chemicals occurring in the soil environment (plant uptake prediction);
2. if soil pore water can be related to mobile water in the soil environment, then soil pore water composition can be used to predict the forms and amounts of chemical that may reach ground and surface water through transport from the soil environment (pollutant fate);
3. if soil pore water approaches a steady state relative to the soil solid phase, then soil pore water composition can be used to predict solid phase components controlling chemical distribution in soil (solid–solution processes).

The validity of these assumptions depends on the way that soil pore water is conceptualized, that is, defined and sampled, and how that concept is translated into an operational method or model whereby soil pore water can be obtained and its composition expressed in a meaningful way. Too often, however, studies skim over a proper definition of soil pore water opting for 'simulating' or bypassing the problem.

2. METHODS FOR SAMPLING SOIL PORE WATER

A range of methods are available for obtaining and analysing 'unaltered' soil pore water, with consideration of ion speciation and complexation, and expression of soil pore water composition in thermodynamic terms. Unfortunately, none of them have been adopted as 'the standard procedure', leading to a rather confused situation. This is particularly disquieting in environmental studies, where standardization is desirable to achieve highly reliable data. One of the main obstacles to standardization, however, is that sampling of soil pore water often presents conceptual ambiguity as well as technical problems, especially if one tries to characterize the liquid phase in terms of its origin within the soil system. In fact, soil porosity generally represents a limiting factor in defining the ratio between the solid, aqueous and gaseous phases of the soil medium (Yaron *et al.*, 1996) because of the open boundaries between these different phases leading to a pattern of continuously changing processes. In addition, solute concentrations in sampled pore water may depend on a number of technical factors, including

o method of extraction,
o imposed tensions,
o flow rate to the sampler, and
o relation of the soil volume sampled versus the scale of heterogeneity in solute concentration (scaling factor).

This situation is made even more complicated by the wide range of methods of pore water extraction used (see Reeder *et al.*, 1998). For this reason a number of pore classifications have been suggested in the past, as demonstrated in Table 10.1. In fact, soil pore water sampling can be approached from different angles, where no single methodology is appropriate to all applications. The choice of method will depend on the particular aim of the study in question. In fact, it is arguable that soil pore water is operationally defined by the methodology employed for its acquisition and subsequent analysis. It is therefore very important to describe the methodology and the assumptions employed. A first distinction to make is between field-based and laboratory-based methods. In general, most field sampling methods have been used to interpret soil pore water chemistry from both static and dynamic perspectives without sufficient consideration of which soil water is being sampled and its chemical reactivity in soils (Wolt, 1994). This possibly makes field-based sampling more suitable than laboratory-based methods for consideration of chemical transport, provided the solutions obtained represent mobile water in the soil environment. By contrast, considerations of biologically important processes relating to plant nutrient availability, phytotoxicity and soil metabolism are probably best related to chemical composition of diffuse soil water as reflected in the composition of displaced soil pore water.

Table 10.1 Examples of soil pore classifications, with description of equivalent soil water phenomena and matric pressures; a brief illustration of the soil system in those conditions is also given; 'd' represents the equivalent diameter of pores and is expressed in μm, unless otherwise stated

Soil system	Functional classification (μm) Greenland and Hayes (1981)	Physical classification Brewer (1964)	Luxmoore (1981)	Predominant water phenomena	'd' (μm)	Equivalent water pressure (kPa)†
Spaces as large as these are commonly formed between the clods of newly ploughed soil. Cracks in dry clay soils can reach widths of this order of magnitude.	Transmission pores: air movement and drainage of excess water $d > 50$	Macropores $d > 1000$	Macropores $d > 1000$	Channel flow through profile from surface ponding and/or perched water table	10000 (1 cm)	−0.015
Pores of about this size and smaller are formed between aggregates of finely tilled soil as for a seed-bed.		Fine macropores $75 < d < 1000$	Mesopores $10 < d < 1000$	Drainage; hysteresis; gravitational driving force for water dynamics	1000 (1 mm)	−0.15
Pores between spherical particles 0.65 mm in diameter in closest packing have this size (Dallavalle, 1948). Roots will not extend into rigid pores smaller than this (Wiersum, 1957).	Storage pores: retention of water against gravity and release to plant roots $0.5 < d < 50$				100	−1.5
Pores larger than about 15 μm (corresponding to 9.8 kPa) are drained in most soils that can be said to be at field capacity (FC).		Mesopores $30 < d < 75$	Micropores $d < 10$	Evapotranspiration; matric pressure gradient for water distribution	10	−15

Description			Physico–chemical classification (adsorption) (IUPAC, 2001)			
Pores down to this size are accessible to bacteria.		Micropores $d <$ 30 ($<$ 0.5)			1	−150
Water in pores of about this size or larger is available to plants in non-saline soil (correspond to 1500 kPa).	Residual pores: retention and diffusion of ions in solution 0.005 $< d <$ 0.5	Ultramicropores 0.5 $< d <$ 0.1	Macropores $d >$ 0.05	Capillary condensation; transport	0.1	−1500 (PWP#)
When micropores are treated as slits between parallel plates, about half the pore space in dried aggregates of clay soil can commonly be attributed to plate separations of 10 nm or less (Sills et al., 1974).			Mesopores 0.002 $< d <$ 0.05		0.01	−15000★
Roughly corresponding to the thickness of three layers of water molecules on a clay surface.	Bonding spaces: support major forces between soil particles $d <$ 0.005	Cryptopores $d <$ 0.1	Micropores $d <$ 0.002	Physisorption; adsorbate–adsorbate and adsorbant–adsorbate interactions	0.001 (1 nm)	−150000★

PWP, permanent wilting point.

2.1. Field-based methods

Field methods for sampling pore water are generally grouped under the general term *lysimetry*. This definition usually comprises a range of types of samplers (Wolt, 1994):

o Monolith—any device using an undisturbed soil block or column;
o Filled-in—devices containing soil where the natural soil structure has been disrupted;
o Tension—also called vacuum, suction, point or mini-lysimeters;
o Passive—also called capillary samplers, zero-tension lysimeters;
o Ebermayer—any lysimeter installation where, by access from a trench, a trough, pan, funnel, plate or wick is placed under undisturbed soil.

Of these, tension and passive samplers are the most widely used and are discussed below.

2.1.1. Tension samplers

In general, the approach employed for extracting pore water *in situ* is to use tension samplers such as porous cups (for a complete review see Grossmann and Udluft, 1991; Litaor, 1988). Porous cups are designed to replicate the function of a plant root, by applying suction to the soil. The method, however, is replete with inadequacies that need to be considered in the acquisition and interpretation of data. Principal limitations are the non-representative sampling of soil water occurring above the capillary fringe and potential artifact effects arising from the reaction of lysimeter materials with the surrounding soil environment. In the first instance, changing the applied vacuum (from 0 to -40 kPa) was observed to generate little effects on the concentrations of chemical species collected (Beier and Hansen, 1992; Beier *et al.*, 1992). In addition to this, water will flow from the soil into the porous cup if the capillary pressure in the cup is lower than that in the soil. With a single pump, a vacuum of -90 kPa can be easily generated and applied to the samplers. Sampling seepage (i.e. slowly percolating) water is, therefore, possible only as long as the capillary pressure in the soil lies above this value. As a result of the low sampling rates at capillary pressures below -70 kPa, the use of this system is limited in the majority of soils (Grossmann and Udluft, 1991).

The other problem connected with the use of these samplers is the sorption of solutes from the pore water. Depending on the cup material [materials used include aluminium oxide, glass sinter, ceramic, teflon, acrylic copolymer with internal nylon support, stainless steel, plastic 'organic' polymers—polyvinylchloride (PVC), PP, PVDF], additional reactions may take place leading to absorption, precipitation or even release of chemical substances, resulting in pronounced effects on the final composition of the water sampled (Litaor, 1988). Several studies have investigated these effects and questioned the validity of results given by these kinds of samplers (e.g. Goyne *et al.*, 2000; Guggenberger and Zech, 1992; Hansen and Harris, 1975; Levin and Jackson, 1977; Nagpal, 1982). Siemens and Kaupenjohann (2003) found that between 0.8 and 63 mg litre^{-1} of dissolved organic carbon (DOC) was released from sealing and glues of pore water samplers. They concluded that samplers should be designed without glues or elastomers, presenting a suction plate entirely made from borosilicate glass that did not release organic C. Interaction with organics may be particularly significant in the case of PTE adsorption on the surface of the samplers (Grossmann *et al.*, 1990; Massee and Maessen, 1981; Shendrikar *et al.*, 1975; Wenzel and Wieshammer, 1995; Wenzel *et al.*, 1997). Different materials have been tested to minimize metals sorption effects. For example, McGuire *et al.* (1992) found that metal adsorption on samplers decreased on porous cups made of materials in the sequence ceramic > stainless steel > fritted glass = poly(tetrafluor-ethene) (PTFE), with PTE being adsorbed in the sequence Zn $\gg$ Co > Cr > Cd. These authors also pointed out the importance of the total metal concentration as well as cleaning method (water vs acid solution) and rinse volumes, which affected the extent of the adsorption. Adsorption was between 2 and 15 times higher for water-cleaned samplers, but also increased on acid–cleaned samplers with decreasing volumes utilized. A later study by Andersen *et al.* (2002) found that cups made of PTFE affected the concentrations of Cd, Cu, Ni and Zn, which were adsorbed at pH > 4.5 for low pore water concentrations. Results on adsorption showed that plastic cups may have some advantage over conventional ceramic cups. With the increasing sample volume, the concentration of a trace metal recorded by the suction cup comes closer to the concentration in the pore water because of the equilibration of the cation exchange surface of the suction cup with the solution. However, the extraction of large sample volumes can cause a significant disturbance of the system (Grossmann and Udluft, 1991). More recently, ceramic cups were found to adsorb PO_4^{3-}, DOC, major and minor cations (Na^+, K^+, Ca^{2+}, Mg^{2+}, Fe^{3+}, Al^{3+}, Mn^{2+} and Zn^{2+}) and SO_4^{2-} and NO_3^- anions. They release H_4SiO_4 and, in addition to this, relative low pH values (5.1–6.2) favoured anion and DOC adsorption, the latter increasing the exchange capacity and cation adsorption of the material (Menendez *et al.*, 2003).

Despite the potential problems, this remains an area of great opportunity for innovation, as illustrated by the development of new types of sampler, such as the soil moisture Rhizon samplers (www.eijkelkamp.com), which will be described later.

2.1.2. Passive samplers

Passive samplers have no tension applied to them. Consequently, they only sample that fraction of the soil water flux occurring under saturated soil conditions or during macropore flow. These devices result in samples of soil water which may represent a combination of bypass water (recent rainfall, irrigation events, i.e. water moving via preferential flow), and 'internal catchment water' (Booltink and Bouma, 1991), that is, water moving by diffusion and/or conduction and exhibiting a range of contact times with the soil matrix. The proportion of bypass water compared to internal catchment water will depend on soil structure, soil moisture conditions before and during percolate sampling and features of the design and operation of the lysimeter. As a result, compositional analysis of 'passive' lysimeter solutions and pore water obtained by laboratory displacement may substantially differ (Zabowsky and Ugolini, 1990).

2.2. Laboratory-based methods

Laboratory-based methods of pore water displacement are designed to approximate diffuse water in quasi-equilibrium with the soil solid phase. Methodologies for obtaining 'unaltered' soil pore water in a laboratory setting may be broadly defined as *displacement techniques* and comprise

- Column displacement (pressure or tension displacement, with or without a displacing head solution);
- Centrifugation (with or without immiscible liquid displacement);
- Saturation extracts (including saturation pastes);
- Water extracts;
- Complexation and exchange techniques (e.g. DGT technique, Hooda *et al.*, 1999); and
- Lysimetric methods (both tension and passive, including Rhizon samplers—see also field methods).

The various column displacement methods are the most widely applicable and reliable techniques, although they require a high degree of operator experience. However, of the methods listed above, four different extraction methods were chosen for a more detailed description and discussion:

1. low (negative-) pressure Rhizon samplers (or 'soil moisture samplers'—SMS);
2. high-speed centrifugation-filtration or drainage centrifugation;
3. high-pressure soil squeezing or 'pressure filtering' (squeezing); and
4. equilibration of dilute soil suspensions.

These methods were selected for different reasons, but mainly for their flexibility and novelty. Rhizon samplers represent the current equivalent of porous cups, widely used in the recent past; centrifugation is possibly the current most widely used method because of the ease and the ready availability of the requisite equipment in most laboratories; squeezing is a novel alternative, since it has been used on soils recently (Di Bonito, 2005) and has the potential to access water contained in small pores; soil suspension or saturation extracts constitute a valid alternative, especially when batch experiments are carried out (Degryse *et al.*, 2003). Furthermore, these methods are capable to perform 'fractionated' extraction on the soil, whereby a combination of the methods can be used to provide soil water originating from a wider range of pores, which can present a variety of interactions with the soil matrix and possibly different chemistry.

3. DESCRIPTION AND DISCUSSION OF SELECTED METHODS

3.1. Rhizon soil moisture samplers

Rhizon samplers are a hybrid device, which can be used in the laboratory, for example, pot experiments, as well as in the field (Cabrera, 1998; Knight *et al.*, 1998). They represent one of the latest developments in terms of tension samplers, where it is necessary to apply a suction to withdraw pore water, either with a syringe, a vacuum tube or a pump. In this chapter, Rhizon samplers obtained from Rhizosphere Research Products (Wageningen, Holland), later acquired by Eijkelkamp (www.eijkelkamp.com), are described. Using this device, a pore water sample is obtained by inserting the sampler into a wet soil, and applying a suction from a vacuum tube or syringe. According to the manufacturer, the yield in water with 100 kPa pressure differential is greater than 1 ml min^{-1}.

For the soil pore water sampling, the procedure described by Knight *et al.* (1998) and Tye *et al.*, (2003) can be normally followed. Samplers are inserted into soil containers and soil pore water extracted by connecting a syringe to each sampler and applying a suction.

3.1.1. Materials

Standard Rhizon samplers (Fig. 10.3) consist of a length of porous, chemically inert hydrophilic polymer plastic (2.5 mm outer diameter, 1.4 mm inner diameter, average pore diameter $\sim$0.1 µm), namely polyethersulphone (PES–G.P.J.P. van Dijk, personal communication). This is capped with nylon at one end, and attached to a 5 or 10 cm length of polyethylene (PE) tubing, with a Luer-Lock (L-L) male connector at the other end. The tubing is double walled, the inner sleeve is PE because it is highly inert, the outer wall (for strength) is PVC. The porous polymer and part of the PVC tube is strengthened by a stainless steel wire. The Bubble Point (BP) of the sampler, that is, the minimum pressure needed to overcome the capillary action of the fluid within the largest pores, which are then emptied,[1] is greater than 200 kPa. The dead volume (or void volume i.e. the total of the volume of the porous material and the inner volume) is relatively low, $\sim$0.5 ml.

Nominal data for physical characteristics and dimensions of the Rhizon samplers are given in Table 10.2.

3.1.2. Theoretical basis of method

When suction is generated within the sampling system, water is sucked inwards through the pores of the sampler until a corresponding capillary pressure occurs in the pores. If the capillary pressure in the sampler is lower than that in the soil, water flows from the soil into the sampler until the capillary pressure in the sampler and in the soil are equal. The maximum capillary pressure in a pore can be calculated by the following equation (Schubert, 1982):

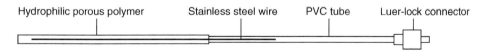

Hydrophilic porous polymer   Stainless steel wire   PVC tube   Luer-lock connector

Figure 10.3 Sketch of a Rhizon sampler (from 'Rhizon soil moisture sampler: operating instructions'; www.eijkelkamp.com).

---

[1] Bubble Point tests are usually carried out to characterize a membrane or porous material consistency or quality; they are also a common procedure to determine the maximum pore size.

Table 10.2 Physical characteristics and dimensions of a Rhizon sampler

q (m^3 s^{-1} (10^8))	BP (kPa)	Porous area (cm^2)	Length (cm)	Internal diameter (cm)	Outer diameter (cm)	Internal volume (cm^3)	Dead volume (ml)
1.7	200	7.90	10	0.14	0.25	0.15	0.5

BP, bubble point; q, volumetric flow rate into the sampler.

$$p_c = \frac{-2\gamma(T)\cos\theta}{rgD_1} \times 10^{-9} \tag{10.1}$$

where p_c is the capillary pressure (MPa), γ is the surface tension (N m^{-1}), T is the temperature, θ is the contact angle, r is the radius of the pore (m), D_1 is the density of the liquid (kg dm^{-3}) and g is the gravitational constant (m s^{-2}).

This equation is valid for pores with a circular cross section. For other shapes, an empirical adjustment factor must be considered. Surface-active substances that are dissolved in the water, for example, humic substances can decrease the surface tension. Materials that are not completely hydrophillic (e.g. plastic) need a smaller pore size.

The time required for sampling depends directly on the actual unsaturated hydraulic conductivity (k) of a soil. Soil pore water will be extracted when $k > 10^{-3}$ m day^{-1} and when there is a good hydraulic contact between the soil and the sampler.

3.1.3. Zone of influence

The zone of influence of the sampler is the zone where sampler installation and operation affect solute flow, the region of the soil from which rhizon water is drawn and the fraction of soil water that is represented in rhizon solutions. According to the supplier (Eijkelkamp, www.eijkelkamp.com), a Rhizon sampler with 10 cm length porous polymer producing a 7-ml sample, will have removed a water cylinder of 1 cm diameter. Following this approach, it was concluded that a generic zone of influence, extending to a radius of 5 cm in all directions from the edge of the porous sampler, should be considered. The assumption of a 5 cm radius of influence, however, is not supported by any consideration of properties such as, hydraulic conductivity or porosity and was only to advise users to space samplers at 5–10 cm distance between each other. Nevertheless, the recharge area of these samplers (the space in which the water flows towards the sampler) will necessarily depend upon the capillary pressure in the soil, the tension applied, the diameter of the sampler and the pore size distribution (PSD) of the soil.

Warrick and Amoozegar-Fard (1977) presented an equation that theoretically described the maximum radius of influence (r_m). The former can be estimated in stationary conditions (steady-state flow) for a point in an infinite medium around the sampler as follows:

$$r_{\mathrm{m}} = \sqrt{\left(\frac{q}{\pi \times K_s}\right) \times 10^{(-\alpha \cdot h_1)}} \qquad (10.2)$$

where q is the volumetric flow rate into the sampler ($\mathrm{cm^3\ s^{-1}}$), K_s is the saturated hydraulic conductivity ($\mathrm{cm\ s^{-1}}$), α is the fitted parameter of hydraulic conductivity function ($\mathrm{kPa^{-1}}$) and h_1 is the pressure head at r_{m} (outside the sphere of influence of sampler, kPa).

An alternative expression for the maximum radius of influence is that given by Morrison and Szecsody (1985), which was derived from Eq. (10.2) but with k (unsaturated hydraulic conductivity) expressed as a function of the hydraulic head (see also Hart et al., 1994 and Hart and Lowery, 1997):

$$k(h) = K_s^{(\alpha \Delta h)} \qquad (10.3)$$

where $\Delta h = h_0 - h_1$, $h_0 =$ suction head at the sampler (kPa).

Therefore, r_{m} will be calculated as:

$$r_{\mathrm{m}} = \sqrt{\left[\frac{4r_0}{\alpha} e^{\alpha r_0/2}\left(1 - e^{\alpha(\Delta h)}\right)\right]} \qquad (10.4)$$

where r_0 is the radius of the sampler.

The constant α is empirically derived and is a measure of the relative importance of gravity and capillarity for water movement in the particular soil (see Bresler, 1978 and Morrison and Szecsody, 1985). Fine soils, where capillarity dominates, have small α values; coarse soils, where gravity effects control water transport, have larger α values (Phillip, 1968).

3.1.4. Uses and limitations for soil pore water extractions

Rhizon samplers are becoming increasingly popular, especially for studies on bio-availability. Knight et al. (1998) were one of the first groups of researchers to apply these devices to extract pore water for metal availability studies (Cd, Zn) on soil pore water. Applications on nutrient properties and distribution in different media are also found (e.g. Cabrera, 1998). The methodology has been compared with centrifugation for microbial ecotoxicity testing on soil amended with Cd and Zn (Tiensing et al., 2001). Luo et al. (2001, 2003) studied Cu and Zn in polluted soil as influenced by γ-radiations, monitoring the changes in metal concentration over time. More recently, the same authors evaluated the use of a multi-layer column device, with installation of Rhizon samplers, to collect soil pore water for study on nitrate leachability in sludge-amended soils. Rhizon samplers were also used by Tye et al. (2002, 2003) to extract pore water to predict As solubility in contaminated soils and to study Cd^{2+} and Zn^{2+} activities in soil pore water on a range of soils.

Figure 10.4 shows r_{m} as calculated by Eq. (10.4), varying the pressure head (Δh) and the type of soil (α value). The real diameter of a Rhizon sampler and optimal

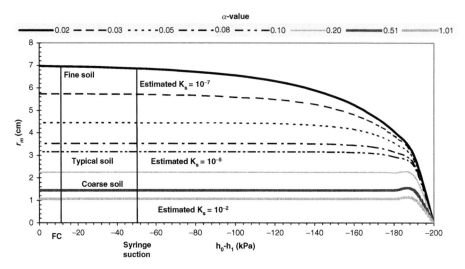

Figure 10.4 Radius of influence of a Rhizon sampler as a function of the pressure head and the soil type.

sampling conditions ($h_0 = -200$ kPa) were assumed for the calculations. The two vertical lines, field capacity (FC) and 'syringe suction', enclose the expected conditions for any soil in these conditions. According to Cabrera (1998), 10 ml plastic syringes applied to Rhizon samplers are likely to generate an average suction of -48.1 ± 0.5 kPa.

As expected, the radius of influence is greater for finer (clayey) soils and smaller for coarser soils. The α values shown in Fig. 10.4, range through most of the published α value for soils ($0.01–1.01$ kPa^{-1}). However, there is a considerable variation in α and K_s, therefore, more accurate r_m calculations can be made if those two parameters are determined experimentally for a particular soil. In general, the calculation showed that the overall axial–radial influence of this type of sampler is very small, confirming the findings of Hart and Lowery (1997). This could result in a limitation of the method, especially when the soil is not homogeneous (or it has not been homogenized) causing preferential flow conditions to prevail.

Samples collected with these devices may inadequately represent the pore water in its natural occurrence because of problems inherent in the technique (Litaor, 1988). This limitation may be additionally influenced by the complex nature of the soil, whose heterogeneity highly affects the chemical concentrations in pore water. Hence, Rhizon samplers, with their small cross-sectional area, may not adequately integrate for spatial variability (Amoozegar-Fard *et al.*, 1982; England, 1974; Haines *et al.*, 1982), and may represent 'point samples' with qualitative rather than quantitative attributes (Biggar and Nielsen, 1976).

Furthermore, as these devices are produced from organic materials (polysulphone fibres), samplers may add some OM to a sample. A recent study (Di Bonito, 2005) posed questions in these areas, and found that after several applications porosity would decrease because of a combination of wearing and organics building up.

In addition, as the PVC tubes contain a plasticizer and stabilizers of which producers do not give information; this needs to be taken into account, particularly, when PVC additives may give problems in analytical methods. In this case, other materials may constitute a better choice. The supplier also informs the users that decaying organic material may influence $N-NH_4$ analysis in auto-analyser systems. Another important consideration on polysulfone fibres application for soil solution sampling is the apparent retention of colloidal Fe at the fibre interface (Jones and Edwards, 1993b) which is not entirely clear and should not to be overlooked, because of the high retention of colloidal Fe and its role in the translocation of PTE in soil.

3.2. Centrifugation

Centrifugal extraction of pore water is a relatively routine and well-established method. Its use started early in the twentieth century (Cameron, 1911), but was little used until its reintroduction by Davies and Davies (1963). Since then few modifications have been applied to the method. Generally, one can distinguish between three main kinds of centrifugation: low-pressure centrifugal displacement or high-pressure centrifugal displacement and centrifugation with immiscible liquid (Fig. 10.5).

The first two types are based on free drainage of the pore water, through a porous plate supporting the sample, into a collecting cup. The third, now mostly in disuse, is based on the displacement of pore water by a dense, immiscible liquid followed by

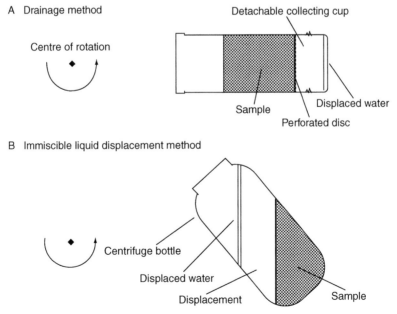

Figure 10.5 Pore water extraction by (A) drainage centrifugation using a swing-out rotor and (B) immiscible liquid displacement using a fixed-angle rotor (after Kinniburgh and Miles, 1983).

subsequent collection of the displaced water after it has floated to the top (Batley and Giles, 1979; Kinniburgh and Miles, 1983; Mubarak and Olsen, 1976; Whelan and Barrow, 1980).

Centrifugation has been widely applied to the extraction of pore waters from various materials including sediments, chalks, sandstones and clayey soils (e.g. Richards and Weaver, 1944; Shaffer *et al.*, 1937). Drainage centrifugation was reported as a method for the removal of fluids from various saturated and partially saturated geological materials in the early soil science literature. These early studies were aimed at measurements of the physical properties of the rocks rather than the characterization of extracted pore waters. Jones *et al.* (1969) and Sholkovits (1973) later reported the use of centrifugation to extract pore water from basin sediments for chemical characterization. Much of the sampling of pore water in hydro-chemical investigations in UK aquifers follows the approach developed and tested in the pioneering work of Edmunds and Bath (1976). The same high-speed centri-fugation technique adopted by these authors was later used by Wheatstone and Gelsthorpe (1982) and others for the extraction of pore waters from Triassic sandstones. The technique gradually became the preferred method in soil science. Adams *et al.* (1980) reported that centrifugal displacement at low pressures (<500 kPa) represented the most widely employed approach to obtaining soil pore water.

The direct centrifugation drainage technique is often preferred as a simple way of obtaining pore water that minimizes risks of contamination (Tyler, 2000). Centri-fugation allows the quick and easy removal of soil water at precise intervals in time at matric suctions greater than 100 kPa, the upper limit of most of the porous ceramic samplers (Jones and Edwards, 1993a). The pressure applied can go up to 1500 kPa with the highest centrifugal speeds.

3.2.1. Materials
Various centrifuges exist on the market to accommodate the vessels required by the method. For example, a Beckman J21C high-speed refrigerated centrifuge, fitted with a 6×500 cm^3 Beckman JA-10 fixed-angle rotor is a common choice. In particular, the study we refer to (Di Bonito, 2005) utilized specially designed polyoxymethylene (Acetal) tubes provided with 316 stainless steel, 20 μm mesh filters, which were manufactured in-house to adapt the rotor available with the centrifuge (Fig. 10.6).

3.2.2. Theoretical basis of method
Although the exact force distribution is difficult to determine, the physics of fluid removal from porous geological materials by drainage centrifugation is fairly well understood (Edmunds and Bath, 1976). Given a column of soil under centrifugation, the tension applied, p_a, can be derived as follows:

$$p_a = \frac{\omega^2}{2g}(r_1^2 - r_2^2) \qquad (10.5)$$

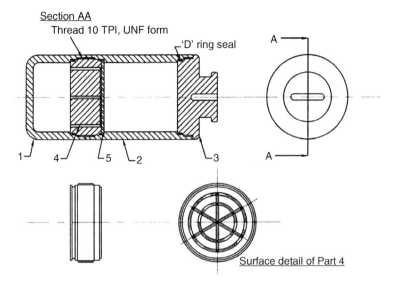

Figure 10.6 Centrifuge tubes for soil separation, designed and manufactured by R&D Workshop BGS, Keyworth; (1) pore water collector cup; (2) upper soil container; (3) screwable acetal top; (4) screwable acetal filter and support and (5) stainless steel filter.

where p_a is the tension applied developed at a generic point r_2 of column (cm water), ω is the angular velocity (rad s^{-1}), g is the gravitational constant (cm s^{-2}) and r_1 is the distance from base of column to centre of rotation (cm).

The applied force is, therefore, a function only of distance from the rotor and the centrifugal speed, that is, it has the same magnitude irrespective of the density and nature of the material tested, and the pattern of water removal will depend on the pore size distribution of the material. At equilibrium, p_a will be everywhere balanced by a capillary pressure, p_c, which can also be expressed as follows (Washburn, 1921):

$$p_c = \frac{2\gamma\cos\theta}{\rho \times r} \tag{10.6}$$

where p_c is the capillary pressure in a pore (N m^{-2}), γ is the surface tension (N m^{-1}), ρ is the specific gravity, θ is the contact angle between porous solid and liquid, and r is the radius of pore (m).

The extent of interstitial water removal is, therefore, a function of the centrifuge dimensions and rotation speed, but it is also governed by the weight of sample used, the degree of initial saturation as well as the material's pore size distribution.

3.2.3. Pore water extraction

The centrifugation drainage procedure described by Edmunds and Bath (1976) and Gooddy *et al.* (1995) can normally be applied for this method. The relationship between the distance to the centre of rotation and speed (hence pressure) applied as well as the optimal choice of centrifuge speed, centrifugation time and sample weight can be established in relation to the particular sample and volume of interstitial water required (see also Kinniburgh and Miles, 1983).

A known quantity of soil is placed into weighed centrifuge buckets. After the samples are spun at the chosen speed, any pore water extracted can be collected using disposable syringes, weighed and filtered through adequate filters for chemical analysis. If only small volumes of pore water were extracted, replicate samples can be used to bulk the extracts. Distilled water blanks are normally passed through the extraction steps to minimize any contamination by materials or handling.

During centrifugation, a soil sample is spun at a specific speed, which corresponds to a relative centrifugal force (RCF) and, according to Eq. (10.5), a corresponding pressure. The pressure on each point within the soil column can be represented, depending on the distance to the axis of rotation and speed. Figure 10.7 shows the variation of the applied pressure within a single bucket, as calculated by using Eq. (10.5). The radius varies between 5 (top of the column) and 9 cm (base of the column) depending on the position of the point inside the bucket, which is inclined at 45° with respect to the axis of rotation. In this situation, we considered the maximum speed achievable as ~7000 rotations per minute (rpm), which was calculated according to a reduction factor for the rotor in use with the centrifuge, in case of materials having density higher than water (Beckman Instruments, 1988).

The mean distance from the axis of rotation is 7 cm, which is the midpoint at which we can calculate the relationship between the varying speed and the pores drained according to Eq. (10.6). Figure 10.8 shows the resulting pressure profile at the midpoint with varying centrifugal speed.

3.2.4. Uses and limitations for soil pore water extractions

Many researchers have tested and reported the yield (defined also as 'extraction efficiency'—Entwisle and Reeder, 1993, see later in this chapter) for different materials and using a range of rotation speeds and lengths of operation. As a general principle, the volume of solution extracted is a function of the initial weight of the

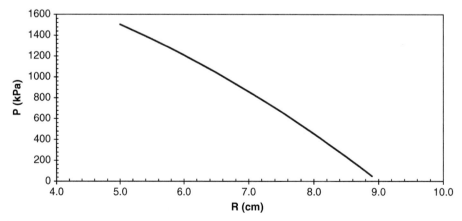

Figure 10.7 Pressure profile depending on distance R from rotation axis at maximum speed (7000 rpm) based on Eq. (10.5).

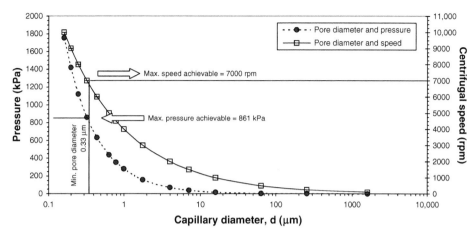

Figure 10.8 Pressure profile depending on the speed at a midpoint of 7 cm from axis of rotation calculated using Eq. (10.6); relationships between speeds of rotation, pressure applied and minimum capillary size drained (radius = 7 cm).

sample, the pore size distribution of the soil, the degree of initial saturation, the centrifuge dimensions and rotational speed (Edmunds and Bath, 1976). In the study of Edmunds and Bath (1976), extraction yields of 20–30% were obtained at low speeds, compared with up to 85–95% of the available pore water using the high-speed centrifuge. In the study of Wheatstone and Gelsthorpe (1982), the moisture content of the samples tested varied between 6 and 15% and the percentage of water extracted increased with the increasing speed from 3000 to 12,000 rpm. At the optimum speed of 12,000 rpm, tests showed that only a marginal increase in the amount of fluid extracted was observed after an initial 30-min operation. In the work of Kinniburgh and Miles (1983), yields were typically 20–50% for soils of moisture content 10–40%, but up to 90% for some chalks with 20% initial moisture content. Di Bonito (2005) found that centrifugation achieved an efficiency of 28% (measured on a single tube), and very little pore water was collected at low speeds (1000 and 2000 rpm). This was due to the initial moisture content of the soil used and the low-pressure differential between the matric potential of the soil (representing the strength at which water is held in the soil), equivalent to FC (-10 kPa), and the corresponding potential applied during these steps (respectively 18 and 70 kPa). When centrifuging, pore water is lost from the sample when positive pressures exerted by centrifugal force are greater than the matric suctions exerted by the solid phase (Jones and Edwards, 1993a). A significant amount of water (up to 40% of the total) remained in the samples at the end of the 5-step centrifugation extraction.

During centrifugation compaction of the soil occurs, the effect being more significant for finer textures (Gamerdinger and Kaplan, 2000). In a recent study (Di Bonito, 2005) from an initial bulk density (ρ_b) of 0.67, a final ρ_b of 1.01 g cm^{-3} was measured, therefore, affecting the nominal pore distribution. Although this is a predictable drawback of the methodology, it could lead to misinterpretation of the possible sources of the water in terms of porosity if not carefully considered.

According to Jones and Edwards (1993a), a moisture content gradient also develops through the sample. As water migrates down through the sample, the base of the soil, from which the solution is released, will be in excess of its water-holding capacity (0 kPa) during part of the centrifugation process, indicating that centrifugation may also yield solution from pores of all sizes at the one time (Lorenz *et al.*, 1994).

Studies comparing different centrifugation methods and other techniques in providing soil pore water have been reported by many authors (e.g. Chapman *et al.*, 1997; Dahlgren, 1993; Giesler *et al.*, 1996; Lorenz *et al.*, 1994; Menzies and Bell, 1988; Sheppard *et al.*, 1992; Zabowsky and Ugolini, 1990). Most studies are fairly consistent in their illustration of the differences among methods in terms of element concentrations in pore water, partly because of different fractions of pore waters considered. Moreover, centrifugation can be used to fractionate the pore water by selecting several centrifugation rates, that is, pore water can be extracted using a number of steps, with an increment of centrifugal speed. When increasing the centrifugal speed, and, therefore, the RCF value, during the various stages of soil centrifugation, less available water may gradually be released and collected (Tyler, 2000), thereby extracting water from a range of pore size distributions. Centrifugal speed has been shown to influence significantly the composition of the extracted soil pore water, which, depending on the specific soil considered, can display an effective increase in metal concentrations (Pérez *et al.*, 2002). In the simplest case of piston and preferential flow (Beven, 1989), water is considered to have a bimodal distribution in velocity, corresponding to 'mobile' and 'immobile' phases (Coats and Smith, 1964). In an unsaturated soil medium with a given degree of heterogeneity, piston flow tends to be dominant at higher water contents and preferential flow at lower water content (Padilla *et al.*, 1999).

3.3. Pressure filtering (squeezing)

The squeezing method represents an approach where it is possible to modify the pressure during the extraction. This technique has been proved to be effective with various structured materials (coherent sediments and rocks) and has also been used for incoherent materials, including peat, clay, till, sand, silt, chalk and sea sediments (Entwisle and Reeder, 1993), but only recently has been used on soils (Di Bonito, 2005).

The squeezing technique was originally developed to obtain pore water samples mainly from unconsolidated marine silts and clays. Manheim (1966) developed a heavy duty squeezer capable of applying a stress up to 150 MPa based on the early designs of a number of Soviet workers, most notably Kriukov (1971). Similar designs have been used by Morgenstern and Balasubramonian (1980), Brightman *et al.* (1985) and Krahn and Fredlund (1972) to evaluate change in salinity with increased pore water extraction and increased pressure. The methodology has been used to extract pore fluid from materials with moisture content slightly below 7%. The method is often unsuccessful on highly cemented, hard material. In comparative studies, squeezing has been found to have a lower potential for contaminations and artifacts, partly because pore water extraction and filtration can be conducted in-line (Bufflap and Allen, 1995b). Very little direct contamination of the pore water resulted from clay studies using this extraction method (Entwisle and Reeder, 1993). In contrast differences in ionic strength and major components were found with increasing pressure applied and decreasing moisture contents (Cave *et al.*, 1998; Reeder *et al.*, 1998).

3.3.1. Materials

The squeezing apparatus described here is the one in use at the British Geological Survey (BGS) and utilizes a hydraulic pump (Wykeham Farrance Engineering Ltd., Slough) which has a maximum output stress of 70 MPa and hydraulic hoses and fittings (Fig. 10.9). The main body of the cell and other metal parts in contact with the test sample or pore water are manufactured from Type 316 stainless steel, selected for its resistance to corrosion and its high tensile strength.

The cell body sample chamber is 75 mm in diameter and 100 mm high (Fig. 10.10). The outside of the cell has a spiral trough through which temperature controlled fluid circulates inside a plastic insulation jacket. Temperature control of the cell is achieved by a heater/chiller, which is capable of temperature control between −10 and 50 °C. The pore water collection pipe screws into the top plate. Pore fluid is collected directly into disposable polypropylene syringes.

Two syringe taps (which can have a 0.45 μm Acrodisc filter in between) are pushed onto the pore water pipe allowing flexibility of pore water collection: taps

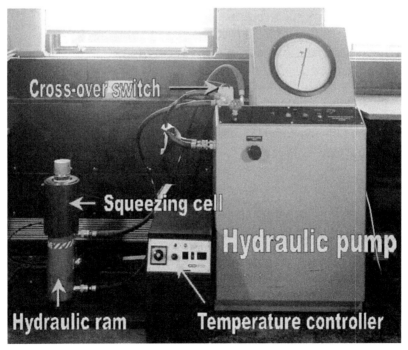

Figure 10.9   Squeezing apparatus, designed and manufactured by R&D Workshop, BGS Keyworth.

Figure 10.10 Clay squeezing cell, designed and manufactured by R&D Workshop, BGS Keyworth.

can be opened and closed when multiple samples are collected and syringes need to be replaced. The metal filter, which has a diameter of 90 mm, is also made of Type 316 stainless steel (see Entwisle and Reeder, 1993 for further details).

3.3.2. Theoretical basis of method

The squeezing process involves the expulsion of pore water from the material being compressed. In general, the material consists of solid particles (mineral phase), and spaces (voids), which in an unsaturated environment such as a soil, contain both air and water. When a squeezing stress is applied to a water-saturated material, its volume decreases by three main mechanisms:

○ compression of the solid phase;
○ compression of the pore water between the solid phase; and
○ escape of water from the voids.

In most circumstances, the compression of the solid and liquid phases is negligible and most of the change in volume is caused by the escape of pore water. This may be illustrated by a hydro-mechanical analogy for load changing and squeezing as shown in Fig. 10.11 (after Lambe and Whitman, 1979). The resistance of the solid phase during compression is represented by a spring and the rate at which the pore fluid flows is dependent upon the size of the valve aperture. In (a) the valve is closed and in equilibrium. When a pressure is added (b) the piston load is apportioned by the water and the spring in relation to the stiffness of each. There is little movement in the piston because the water is relatively incompressible. Most of the load is carried by the water and this increases the water pressure. If the valve is now opened (c) the excess pore pressure dissipates by water escaping through the valve (d). The piston

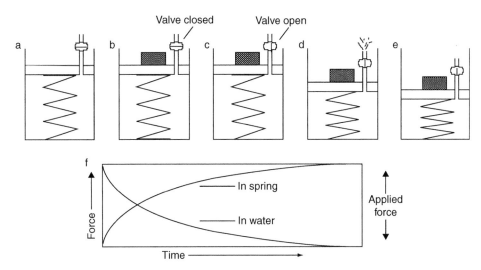

Figure 10.11 A hydro-mechanical analogy for load changes during squeezing (after Lambe and Whitman, 1979).

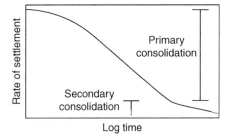

Figure 10.12 The rate of settlement for increasing stress (modified from Entwisle and Reeder, 1993).

drops and the volume of the chamber decreases until there is a new equilibrium when the applied load is carried by the spring and the water pressure has returned to the original hydrostatic condition (e). The gradual transfer of load from the water to the spring is shown in (f). The dissipation of the pore water is called primary consolidation.

The rate at which the pore fluid is expelled is related to the length of the sample and the pore size. A typical graph of the rate of settlement, and, therefore, pore fluid extraction after the addition of a load (Fig. 10.12) shows both primary and secondary consolidation.

Most of the excess pore pressure dissipates during primary consolidation. Secondary consolidation involves the movement of particles as they adjust to the increase in effective pressure and the dissipation of excess pore pressure from very small pores. The pore water extracted during squeezing is mainly because of primary consolidation.

3.3.3. Pore water extraction

Soil is placed into the clean dry cell. A 90-mm diameter Whatman filter paper is then placed on to the shoulder of the sample chamber and a clean steel filter placed on top of the filter paper. The top plate is screwed into the cell to contact the metal filter; the temperature control unit cooler is then switched on. A small nominal stress (<1 MPa) needs to be applied to remove most of the air from the cell and to allow the sample and the components to bed in. When the selected temperature (± 2 C) is attained, the pressure can be increased and water collected.

The squeezing test may take from 1 h to in excess of 2 or more weeks, depending on the set up and the physical properties of the material, producing either a single bulk sample or a number of 'sequential' water samples (Ross *et al.*, 1989). When sufficient volume of pore water had been obtained ($\sim$15 ml to allow chemical analysis), the syringe is removed and the sample filtered as soon as collected through appropriate filters and treated for analysis. This process can be repeated by continuing squeezing, using a new syringe assemblies as necessary, until no further pore water is obtained (Cave *et al.*, 1998).

After the extraction is completed, typically the specimen is removed and measured to calculate its volume, weighed and oven dried at 105 °C for density and moisture content determinations.

3.3.4. Extraction efficiency

The percentage of the available pore water extracted, E, (Entwisle and Reeder, 1993) is determined as follows:

$$E = \frac{W_{\mathrm{p}}}{W_{\mathrm{si}} - W_{\mathrm{sd}}} \times 100 \qquad (10.7)$$

where W_{p} is the weight of pore water collected, W_{s} is the weight of sample initially tested and W_{sd} is the weight of sample post squeezing (centrifuging) after oven drying.

This can be written as follows:

$$E = \frac{W_{\mathrm{p}}}{W_{\mathrm{s}} \times \theta_{\mathrm{w}}} \times 1000 \qquad (10.8)$$

where θ_{w} is the moisture content with respect to initial wet sample weight. This concept is applied both to centrifugation and squeezing, where moisture contents are normally reported with respect to the dry weight. A known mass (at least 50 g, in triplicates) of the original sample is tested by determining its weight before and after heating at $\sim$105–110 C for a minimum of 24 h.

The percentage moisture content with respect to the initial wet sample weight, θ_{w}, can be determined as follows:

$$\theta_{\mathrm{w}} = \frac{W_{\mathrm{w}} - W_{\mathrm{d}}}{W_{\mathrm{d}}} \times 100 \qquad (10.9)$$

where W_{w} is the wet sample weight and W_{d} is the dry sample weight.

3.3.5. Uses and limitations for soil pore water extractions

This method was recently tested on three different soil types (Di Bonito, 2005): a Brown sand from the Newport region, a Calcareous pelosol from the Hanslope region and a sandy silt with high OM content (LOI 27.1%, OC 15.7%) from the Nottingham region. Extraction efficiency, %E, for the three soils were calculated and resulted to be 27.4% for a sandy soil. This is due to the initial low moisture content (11.1%) and the sandy texture of the soil. Previous tests on different materials (Entwisle and Reeder, 1993) showed that samples with an initial moisture content of less than 10% present a low extract efficiency and that there is an apparent cut-off of about 7–8% below which no water is collected. Furthermore, the sandy texture of the soil suggests that solid particle repacking, which is one of the main mechanisms through which pore water is displaced, will be limited during squeezing. The rate of settlement and, therefore, the pore water extraction by primary consolidation are also related to the change in volume (i.e. voids ratio) that the sample suffers during the process; this is related to the texture of the sample, with sand giving lower porosity. Extraction efficiency, %E, for the other two soils, organic and clayey, were 68.3% and 67.4%, respectively.

The time to complete the extractions can vary, depending on the approach used and can be of weeks if the aim of the extraction is to go through states of equilibrium with the pressure applied. In general, stiffer materials of low permeability require longer periods of squeezing.

Figure 10.13 shows the cumulative release of pore water for three soils tested with increasing pressure (Di Bonito, 2005). The dotted line represents the pressure corresponding to the hygroscopic coefficient, that is, the upper limit for the capillary water. This illustrates the capability of squeezing to reach a wider range of pores. This pressure value should also correspond to most of the available water in the soil.

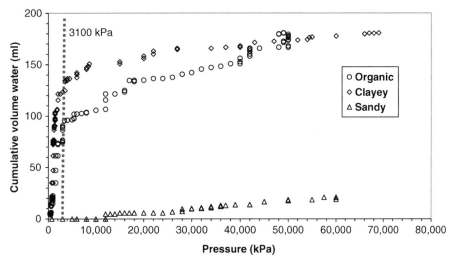

Figure 10.13 Cumulative volume of pore water extracted for the three soils tested with increasing pressure; the dotted line represents the pressure corresponding to the hygroscopic coefficient, that is, the upper limit for the capillary water.

In fact, by this stage almost 75% of the total water was extracted for the clayey soil and 60% for the organic soil.

Soil compaction was clearly very marked for the clayey and the organic soils, and to a lesser extent for the sandy one, because of the soils intrinsic characteristics.

3.4. Soil suspensions

Under this umbrella, a number of different applications lie, amongst others the batch equilibration and equilibrium soil solution methods. According to the USEPA (1999), the former represents the most common laboratory method for determining partition coefficients—normally defined as K_d—both for contaminated sites studies and for predictions of chemicals behaviour in soils (OECD, 2002). The batch equilibration method consists of mixing a soil with a known amount of liquid (background electrolyte), which is then shaken into a slurry and allowed to equilibrate for an adequate time. The solution will be separated from the solids by centrifuging the slurry, resulting in a supernatant and a separated solid phase. The supernatant will, therefore, be removed, filtered and analysed.

3.4.1. Materials

There is a multitude of choices for shakers (reciprocal or not) and centrifuges with annexed spares, with several variations to the above described general procedure. Widely different soil:solution ratios (R, kg litre^{-1}) have been used to investigate the adsorption and desorption behaviour of trace-metal ions by soils, from 1:2 up to 1:200 soil:solution dilutions (Table 10.3). The background electrolyte solution and

Table 10.3 Some example of the background electrolyte composition and soil:solution ratios used in the literature, ordered by decreasing soil:solution ratio

Potentially toxic elements studied	R (kg litre^{-1})	Background electrolyte	Ionic strength (mmol litre^{-1})
Cr	1:1; 1:20	$NaNO_3$	100
Cd, Cu	1:4	$CaCl_2$	15–150
Cd	1:10	$(Ca,Na)(Cl,NO_3)$	30
Zn	1:10	$CaCl_2$, $Na(Cl,NO_3,SO_4)$	5–300
Cu, Zn	1:10	$Mg(NO_3)_2$	10
Cd	1:20	$NaNO_3$, $Ca(NO_3)_2$	30
Cd, Cr, Cu, Ni, Pb, Zn	1:30	Na-acetate	10
Cd	1:40	H_2O	–
Cd	1:40	$Ca(NO_3)_2$	10
Co, Cu, Ni	1:100	$CaCl_2$	0.5
Cu, Ni, Pb, Zn	1:100	H_2O	–
Cu	1:200	$Ca(NO_3)_2$	30

Source: Harter and Naidu (modified), 2001 and references thereafter.

ionic strength also vary extensively between reported studies. It is commonplace in many soil and environmental studies to utilize 5 mmol litre^{-1} (0.01 N) Ca^{2+} solutions (where the counter ion is NO$_3^-$, Cl$^-$, or occasionally, SO$_4^{2-}$) for batch equilibration with soil to mimic soil pore water ionic strength and composition (Wolt, 1994). This arises from a historical precedent originating from earlier studies of soil exchangeable cations and dissolved salt content, but is still largely used in many studies (Table 10.3). However, depending on the value of R, the equilibrating solution may reflect the ionic strength and ion composition of the added electrolyte more than it does that of the *in situ* pore water. Some studies have not made provision for background electrolyte composition that would be comparable to the pore water, sometimes even having an ionic strength (I) that approaches zero (Table 10.3). However, it is well known that solid:solution equilibria in soils depend significantly on ionic strength (I).

Time is another variable for this methodology, where a 24–48-h extraction time is generally typical for many batch extractions (e.g. Anderson and Christensen, 1988; Yin *et al.*, 2002) but it may also vary and extend into several weeks according to the aim of the study (Jopony and Young, 1994). Longer equilibration times are normally selected to try to take into account the slow kinetics of the organic-metal desorption.

3.4.2. Theoretical basis of method
The method of soil suspensions extracts is based on metal desorption/dissolution processes, which primarily depend on the physico-chemical characteristics of the metals, selected soil properties and environmental conditions. Metal adsorption/desorption and solubility studies are important in the characterization of metal mobility and availability in soils. Metals are, in fact, present within the soil system in different 'pools' and can follow either adsorption and precipitation reactions or desorption and dissolution reactions (Selim and Sparks, 2001). The main factors affecting the relationship between the soluble/mobile and immobile metal pools are soil pH, redox potential, adsorption and exchange capacity, the ionic strength of soil pore water, competing ions and kinetic effects (e.g. contact time) (Evans, 1989; Impellitteri *et al.*, 2001; McBride, 1994; Sparks, 1995).

Adsorption equilibria are often summarized by the *distribution or partition coefficient*, K_d, which expresses the relative affinity for a sorbate in solution to be absorbed or desorbed (Oscarson and Hume, 1998). The distribution coefficient, K_d, is usually defined as the ratio of concentrations in the adsorbed and liquid phase, thus:

$$K_d = \frac{[M_{soil}]}{[M_{solution}]} \tag{10.10}$$

usually expressed in litre kg^{-1}. However, when reporting K_d values for soils, it is very important that the definitions of [M_{soil}] and [$M_{solution}$] are given, identifying what form of metals is described. For example, for estimating immediately bioavailable metal in soil pore water, the free-ion activity estimated by speciation programmes can be used (Jopony and Young, 1994); conversely, if the study focuses on metal transportation to groundwater, it may be more appropriate (pragmatically) to include all species of metals in solution (Impellitteri *et al.*, 2001).

The extent to which a chemical is adsorbed or desorbed can generally be rationalized within the frameworks of solution and surface chemistry. *Internal* and *external* factors affecting sorption and/or desorption can be distinguished (Harter and Naidu, 2001). Examples of the first category are:

○ Ionic strength
○ Cation, anion and organic ligands
○ pH
○ Total metal concentration

Examples of the second category are:

○ Pressure
○ Temperature
○ Soil:solution ratio (R)
○ Experiment technique and sample storage conditions

3.4.3. Uses and limitations for soil pore water extractions

The main advantage of the use of this method is related to the costs, as there is often no need for extra equipment in the laboratory, no chemicals are involved unless trying to reproduce the composition of the natural background electrolyte. The preparation is easy and the use is widespread, allowing comparison with results from other studies.

Nevertheless, variations in R have been found to influence the aqueous-phase chemistry of trace-metal ions, and thereby affect sorption and desorption processes (Celorie *et al.*, 1989; Di Toro, 1985; O'Conner and Connolly, 1980; Voice *et al.*, 1983). Generally, K_d values tend to decrease with the increasing soil concentration (high R), suggesting increased metal solubility. Grover and Hance (1970) suggested that this effect is predominantly caused by higher surface area exposure at high R. Another explanation would be that there are simply more particles that pass through a given filter at higher solid concentrations. Therefore, more particles transporting bound metal are erroneously analysed as 'soluble' or desorbed metals in a supernatant, yielding lower K_d values (Van Benschoten *et al.*, 1998; Voice and Weber, 1985). A third and more plausible explanation is that K_d increases as labile metal is depleted (stronger sites accessed) and so a small value of R (substantial depletion) will produce a greater K_d value. Early studies on this topic revealed that total concentrations of soluble salts generally decreased with an increase in the soil moisture content (increasing dilution) (Khasawneh and Adams, 1967; Larsen and Widdowson, 1968; Reitemeier, 1946). More recently, Yin *et al.* (2002) used a series of R values (from 0.5 to 0.02) in batch experiments to quantify the solubility of Cu, Ni and Zn in soils. Again, greater concentrations arose from larger values of R.

The effect of R on metal concentrations in soil pore water is of interest because it can characterize soil pore water at different water content, and also to explain the effect of different values of R in procedures for extracting equilibrium soil pore water. Water extracts are often used to measure pore water chemistry and thus extrapolating its equivalent composition at FC. However, varying soil:solution ratios showed that this could be an ambiguous exercise if not corroborated by sound speciation models. A major part in the final result can be played by OM in the soil, where desorption of OM can render unreliable or inconsistent values of the metal released with varying R (Di Bonito, 2005). On the other hand, there is good evidence that solubility/adsorption equilibria of free divalent metals are maintained reasonably well (relatively constant $[M^{2+}]$ and very limited depletion of labile pool). The significance of this is that it is possible to attain a fairly reliable estimate of bioavailable metal (M^{2+}) even in strongly dilute suspension but poor estimates of M_{soln} (e.g. for leaching studies) because they are hopelessly dependent on operational aspects of the suspension/extraction scheme. USEPA (1999) and Yin *et al.* (2002) drew the conclusions that values of metal dissolved based on this technique conducted with a soil:solution ratio significantly less than those existing in the field would overestimate metal sorption and underestimate metal migration (see also Strandberg and Fortkamp, 2005).

4. CONCLUSIONS AND RECOMMENDATIONS

Rhizon soil moisture samplers, centrifugation, high-pressure squeezing and soil water suspensions constitute good laboratory techniques for the recovery of pore water. The functionality, benefits and limitations of these methods can be summarized as follows:

Rhizon samplers should be used preferably as a disposable device, to avoid material modification with time, decrease in porosity and possibly significant variations in solute chemistry; this will inevitably increase the experimental costs in the case of larger designs. Given the nature of the materials, results should be treated cautiously in cases where humic substances are present in the soil. There is no equipment required except syringes (or vacuum tubes), ease of handling/deployment and straightforward use constitute major advantages. There are no 'side effects' (provided time is allowed for equilibration with the soil) if M^{2+} (free-ion activity) were needed as opposed to M_{soln} (total metal in pore water), as often required in environmental studies. The overall geometric mean axial–radial influence of these samplers is very small, suggesting that careful consideration of the samplers' placement and experimental designs as a function of the characteristics of the soil under investigation is required. Rhizon samplers only function when the matric potential is greater than 10 kPa (above FC) otherwise the potential gradient and the hydraulic conductivity are too low to obtain a sample. Rates of solution accumulation by tension samplers can vary in a given soil because of heterogeneity of moisture content and solution flow pathways. Increasing the applied tension increases the non-uniformity of sampling. Hence, the more uniform the particle size the more uniform is the sampling rate.

Centrifugation presents fewer material problems compared to Rhizon samplers; however, tubes and other spare parts are not commercially available and often have to be built in-house. The method is optimal for bulk solution studies, or when homogenization represents a key experimental point; targeted studies are also possible but would necessitate prior examination of the soil water-holding capacity and release under varying centrifugal speeds, which could be demanding in terms of time and effort.

Centrifugation covers a wider range of pore sizes compared to Rhizon sampling, but nevertheless will only yield a fraction of the total pore fluids. A consideration when using this methodology is that compaction and pore size reduction may occur during centrifugation. As this effect is more significant for finer textures, therefore, these soils would need to be tested for compaction before experimentation. Furthermore, a moisture content gradient also develops through the sample. As the water migrates down through the sample, the base of the soil will be in excess of its water-holding capacity (0 kPa) during part of the centrifugation process, resulting in saturation of the sample at its base. This could cause alteration of the chemical composition of the extracted pore water, caused by a mixing process, which would homogenize the composition compared to the heterogeneous *in situ* conditions. Some studies reached similar conclusions, indicating that centrifugation yielded solution from pores of all sizes at the one time. This effect would, however, be much more evident at soil matric potential increasingly above FC, where preferential flow conditions and gravitational water prevail with respect to capillary water, and for single extractions. In fact, as this method is destructive and produces compaction of the soil sample, capillary conditions are most likely to be established as the extraction would proceed to later fractions, because of the reduction of the total nominal porosity.

Soil squeezing is not as accessible as the other three, at least with the present apparatus required, but there is a space for further development and improvement (i.e. simplification of the design and/or reduction of labour required). The rate of pore water collected during squeezing tests shows that sample texture and OM content are important variables to consider before extraction. In the cases of sandy soils, exceeding FC would be advisable to improve the release of pore water. Conversely, because of the higher range of pressures exerted compared to Rhizon samplers and centrifugation, a greater amount of the total available water was generally extracted. Efficiencies observed for a clayey and a sandy silt with high OM were 67% and 68%, respectively. The principal constraint of this methodology is the time involved: if the extractions require 'multiple-step at quasi-equilibrium' protocol then a much longer duration, compared to centrifugation and Rhizon samplers, is needed for a complete test. The duration of the extraction constitutes a major limitation, especially when OM is a key component of the soil studied. Decomposition of OM and anaerobism can lead to massive variations in chemical composition and speciation. Careful consideration should, therefore, be made in those circumstances. However, 'fast' extractions with high initial pressures may produce good results and minimize the alterations produced by anaerobism.

Batch extractions are very accessible and easy to reproduce. Costs and ease represent the main advantages of this method: often no extra equipment and no chemicals are necessary, unless the aim is to reproduce the composition of the natural background electrolyte. One of the main drawbacks is related to the actual comparability of results, given the wide range of operating factors used in the literature. In particular, variations in soil:solution ratios (R) have been found to influence the aqueous-phase chemistry of trace-metal ions, and thereby affect sorption and desorption processes. Varying R can show to be an ambiguous exercise if not corroborated by sound speciation models, particularly, where a major part in the final result can be played by OM in the soil. Batch extractions have proven to be very consistent in estimating the bioavailable fraction of the metal (M^{2+}) even in strongly dilute suspension, whereas they are not always reliable for estimates of M_{soln} (e.g. for leaching studies) because they are hopelessly dependent on operational aspects of the suspension/extraction scheme.

One important conclusion drawn in the literature is that values of metal dissolved based on this technique conducted with an R significantly less than those existing in the field would overestimate metal sorption and underestimate metal migration. As a consequence, suspensions with high solids concentrations should be used for batch experiment extractions to approach more closely natural conditions, unless only estimates of free metal ion activity (M^{2+}) are needed.

In summary, the importance of the method employed for soil pore water extraction should not be underestimated. Experimental design and performance should be chosen to reflect the particular aim of the study, reported in sufficient detail to allow others to make appropriate comparisons and the parameters operationally defined as a function of the method employed.

REFERENCES

Adams, F., Burmester, C., Hue, N. V., and Long, F. L. (1980). Comparison of column-displacement and centrifuge methods for obtaining soil solution. *Soil Sci. Soc. Am. J.* **44**, 733–735.

Amoozegar-Fard, A. D., Nielsen, D. R., and Warrick, A. W. (1982). Soil solute concentration distribution for spatially varying pore water velocities and apparent diffusion coefficients. *Soil Sci. Soc. Am. J.* **46**, 3–9.

Andersen, M. K., Raulund-Rasmussen, K., Strobel, B. W., and Hansen, H. C. B. (2002). Adsorption of cadmium, copper, nickel, and zinc to a poly(tetrafluorethene) porous soil solution sampler. *J. Environ. Qual.* **31**(1), 168–175.

Anderson, P. R., and Christensen, T. H. (1988). Distribution coefficients of Cd, Co, Ni, and Zn in soils. *J. Soil Sci.* **39**, 15–22.

Barber, S. A. (1984). "Soil Nutrient Bioavailability: A Mechanistic Approach." John Wiley, New York.

Basta, N. T., and Tabatabai, M. A. (1992). Effect of cropping systems on adsorption of metals by soils. II. Effect of pH. *Soil Sci.* **153**, 195–204.

Batley, G. E., and Giles, M. S. (1979). Solvent displacement of sediment interstitial waters before trace-metal analysis. *Water Res.* **13**(9), 879–886.

Beckman Instruments. (1988). "Instructions for Using the JA-10 Fixed Angle Rotor. Beckman J-21 Series, J-6, and J-6B." Spinco Division of Beckman Instruments, INC., Palo Alto, CA.

Beier, C., and Hansen, K. (1992). Evaluation of porous cup soil-water samplers under controlled field conditions: Comparison of ceramic and PTFE cups. *J. Soil Sci.* **43**, 261–271.

Beier, C., Hansen, K., Gundersen, P., and Andersen, B. R. (1992). Long-term field comparison of ceramic and poly(tetrafluoroethene) porous cup soil water samplers. *Environ. Sci. Technol.* **26**, 2005–2011.

Benedetti, M. F., VanRiemsdijk, W. H., Koopal, L. K., Kinniburgh, D. G., Gooddy, D. C., and Milne, C. J. (1996a). Metal ion binding by natural organic matter: From the model to the field. *Geochim. Cosmochim. Acta* **60**(14), 2503–2513.

Benedetti, M. F., VanRiemsdik, W. H., and Koopal, L. K. (1996b). Humic substances considered as a heterogeneous donnan gel phase. *Environ. Sci. Technol.* **30**(6), 1805–1813.

Beven, K. (1989). Changing ideas in hydrology – the case of physically-based models. *J. Hydrol.* **105**(1–2), 157–172.

Biggar, J. W., and Nielsen, D. R. (1976). Spatial variability of leaching characteristics of a field soil. *Water Resour. Res.* **12**(1), 78–84.

Booltink, H. W. G., and Bouma, J. (1991). Morphological characterization of bypass flow in well-structured clay soil. *Soil Sci. Soc. Am. J.* **55**, 1249–1254.

Bresler, E. (1978). Analysis of trickle irrigation with application to design problems. *Irrig. Sci.* **1**, 3–17.

Brewer, R. (1964). "Fabric and Mineral Analysis of Soil." John Wiley, New York.

Brightman, M. A., Bath, A. H., Cave, M. R., and Darling, W. G. (1985). "Pore Fluids from the Argillaceous Rocks of the Harwell Region 85–86." British Geological Survey, Keyworth, Nottingham.

Bufflap, S. E., and Allen, H. E. (1995b). Comparison of pore water sampling techniques for trace metals. *Water Res.* **29**(9), 2051–2054.

Cabrera, R. I. (1998). Monitoring chemical properties of container growing media with small soil solution samplers. *Sci. Hortic.* **75**, 113–119.

Cameron, F. K. (1911). "The Soil Solution: The Nutrient Medium for Plant Growth." Chemical Publishing, Easton, PA.

Cances, B., Ponthieu, M., Castrec-Rouelle, M., Aubry, E., and Benedetti, M. F. (2003). Metal ions speciation in a soil and its solution: Experimental data and model results. *Geoderma* **113**(3–4), 341–355.

Cavallaro, N., and McBride, M. B. (1980). Activities of Cu^{2+} and Cd^{2+} in soil solutions as affected by pH. *Soil Sci. Soc. Am. J.* **44**, 729–732.

Cavallaro, N., and McBride, M. B. (1984). Zinc and copper sorption and fixation by an acid soil clay: Effect of selective dissolutions. *Soil Sci. Soc. Am. J.* **48**, 1050–1054.

Cave, M. R., Reeder, S., Entwisle, D. C., Blackwell, P. A., Trick, J. K., Wragg, J., and Burden, S. R. (1998). "Extraction and Analysis of Pore Waters from Bentonite Clay. *WI/98/9C*, British Geological Survey." Keyworth, Nottingham.

Celardin, F. (1999). Semi-empirical correlations between so-called mobilizable and mobile concentrations of heavy metals in soils. *Commun. Soil Sci. Plant Anal.* **30**(5–6), 843–854.

Celorie, J. A., Woods, S. L., Vinson, T. S., and Istok, J. D. (1989). A comparison of sorption equilibrium distribution coefficients using batch and centrifugation methods. *J. Environ. Qual.* **18** (3), 307–313.

Chapman, P. J., Edwards, A. C., and Shand, C. A. (1997). The phosphorus composition of soil solutions and soil leachates: Influence of soil:solution ratio. *Eur. J. Soil Sci.* **48**, 703–710.

Coats, K. H., and Smith, B. D. (1964). Dead-end pore volume and dispersion in porous media. *Soc. Pet. Eng. J.* **4**(1), 73–84.

Dahlgren, R. A. (1993). Comparison of soil solution extraction procedures: Effect on solute chemistry. *Commun. Soil Sci. Plant Anal.* **24**(15–15), 1783–1794.

Davies, B. E., and Davies, R. I. (1963). A simple centrifugation method for obtaining small samples of soil solution. *Nature* **198**, 216–217.

Davranche, M., and Bollinger, J. C. (2001). A desorption-dissolution model for metal release from polluted soil under reductive conditions. *J. Environ. Qual.* **30**(5), 1581–1586.

Davranche, M., Bollinger, J. C., and Bril, H. (2003). Effect of reductive conditions on metal mobility from wasteland solids: An example from the Mortagne-du-Nord site (France). *Appl. Geochem.* **18**(3), 383–394.

Degryse, F., Broos, K., Smolders, E., and Merckx, R. (2003). Soil solution concentration of Cd and Zn can be predicted with a $CaCl_2$ soil extract. *Eur. J. Soil Sci.* **54**(1), 149–157.

Di Bonito, M. (2005). "Trace Elements in Soil Pore Water: A Comparison of Sampling Methods." Ph.D. thesis, University of Nottingham, Nottingham (http://etheses nottingham.ac.uk/archive/00000123/).

Di Toro, D. M. (1985). A particle interaction model of reversible organic chemical sorption. *Chemosphere* **14**, 1503–1538.

Dallavalle, J. M. (1948). Micromeritics: The technology of fine particles. Pitman, New York.

Dumestre, A., Sauve, S., McBride, M., Baveye, P., and Berthelin, J. (1999). Copper speciation and microbial activity in long-term contaminated soils. *Arch. Environ. Contam. Toxicol.* **36**(2), 124–131.

Dzombak, D. A., and Morel, F. M. M. (1987). Adsorption of inorganic pollutants in aquatic systems. *J. Hydraul. Engng.-ASCE* **113**(4), 430–475.

Edmunds, W. M., and Bath, A. H. (1976). Centrifuge extraction and chemical analysis of interstitial waters. *Environ. Sci. Technol.* **10**, 467–472.

Elliot, H. A., Liberati, M. R., and Huang, C. P. (1986). Competitive adsorption of heavy metals by soils. *J. Environ. Qual.* **15**, 214–219.

England, C. B. (1974). Comments on "A technique using porous cup for water sampling at any depth in the unsaturated zone" by W. W. Wood. *Water Resour. Res.* **10**, 1049.

Entwisle, D. C., and Reeder, S. (1993). New apparatus for pore fluid extraction from mudrocks for geochemical analysis. *In* "Geochemistry of Clay-Pore Fluid Interaction" (D. A. C. Manning, P. L. Hall, and C. R. Hughes, eds.), pp. 365–388. Chapman & Hall, London.

Evans, L. J. (1989). Chemistry of metal retention by soils. *Environ. Sci. Technol.* **23**, 1046–1056.

Gamerdinger, A. P., and Kaplan, D. I. (2000). Application of a continuous-flow centrifugation method for solute transport in disturbed, unsaturated sediments and illustration of mobile-immobile water. *Water Resour. Res.* **36**(7), 1747–1755.

Garcia-Miragaya, J. (1984). Levels, chemical fractionation, and solubility of lead in roadside soils of Caracas, Venezuela. *Soil Sci.* **138**, 147–152.

Gerritse, R. G., and Vandriel, W. (1984). The relationship between adsorption of trace metals, organic matter, and pH in temperate soils. *J. Environ. Qual.* **13**(2), 197–204.

Gerritse, R. G., Adeney, J. A., Baird, G., and Colquhoun, I. (1992). The reaction of copper ions and hypochlorite with minesite soils in relation to fungicidal activity. *Aust. J. Soil Res.* **30**(5), 723–735.

Giesler, R., Lundstrom, U. S., and Grip, H. (1996). Comparison of soil solution chemistry assessment using zero-tension lysimeters or centrifugation. *Eur. J. Soil Sci.* **47**(3), 395–405.

Gooddy, D. C., Shand, P., Kinnigurgh, D. G., and Van Riemsdijk, W. H. (1995). Field-based partition coefficients for trace elements in soil solutions. *Eur. J. Soil Sci.* **46**, 265–285.

Goyne, K. W., Day, R. L., and Chorover, J. (2000). Artifacts caused by collection of soil solution with passive capillary samplers. *Soil Sci. Soc. Am. J.* **64**(4), 1330–1336.

Green, C. H., Heil, D. M., Cardon, G. E., Butters, G. L., and Kelly, E. F. (2003). Solubilization of manganese and trace metals in soils affected by acid mine runoff. *J. Environ. Qual.* **32**(4), 1323–1334.

Greenland, D. J., and Hayes, M. H. B. (1981). "The Chemistry of Soil Processes." John Wiley & Sons, Chichester.

Grossmann, J., and Udluft, P. (1991). The extraction of soil water by the suction-cup method: A review. *J. Soil Sci.* **42**, 83–93.

Grossmann, J., Bredemeier, M., and Udluft, P. (1990). Sorption of trace metals by suction cups. *Zeitschrift für Pflanzenernahrung und Bodenkunde* **153**, 359–364.

Grover, R., and Hance, R. J. (1970). Effect of ratio of soil to water on adsorption of linuron and atrazine. *Soil Sci.* **100**, 136–138.

Guggenberger, G., and Zech, W. (1992). Sorption of dissolved organic-carbon by ceramic P-80 suction cups. *Zeitschrift Fur Pflanzenernahrung Und Bodenkunde* **155**(2), 151–155.

Haines, B. L., Waide, J. B., and Todd, R. L. (1982). Soil solution nutrient concentrations sampled with tension and zero-tension lysimeters: Report of discrepancies. *Soil Sci. Soc. Am. J.* **46**, 658–661.

Hani, H. (1996). Soil analysis as a tool to predict effects on the environment. *Commun. Soil Sci. Plant Anal.* **27**(3–4), 289–306.

Hansen, E. A., and Harris, A. R. (1975). Validity of soil-water samples collected with porous ceramics cups. *Soil Sci. Soc. Am. J.* **39**, 528–536.

Hart, G. L., and Lowery, B. (1997). Axial-radial influence of porous cup soil solution samplers in a sandy soil. *Soil Sci. Soc. Am. J.* **61**(6), 1765–1773.

Hart, G. L., Lowery, B., McSweeney, K., and Fermanich, K. J. (1994). *In situ* characterization of hydrologic properties of Sparta sand: Relation to solute travel using time domain reflectometry. *Geoderma* **64**, 41–55.

Harter, R. D. (1983). Effect of soil pH on adsorption of lead, copper, zinc, and nickel. *Soil Sci. Soc. Am. J.* **47**(1), 47–51.

Harter, R. D., and Naidu, R. (2001). An assessment of environmental and solution parameter impact on trace-metal sorption by soils. *Soil Sci. Soc. Am. J.* **65**(3), 597–612.

Haworth, A. (1990). A review of the modelling of sorption from aqueous-solution. *Adv. Colloid Interface Sci.* **32**(1), 43–78.

Hesterberg, D. (1998). Biogeochemical cycles and processes leading to changes in mobility of chemical in soils. *Agric. Ecosyst. Environ.* **67**, 121–133.

Hooda, P. S., Zhang, H., Davison, W., and Edwards, A. C. (1999). Measuring bioavailable trace metals by diffusive gradients in thin films (DGT): Soil moisture effects on its performance in soils. *Eur. J. Soil Sci.* **50**(2), 285–294.

Impellitteri, C. A., Allen, H. E., Yin, Y., You, S., and Saxe, J. K. (2001). Soil properties controlling metal partitioning. *In* "Heavy Metals Release in Soils" (H. M. Selim and D. L. Sparks, eds.), pp. 149–165. CRC Press LLC, Lewis Publishers, Boca Raton, Florida.

Impellitteri, C. A., Saxe, J. K., Cochran, M., Janssen, G., and Allen, H. E. (2003). Predicting the bioavailability of copper and zinc in soils: Modeling the partitioning of potentially bioavailable copper and zinc from soil solid to soil solution. *Environ. Toxicol. Chem.* **22**(6), 1380–1386.

IUPAC. (2001). Appendix II. Definitions, Terminology and Symbols in Colloid and Surface Chemistry, part I. 57, International Union of Pure and Applied Chemistry, Commission on Colloid and Surface Chemistry Including Catalysis, Washington, DC; available on internet at the URL: http://www.iupac.org/reports/2001/colloid_2001/manual_of_s_and_t.pdf.

Jenne, E. A. (1968). Controls on Mn, Fe, Co, Ni, Cu, and Zi concentrations in soils and water: The significant role of hydrous Mn and Fe oxides. *Adv. Chem. Phys. Ser.* **73**, 337–387.

Jones, D. L., and Edwards, A. C. (1993a). Effect of moisture content and preparation technique on the composition of soil solution obtained by centrifugation. *Commun. Soil Sci. Plant Anal.* **24**(1–2), 171–186.

Jones, D. L., and Edwards, A. C. (1993b). Evaluation of polysulfone hollow fibres and ceramic suction samplers as devices for the *in situ* extraction of soil solution. *Plant Soil* **150**(2), 157–175.

Jones, B. F., Vandenburgh, A. S., Truesdell, A. H., and Rettig, S. L. (1969). Interstitial brines in playa sediments. *Chem. Geol.* **4**, 253–262.

Jopony, M., and Young, S. D. (1994). The solid–solution equilibria of lead and cadmium in polluted soils. *Eur. J. Soil Sci.* **45**(1), 59–70.

Kashem, M. A., and Singh, B. R. (2001). Metal availability in contaminated soils: I. Effects of flooding and organic matter on changes in Eh, pH and solubility of Cd, Ni and Zn. *Nutr. Cycl. Agroecosyst.* **61**(3), 247–255.

Khasawneh, F. E., and Adams, F. (1967). Effect of dilution on calcium and potassium contents of soil solutions. *Soil Sci. Soc. Am. J.* **31**, 172–176.

Kinniburgh, D. G., and Miles, D. L. (1983). Extraction and chemical analysis of interstitial water from soils and rocks. *Environ. Sci. Technol.* **17**(6), 362–368.

Kinniburgh, D. G., Jackson, M. L., and Syers, J. K. (1976). Adsorption of alkaline earth, transition, and heavy metal cations by hydrous oxide gels of iron and aluminium. *Soil Sci. Soc. Am. J.* **40**, 796–799.

Kinniburgh, D. G., van Riemsdijk, W. H., Koopal, L. K., Borkovec, M., Benedetti, M. F., and Avena, M. J. (1999). Ion binding to natural organic matter: Competition, heterogeneity, stoichiometry and thermodynamic consistency. *Colloids Surf., A—Physicochem. Eng. Aspects* **151**(1–2), 147–166.

Knight, B. P., Chaudri, A. M., McGrath, S. P., and Giller, K. E. (1998). Determination of chemical availability of cadmium and zinc in soils using inert soil moisture samplers. *Environ. Pollut.* **99**, 293–298.

Krahn, J., and Fredlund, D. G. (1972). On total, matric and osmotic suction. *Soil Sci.* **114**(5), 339–348.

Kriukov, P. A. (1971). "Gornye, pochverrye i Ilovye Rastvory (Interstitial waters of soils, rocks and sediments)." Izdatel, stvo Nauka, Moskow.

Kuo, S. (1986). Concurrent sorption of phosphate and zinc, cadmium, or calcium by a hydrous ferric oxide. *Soil Sci. Soc. Am. J.* **50**, 1412–1419.

Lambe, T. W., and Whitman, R. V. (1979). "Soil Mechanics, SI Version." John Wiley & Sons, New York.

Larsen, S., and Widdowson, A. E. (1968). Chemical composition of soil solution. *J. Sci. Food Agric.* **19**, 693–695.

Levin, M. J., and Jackson, D. R. (1977). A comparison of *in situ* extractors for sampling soil water. *Soil Sci. Soc. Am. J.* **41**, 535–536.

Lindroos, A. J., Brugger, T., Derome, J., and Derome, K. (2003). The weathering of mineral soil by natural soil solutions. *Water, Air, Soil Pollut.* **149**(1–4), 269–279.

Lindsay, W. L. (1979). "Chemical Equilibria in Soils." John Wiley, New York.

Litaor, M. I. (1988). Review of soil solution samplers. *Water Resour. Res.* **24**(5), 727–733.

Lorenz, S. E., Hamon, R. E., and McGrath, S. P. (1994). Differences between soil solutions obtained from rhizosphere and non-rhizosphere soils by water displacement and soil centrifugation. *Eur. J. Soil Sci.* **45**, 431–438.

Luo, Y. M., Yan, W. D., and Christie, P. (2001). Soil solution dynamics of Cu and Zn in a Cu- and Zn-polluted soil as influenced by gamma-irradiation and Cu-Zn interaction. *Chemosphere* **42**(2), 179–184.

Luo, Y. M., Qiao, X. L., Song, J., Christie, P., and Wong, M. H. (2003). Use of a multi-layer column device for study on leachability of nitrate in sludge-amended soils. *Chemosphere* **52**(9), 1483–1488.

Luxmoore, R. J. (1981). Micro-, meso- and macroporosity of soil. *Soil Sci. Soc. Am. J.* **45**, 671.

Manheim, F. T. (1966). A hydraulic squeezer for obtaining interstitial waters from consolidated and unconsolidated sediments. pp. 550–556. USGS, Washington.

Massee, R., and Maessen, F. J. M. J. (1981). Losses of silver, arsenic, cadmium, selenium, and zinc traces from distilled water and artificial sea-water by sorption on various container surface. *Anal. Chim. Acta* **127**, 206–210.

Mattigod, S. V., Sposito, G., and Page, A. L. (1981). Factors affecting the solubilities of trace metals in soils. *In* "Chemistry in the Soil Environment" (M. Stelly, ed.), pp. 203–221. American Society of Agronomy - Soil Science Society of America, Madison, WI.

McBride, M. B. (1994). Environmental chemistry of soils. Oxford University Press, Inc., New York.

McBride, M. B., and Blasiak, J. J. (1979). Zinc and copper solubility as a function of pH in an acid soil. *Soil Sci. Soc. Am. J.* **43**, 866–870.

McBride, M. B., Sauve, S., and Hendershot, W. (1997). Solubility control of Cu, Zn, Cd and Pb in contaminated soils. *Eur. J. Soil Sci.* **48**(2), 337–346.

McGuire, P. E., Lowery, B., and Helmke, P. A. (1992). Potential sampling error—Trace metal adsorption on vacuum porous cup samplers. *Soil Sci. Soc. Am. J.* **56**(1), 74–82.

Menendez, I., Gallardo, J. F., and Vicente, M. A. (2003). Functional and chemical calibrates of ceramic cup water samplers in forest soils. *Commun. Soil Sci. Plant Anal.* **34**(7–8), 1153–1175.

Menzies, N. W., and Bell, L. C. (1988). Evaluation of the influence of sample preparation and extraction technique on soil solution composition. *Aust. J. Soil Res.* **26**(3), 451–464.

Morgenstern, N. R., and Balasubramonian, B. I. (1980). Effects of pore fluid on the swelling of clay shale. *In* "4th International Conference on Expansive Soils," pp. 190–205. Denver.

Morrison, R., and Szecsody, J. (1985). Sleeve and casing lysimeters for soil pore water sampling. *Soil Sci.* **139**(5), 446–451.

Mubarak, A., and Olsen, R. A. (1976). Immiscible displacement of soil solution by centrifugation. *Soil Sci. Soc. Am. J.* **40**(2), 329–331.

Nagpal, N. K. (1982). Comparison among, and evaluation of, ceramic porous cup soil water samplers for nutrient transport studies. *Can. J. Soil Sci.* **62**, 685–694.

OECD (2002). OECD Guidelines for the Testing of Chemicals Revised, Proposal for a New Guideline 312, Leaching in Soil Columns.

O'Conner, D. J., and Connolly, J. P. (1980). The effect of concentration of adsorbing solids on the partition coefficient. *Water Res.* **14**, 1517–1523.

Oscarson, D. W., and Hume, H. B. (1998). Effect of solid:liquid ratio on the sorption of Sr^{2+} and Cs^+ on bentonite. *In* "Adsorption of Metals by Geomedia" (E. A. Jenne, ed.), pp. 277–289. Academic Press, London.

Padilla, I. Y., Yeh, T. C. J., and Conklin, M. H. (1999). The effect of water content on solute transport in unsaturated porous media. *Water Resour. Res.* **35**(11), 3303–3313.

Percival, H. J. (2003). Soil and soil solution chemistry of a New Zealand pasture soil amended with heavy metal-containing sewage sludge. *Aust. J. Soil Res.* **41**(1), 1–17.

Pérez, D. V., de Campos, R. C., and Novaes, H. B. (2002). Soil solution charge balance for defining the speed and time of centrifugation of two Brazilian soils. *Commun. Soil Sci. Plant Anal.* **33**(13–14), 2021–2036.

Phillip, J. (1968). Steady infiltration from buried point sources and spherical cavities. *Water Resour. Res.* **4**(5), 1039–1047.

Prokop, Z., Cupr, P., Zlevorova-Zlamalikova, V., Komarek, J., Dusek, L., and Holoubek, I. (2003). Mobility, bioavailability, and toxic effects of cadmium in soil samples. *Environ. Res.* **91**(2), 119–126.

Qafoku, N. P., Ainsworth, C. C., Szecsody, J. E., Qafoku, O. S., and Heald, S. M. (2003). Effect of coupled dissolution and redox reactions on Cr(VI)(aq) attenuation during transport in the sediments under hyperalkaline conditions. *Environ. Sci. Technol.* **37**(16), 3640–3646.

Reeder, S., Cave, M. R., Entwisle, D. C., and Trick, J. K. (1998). "Extraction of Water and Solutes from Clayey Material: A Review and Critical Discussion of Available Techniques. *WI/98/4C*, British Geological Survey." Keyworth, Nottingham.

Reitemeier, R. F. (1946). Effects of moisture content on the dissolved and exchangeable ions of soils of arid regions. *Soil Sci.* **61**, 195–214.

Richards, L. A., and Weaver, L. R. (1944). Moisture retention by some irrigated soils as related to soil moisture tension. *J. Agric. Res.* **69**, 215–235.

Robb, F., and Young, S. D. (1999). Addition of calcareous metal wastes to acid soils: Consequences for metal solubility. *Water, Air, Soil Pollut.* **111**(1–4), 201–214.

Ross, C. A. M., Bath, A. H., Entwisle, D. C., Cave, M. R., Fry, M. B., Green, K. A., and Reeder, S. (1989). "Hydrochemistry of Pore Waters from Jurassic Oxford Clay, Kellaways Beds, Upper Estuarine and Upper Lias Formations at the Elstow Site, Bedfordshire. *WE/89/28*, British Geological Survey." Keyworth, Nottingham.

Sauvé, S., Dumestre, A., McBride, M., and Hendershot, W. (1998). Derivation of soil quality criteria using predicted chemical speciation of Pb^{2+} and Cu^{2+}. *Environ. Toxicol. Chem.* **17**(8), 1481–1489.

Sauvé, S., Hendershot, W., and Allen, H. E. (2000). Solid-solution partitioning of metals in contaminated soils: Dependence on pH, total metal burden, and organic matter. *Environ. Sci. Technol.* **34**(7), 1125–1131.

Schubert, H. (1982). "Kapillarität in porösen feststoffsystemen." Springer-Verlag, Berlin.

Selim, H. M., and Sparks, D. L. (2001). Heavy metals release in soils. CRC Press LLC, Lewis Publishers, Boca Raton, Florida.

Shaffer, R. J., Wallace, J., and Garwood, F. (1937). The centrifuge method of investigating the variation of hydrostatic pressure with water content in porous material. *Trans. Faraday Soc.* **33**, 723–734.

Shan, X. Q., Wang, Z. W., Wang, W. S., Zhang, S. Z., and Wen, B. (2003). Labile rhizosphere soil solution fraction for prediction of bioavailability of heavy metals and rare earth elements to plants. *Anal. Bioanal. Chem.* **375**(5), 400–407.

Shaw, E. M. (1993). "Hydrology in Practice." Chapman & Hall, London.

Shendrikar, A. D., Dharmarajan, V., Walker-Merrick, H., and West, P. W. (1975). Adsorption characteristics of traces of barium, beryllium, cadmium, manganese, lead, and zinc on selected surfaces. *Anal. Chim. Acta* **84**, 409–417.

Sheppard, M. I., Thibault, D. H., and Smith, P. A. (1992). Effect of extraction techniques on soil pore-water chemistry. *Commun. Soil Sci. Plant Anal.* **23**(13–14), 1643–1662.

Sholkovits, E. (1973). Interstitial water chemistry of the Santa Barbara basin sediments. *Geochim. Cosmochim. Acta* **37**, 2043–2078.

Siemens, J., and Kaupenjohann, M. (2003). Dissolved organic carbon is released from sealings and glues of pore-water samplers. *Soil Sci. Soc. Am. J.* **67**(3), 795–797.

Sills, I. D., Aylmore, L. A. G., and Quirk, J. P. (1974). Relationship between pore size distribution and physical properties of a clay soil. *Aust. J. Soil Res.* **12**, 107–117.

Sparks, D. L. (1995). Environmental soil chemistry. Academic Press, Inc., San Diego, CA.

Sposito, G. (1989). "The Chemistry of Soils." Oxford University Press, New York.

Strandberg, J., and Fortkamp, U. (2005). Investigations on methods for site specific, determination of the partition coefficient—K_d, for contaminants in soils. IVL report B1619, available online at: http://www.ivl.se/rapporter/pdf/B1619.pdf.

Tack, F. M. G., and Verloo, M. G. (1995). Chemical speciation and fractionation in soil and sediment heavy metal analysis: A review. *Int. J. Environ. Anal. Chem.* **59**(2–4), 225.

Tiensing, T., Preston, S., Strachan, N., and Paton, G. I. (2001). Soil solution extraction techniques for microbial ecotoxicity testing: A comparative evaluation. *J. Environ. Monit.* **3**(1), 91–96.

Tiller, K. G., Hogson, J. F., and Peech, M. (1963). Specific sorption of cobalt by soil clays. *Soil Sci.* **15**, 392–399.

Tye, A. M., Young, S. D., Crout, N. M. J., Zhang, H., Preston, S., Bailey, E. H., Davison, W., McGrath, S. P., Paton, G. I., and Kilham, K. (2002). Predicting arsenic solubility in contaminated soils using isotopic dilution techniques. *Environ. Sci. Technol.* **36**(5), 982–988.

Tye, A. M., Young, S. D., Crout, N. M. J., Zhang, H., Preston, S., Barbosa-Jefferson, V. L., Davison, W., McGrath, S. P., Paton, G. I., Kilham, K., and Resende, L. (2003). Predicting the activity of Cd^{2+} and Zn^{2+} in soil pore water from the radio-labile metal fraction. *Geochim. Cosmochim. Acta* **67**(3), 375–385.

Tyler, G. (2000). Effects of sample pretreatment and sequential fractionation by centrifuge drainage on concentrations of minerals in a calcareous soil solution. *Geoderma* **94**, 59–70.

USEPA, Office of Air and Radiation (1999). Understanding variation in partition coefficient, K_d values—Volume 1. The K_d Model, Methods of Measurements, and Applications of Chemical Reactions Codes. EPA 402-R-99–004A.

Van Benschoten, J. E., Young, W. H., Matsumoto, M. R., and Reed, B. E. (1998). A nonelectrostatic surface complexation model for lead sorption on soils and mineral surfaces. *J. Environ. Qual.* **27**, 24–30.

Vig, K., Megharaj, M., Sethunathan, N., and Naidu, R. (2003). Bioavailability and toxicity of cadmium to microorganisms and their activities in soil: A review. *Adv. Environ. Res.* **8**(1), 121–135.

Voice, T. C., and Weber, J. W. J. (1985). Sorbent concentration effects in liquid/solid partitioning. *Environ. Sci. Technol.* **19**, 789–796.

Voice, T. C., Rice, C. P., and Weber, J. W. J. (1983). Effects of solids concentration on the sorptive partitioning of hydrophobic pollutants in aquatic systems. *Environ. Sci. Technol.* **17**, 513–518.

Warren, L. A., and Haack, E. A. (2001). Biogeochemical controls on metal behaviour in freshwater environments. *Earth-Sci. Rev.* **54**(4), 261–320.

Warrick, A. W., and Amoozegar-Fard, A. (1977). Soil-water regimes near porous cup water samplers. *Water Resour. Res.* **13**(1), 203–207.

Washburn, E. W. (1921). The dynamics of capillary flow. *Am. Phys. Soc. 2nd series* **17**, 374–375.

Weng, L. P., Temminghoff, E. J. M., Lofts, S., Tipping, E., and Van Riemdijk, W. H. (2002). Complexation with dissolved organic matter and solubility control of heavy metals in a sandy soil. *Environ. Sci. Technol.* **36**(22), 4804–4810.

Wenzel, W. W., and Wieshammer, G. (1995). Suction cup materials and their potential to bias trace-metal analyses of soil solutions—A review. *Int. J. Environ. Anal. Chem.* **59**(2–4), 277–290.

Wenzel, W. W., Sletten, R. S., Brandstetter, A., Wieshammer, G., and Stingeder, G. (1997). Adsorption of trace metals by tension lysimeters: Nylon membrane vs. porous ceramic cup. *J. Environ. Qual.* **26**(5), 1430–1434.

Wheatstone, K. C., and Gelsthorpe, D. (1982). Extraction and analysis of interstitial water from sandstone. *Analyst* **107**, 731–736.

Whelan, B. R., and Barrow, N. J. (1980). A study of a method for displacing soil solution by centrifuging with an immiscible liquid. *J. Environ. Qual.* **9**(2), 315–319.

Wiersum, L. K. (1957). The relationship of the size and structural rigidity of pores to their penetration by roots. *Plant Soil* **9**, 75–85.

Wolt, J. (1994). "Soil Solution Chemistry: Applications to Environmental Science and Agriculture." John Wiley & Sons, New York.

Yaron, B., Calvet, R., and Prost, R. (1996). "Soil Pollution—Processes and Dynamics." Springer-Verlag, Berlin.

Yin, Y. J., Impellitteri, C. A., You, S. J., and Allen, H. E. (2002). The importance of organic matter distribution and extract soil: Solution ratio on the desorption of heavy metals from soils. *Sci. Total Environ.* **287**(1–2), 107–119.

Zabowsky, D., and Ugolini, F. C. (1990). Lysimeter and centrifuge soil solutions: Seasonal differences between methods. *Soil Sci. Soc. Am. J.* **54**, 1130–1135.

Ziper, C., Komarneni, S., and Baker, D. E. (1988). Specific cadmium sorption in relation to the crystal of clay minerals. *Soil Sci. Soc. Am. J.* **52**, 49–53.

CHAPTER ELEVEN

SEWAGE SLUDGE IN EUROPE AND IN THE UK: ENVIRONMENTAL IMPACT AND IMPROVED STANDARDS FOR RECYCLING AND RECOVERY TO LAND

Marcello Di Bonito*

Contents

Abstract

Sewage treatment and sewage sludge (also called biosolids) represent a serious environmental issue that has affected modern society for the past century. In response to this, an increasing number of controls and resulting regulations have been introduced to avoid polluting our rivers and seas with pathogens, oxygen-demanding organic debris, potentially toxic elements (PTEs) and eutrophying nutrients. Over the past few decades, improved technology and more stringent regulations, driven in Europe by the increasingly precautionary European Community (EC) legislation, have worked together to achieve a net decrease of the amount of some of the pollutants, both through treatment of the wastewater and by cutting or forbidding the intake of contaminants at source. The growing importance of cleaning up wastewater before returning it to natural waters has

* Environment Agency, Trentside Offices, West Bridgford, Nottingham NG2 5FA, UK

Environmental Geochemistry
DOI: 10.1016/B978-0-444-53159-9.00011-5

© 2008 Elsevier B.V.
All rights reserved.

led to (1) a vast increase of the quantity and the quality of the sludge resulting from the treatment of wastewater, with a greater amount of pollutants removed during the treatment process; (2) a greater effort towards the reuse and recycling of the sludge as opposed to disposal (e.g. landfilling, incineration). One of the consequences of this situation is a rising interest in using soils (agricultural or not) to address the latter through various applications. In fact, as the sludge has to be disposed off safely, soils can be used as a system of assimilating, recycling or disposing off the sewage sludge. Science and legislation are trying to provide the safest possible route to accomplish these targets, but the subject is not free from controversy, often causing hot debates between the interested parties. This chapter aims to review the improved standards achieved with sewage sludge, touching in particular the British experience in the field of regulating the disposal and reuse of these materials.

1. WASTEWATER AND SLUDGE: DEFINITIONS AND TREATMENT

The terminology that refers to wastewater is often cause of confusion and ambiguity. The term 'sewage' refers to the wastewater produced by residential and commercial establishments and discharged into sewers, specifically the liquid waste from toilets, baths, showers, kitchens, etc., that is disposed off via sewers. In many areas, sewage also includes some liquid waste from industry and commerce. However, it should be noted that specific industrial sources of wastewater often require specialised treatment processes (e.g. pulp and papermill, chemical or petroleum manufacturer wastewaters). In the UK, the waste from toilets is called 'foul waste', the waste from basins, baths, kitchens is termed 'sullage water' and the industrial and commercial waste is termed 'trade waste'. Furthermore, the division of household water drains into greywater and blackwater is becoming more common in the developed world, with greywater being permitted to be used for watering plants or recycled for flushing toilets. Much sewage also includes some surface water from roofs or hard-standing areas. Municipal wastewater therefore includes residential, commercial and industrial liquid waste discharges, and may include stormwater run-off.

Municipal or urban wastewaters (UWW) are treated at wastewater treatment plants (WWTPs) principally removing the impurities and the contaminants so that the remaining wastewater can be safely returned to the river or sea and become part of the natural water cycle again.

A WWTP separates solids from liquids by physical processes and purifies the liquid by biological processes. Processes may vary but the following waste stream is typical. In the preliminary treatment, solids like wood, paper, rags and plastic are removed by screens, washed, dried and taken away for safe disposal at a licensed waste tip. Grit and sand, which would damage pumps, are also removed and disposed off in a similar way. In the primary treatment, the remaining solids are separated from the liquid by passing the sewage through large settlement tanks, where most of the solid material sinks to the bottom. About 70% of solids settle down at this stage and are referred to as sewage sludge or 'biosolids'. The remaining liquid portion undergoes the secondary treatment, a biological process which relies on naturally occurring micro-organisms acting to break down organic material and purify the liquid. In a simple sewage treatment process, micro-organisms are encouraged to grow on stones over which the sewage is trickled. The micro-organisms, which need oxygen to thrive, feed on the bacteria in the sewage and purify the water. These treatment units are called percolating filters. This process can be speeded up by blowing air into tanks of sewage where the micro-organisms float freely and feed on the bacteria. These treatment units are called aeration tanks. Following either form of secondary treatment, the wastewater is settled in tanks to separate the biological sludge from the purified wastewater. Sometimes, extra treatment is needed to give the wastewater a final 'polish'. This is known as tertiary treatment. Various methods may be used, including sand filters, reed beds or grass plots. Disinfection, using ultraviolet light to kill bacteria, is another method often used at a number of coastal sewage treatment schemes.

1.1. Wastewater treatment in Europe

The figures for water supply (litres/inhabitant/day), consumption and treated wastewater very much vary across Europe. The water supply fluctuates greatly across the European urban areas (EC, 2001b; Ginés, 1997). Not all collected wastewater is treated; the percentage of urban wastewater not receiving treatment ranges from 3% in Germany up to 77% in Greece. This may include unplanned wastewater collection, such as septic systems or leakage of urban wastewater collection systems. Cities such as Milan and Brussels do not yet have a centralised WWTP. According to the European Waste Water Group (European Waste Water Group (EWWG), 1997) report on urban wastewater treatment in the EU and accession countries, there are marked differences between various European regions in terms of primary, secondary and tertiary treatment (Table 11.1). In the Western Region (UK, Ireland, France and the Benelux countries, Belgium, the Netherlands and Luxembourg), wastewater treatment is among the most advanced in the EU. In the UK, there are over 300,000 km sewers and 7600 WWTPs. The collection of sewage, its treatment and disposal of effluent and sludge are the responsibility of privately financed water service companies in England and Wales, public water and sewerage authorities in Scotland and the Water Service of the Department of Environment in Northern Ireland (European Waste Water Group (EWWG), 1997). In the Republic of Ireland, the urban wastewater collection and treatment is the responsibility of local councils. Private–public partnerships (PPP) schemes are in place for modernising and upgrading the WWTPs (Dept. of Environment and Local Govt., Ireland, 1999).

Sewage treatment and sludge characteristics have evolved over the past century in response to society's need and the resulting regulations in order to avoid polluting our rivers and seas with pathogens, oxygen-demanding organic debris, potentially toxic elements (PTEs) and eutrophying nutrients. The growing importance of cleaning up wastewater before returning it to natural waters has led to (1) a vast increase of the amount of material 'removed' from the wastewater during the treatment process and (2) a greater effort towards an advanced treatment by removing or reducing nutrients (mainly phosphorus, but also nitrogen) from the effluent (Brady and Weil, 1999). This increase is mainly due to the practical implementation

Table 11.1 Population and household access to sewerage and wastewater treatment facilities (modified from EC, 2001a)

Country	Population (1998) thousands	Percentage of population with access to sewerage and public wastewater treatment (WWT) facilities					
		Access to sewerage (%)	No treatment (%)	P (%)	P + S (%)	P + S + T (%)	Year
Austria	8075	76	1	1	39	35	1995
Belgium	10,192	78				37		1997
Denmark	5295	87	0				18	1996
Finland	5147	78	0	7		71	1997
France	58,727	81	4	0	0	77	1994
Germany	82,057	92			>31	>47	1995
Greece	10,511	70	16	0	0	54	1996
Ireland	3694	68	32	23	12	1	1995
Italy	57,563	84	16	3	38	26	1996
Luxembourg	424	88	0	19	57	11	1995
The Netherlands	15,654	98	2	0	68	28	1994
Portugal	9957	55	34	9	11	0	1990
Spain	39,348	62	13	11	34	3	1995
Sweden	8848	86	0	0	5	81	1995
UK	59,090	96	9	9	64	14	1996
England & Wales	—	96	10	0	0	86	1995
N. Ireland	—	83					1996
Scotland	—	94					1996
Total EU	374,582	84				48	

P = primary, S = secondary and T = tertiary treatment.

of a number of measures that have been attempted at various levels, with varying success, to reduce UWW pollution closer to source. These instruments are consistent with an overall strategy of waste minimisation, polluter pays and reduction at source, and include individual regulatory, economic and voluntary and educational instruments and mainly stem from European Directives. The Directive sets more and more stringent provisions for agglomerations discharging into sensitive areas such as fresh waters or estuaries and resulted in a slow but constant rise in the number of households connected to sewers and the increase in the level of treatment (up to tertiary treatment with removal of nutrients in some Member States). All municipal WWTPs produce sewage sludge which invariably needs to be reused or disposed off appropriately. As a consequence, there is a rising interest in using soils to address these two issues as soils have proved to be used as a system of assimilating, recycling or disposing off the sewage sludge. In addition, the interaction of the effluent with a soil–plant system can be used as a means of carrying out the final removal of nutrients and organics from the liquid effluent.

2. WASTEWATER AND SLUDGE COMPOSITION

At the beginning of the treatment, municipal wastewater is mainly comprised of water (99.9%) together with relatively small concentrations of suspended and dissolved organic and inorganic solids. Among the organic substances present in sewage are carbohydrates, lignin, fats, soaps, synthetic detergents, proteins and their decomposition products, as well as various natural and synthetic organic chemicals from the process industries. Table 11.2 shows the levels of the major constituents of

Table 11.2 Major constituents of typical domestic wastewater (modified from FAO, 1992 and references therein)

	Concentration mg L^{-1}		
Constituent	Strong	Medium	Weak
Total solids	1200	700	350
Dissolved solids (TDS)[a]	850	500	250
Suspended solids	350	200	100
Nitrogen (as N)	85	40	20
Phosphorus (as P)	20	10	6
Chloride[a]	100	50	30
Alkalinity (as $CaCO_3$)	200	100	50
Grease	150	100	50
BOD_5[b]	300	200	100

[a] The amounts of TDS and chloride should be increased by the concentrations of these constituents in the carriage water.
[b] BOD_5 is the biochemical oxygen demand at 20 °C over 5 days and is a measure of the biodegradable organic matter in the wastewater.

strong, medium and weak domestic wastewaters. In arid and semi-arid countries, water use is often fairly low and sewage tends to be very strong.

Pathogenic viruses, bacteria, protozoa and helminths may be present in raw municipal wastewater at the levels indicated in Table 11.3 and will survive in the environment for long periods. Pathogenic bacteria will be present in wastewater at much lower levels than that of the coliform group of bacteria, which are much easier to identify and enumerate (as total coliforms/100 ml). *Escherichia coli* are the most widely adopted indicator of faecal pollution and they can also be isolated and identified fairly simply, with their numbers usually being given in the form of faecal coliforms (FC)/100 ml of wastewater.

Municipal wastewater also contains a variety of organic and inorganic substances (Table 11.4) from domestic and industrial sources, including a number of PTEs such as arsenic (As), cadmium (Cd), chromium (Cr), copper (Cu), lead (Pb), mercury (Hg), zinc (Zn), etc., which constitute a major issue in terms of abatement and/or treatment (for a review of organic chemicals in sewage sludge, see EC, 2001a; Harrison *et al.*, 2006).

All the potential contaminants are dealt with during the various phases of the wastewater treatment at WWTPs, with the primary goal of protecting the receiving waters from pollution by harmful effluent. UWW treatment, however, is not sufficient to deal with all the pollution. Although different contaminants end up to varying degrees in effluent from WWTPs, most of the pollutants are concentrated in

Table 11.3 Possible levels of pathogens in wastewater (modified from Feachem *et al.*, 1983)

Type of pathogen	Possible concentration per litre in municipal wastewater[a]
Viruses	
Enteroviruses[b]	5000
Bacteria	
Pathogenic *E. coli*[c]	Uncertain
Salmonella spp.	7000
Shigella spp.	7000
Vibrio cholerae	1000
Protozoa	
Entamoeba histolytica	4500
Helminths	
Ascaris lumbricoides	600
Hookworms[d]	32
Schistosoma mansoni	1
Taenia saginata	10
Trichuris trichiura	120

[a] Based on 100 L per capita per day (lpcd) of municipal sewage and 90% inactivation of excreted pathogens.
[b] Includes polio-, echo- and coxsackie viruses.
[c] Includes enterotoxigenic, enteroinvasive and enteropathogenic *E. coli*.
[d] *Anglostoma duedenale* and *Necator americanus*.

Table 11.4 Organic and inorganic constituents of drinking water of health significance (modified from World Health Organization, 1984)

Organic	Inorganic
Aldrin and dieldrin	Arsenic
Benzene	Cadmium
Benzo-*a*-pyrene	Chromium
Carbon tetrachloride	Cobalt
Chlordane	Copper
Chloroform	Cyanide
2,4 D	Fluoride
DDT	Lead
1,2 Dichloroethane	Mercury
1,1 Dichlorethylene	Nickel
Heptachlor and heptachlor epoxide	Nitrate
Hexachlorobenzene	Selenium
Lindane	Zinc
Methoxychlor	
Pentachlorophenol	
Tetrachlorethylene	
2, 4, 6 Trichloroethylene	
Trichlorophenol	

sewage sludge produced at the end of the primary treatment. Sewage sludge can then be contaminated with PTEs (Koch and Rotard, 2001; Koch *et al.*, 2001, Sörme and Lagerkvist, 2002), bacteria and viruses and a number of organic substances.[1] Concentrating pollutants in sewage sludge, while it may protect receiving waters, presents a number of problems for its disposal or use on agricultural land. Therefore, sludge needs additional treatment to make it suitable for reusing. One of the standard treatments occurs through a digestion and takes place in large, enclosed tanks in anaerobic conditions. The digestion process is speeded up by heating the sludge to a temperature where naturally occurring bacteria respond to these comfortable conditions and feed on other bacteria. On cooling, the well-fed bacteria die off, and the sludge is suitable for use on agricultural land. A by-product of the sludge digestion process is methane gas, which can be burned to produce electricity. The electricity can be used to heat more sludge or to provide heat and light for the treatment works. Further treatments are available to reduce pollutants from sewage sludge, particularly PTE, due to their persistence and connected health risk during land application. For a review of the various methodologies available for PTE removal from sewage sludge, see Babel and del Mundo Dacera (2006).

---

[1] http://www.fwr.org/defrawqd/wqd0001.htm.

2.1. Potential hazards of chemicals in sewage sludge

From the point of view of health, which is a very important consideration in agricultural use of sewage sludge, some of the contaminants of greatest concern relate to the pathogenic micro- and macro-organisms. However, despite continuing to be a risk, with the recent improvement in sewage treatment and regulation, this aspect represents only a relatively minor issue where tertiary and sludge treatment are undertaken properly.

The principal health hazards is instead associated with the chemical constituents of the initial wastewaters which can therefore produce contamination of crops or groundwaters. Cu, Cr, Zn and Se are essential trace elements; however, they are PTEs, and above certain concentrations may interfere with or inhibit the actions of cellular enzymes. After the treatment, even if toxic materials are not present in the sludge in concentrations likely to affect humans, they might well be at phytotoxic levels, which would limit their agricultural use (Table 11.5). Furthermore, Hillman (1988) has drawn attention to the particular concern attached to the cumulative poisons, principally PTEs, and carcinogens, mainly organic chemicals. World Health

Table 11.5 Threshold levels of potentially toxic elements (PTEs) in Water for crop production (modified from Pescod, 1992 and references therein)

Element	Recommended maximum concentration[a] (mg L^{-1})	Remarks
Al	5.0	Can cause non-productivity in acid soils (pH < 5.5), but more alkaline soils at pH > 7.0 will precipitate the ion and eliminate any toxicity
As	0.10	Toxicity to plants varies widely, ranging from 12 mg L^{-1} for Sudan grass to less than 0.05 mg L^{-1} for rice
Be	0.10	Toxicity to plants varies widely, ranging from 5 mg L^{-1} for kale to 0.5 mg L^{-1} for bush beans
Cd	0.01	Toxic to beans, beets and turnips at concentrations as low as 0.1 mg L^{-1} in nutrient solutions. Conservative limits recommended due to its potential for accumulation in plants and soils to concentrations that may be harmful to humans
Co	0.05	Toxic to tomato plants at 0.1 mg L^{-1} in nutrient solution. Tends to be inactivated by neutral and alkaline soils
Cr	0.10	Not generally recognised as an essential growth element. Conservative limits recommended due to lack of knowledge on its toxicity to plants
Cu	0.20	Toxic to a number of plants at 0.1–1.0 mg L^{-1} in nutrient solutions
F	1.0	Inactivated by neutral and alkaline soils

Table 11.5 *(Continued)*

Element	Recommended maximum concentration[a] (mg L^{-1})	Remarks
Fe	5.0	Not toxic to plants in aerated soils, but can contribute to soil acidification and loss of availability of essential phosphorus and molybdenum. Overhead sprinkling may result in unsightly deposits on plants, equipment and buildings
Li	2.5	Tolerated by most crops up to 5 mg L^{-1}; mobile in soil. Toxic to citrus at low concentrations (<0.075 mg L^{-1}). Acts similarly to boron
Mn	0.20	Toxic to a number of crops at a few tenths to a few mg L^{-1}, but usually only in acid soils
Mo	0.01	Not toxic to plants at normal concentrations in soil and water. Can be toxic to livestock if forage is grown in soils with high concentrations of available molybdenum
Ni	0.20	Toxic to a number of plants at 0.5–1.0 mg L^{-1}; reduced toxicity at neutral or alkaline pH
Pd	5.0	Can inhibit plant cell growth at very high concentrations
Se	0.02	Toxic to plants at concentrations as low as 0.025 mg L^{-1} and toxic to livestock if forage is grown in soils with relatively high levels of added selenium. As essential element to animals but in very low concentrations
Ti	—	Effectively excluded by plants; specific tolerance unknown
V	0.10	Toxic to many plants at relatively low concentrations
Zn	2.0	Toxic to many plants at widely varying concentrations; reduced toxicity at pH > 6.0 and in fine-textured or organic soils

[a] The maximum concentration is based on a water application rate which is consistent with good irrigation practices (10,000 m^3 ha^{-1} year^{-1}). If the water application rate greatly exceeds this, the maximum concentrations should be adjusted downward accordingly. No adjustment should be made for application rates less than 10,000 m^3 ha^{-1} year^{-1}. The values given are for water used on a continuous basis at one site.

Organization guidelines for drinking water quality (World Health Organization, 1984) include limit values for the organic and toxic substances (see Table 11.6), based on acceptable daily intakes (ADI). These can be adopted directly for groundwater protection purposes, but in view of the possible accumulation of certain toxic elements in plants (e.g. cadmium and selenium), the intake of toxic materials through eating the crops irrigated with contaminated sludge or wastewater must be carefully assessed. In fact, a significant source of metal contamination of crops in the UK arises from the disposal of sewage sludge on arable land (Alloway, 1995; Reilly, 1991). Here, the availability of agricultural land in the vicinity of WWTPs has been the primary factor determining disposal routes, resulting in a historical legacy of heavily contaminated land (Gendebien *et al.*, 1999). Furthermore, in view of the very long residence periods of PTEs in most soils, it is important to be aware of the existing legacy from metals in sludges applied to land in the past.

Effects of PTE are summarised in Table 11.6, together with the maximum concentration limits (MCL) in drinking water (*source*: EPA and WHO). PTE at high concentrations are acutely toxic to humans. High concentrations are rare in urban wastewater, but could possibly result from accidental spills, although there is limited exposure from this route. The major concern is exposure to low concentrations over longer time periods. This is chronic exposure and may have more subtle effects.

The chemical form and corresponding bioavailability of PTEs to plants, fungi, micro-organisms and animals are also important, therefore affecting the regulation regarding the specific compound. For example, pH and redox conditions can influence the form in which chemicals are in the environment, and therefore their capacity to bind to other compounds or to be bioavailable (for plants and animals). As a consequence, these parameters need to be considered when deciding adequate ranges of acceptability. Furthermore, complex dietary interactions are inherently protective limiting the accumulation of metals in animal body tissues. The phytotoxic concentrations of Zn and Cu in plant leaves less than the zootoxic concentration of these elements in plant produce consumed by animals and humans. Technically based, precautionary soil limit values for these elements can protect crop yields and the human diet from the potentially toxic effects of metals in sewage sludge, which is why both the EU and national regulations set limits for contaminant concentrations to protect the soil and humans from pollution.

3. THE LEGISLATIVE DEBATE AND REGULATIVE TOOLS IN EUROPE AND IN THE UK

3.1. Regulatory instruments in Europe

The existing legislation related to sludge treatment, disposal and recycling focuses on the specific legal requirements necessary principally for its application in agriculture. For the moment, other uses or disposal routes for sludge generally fall under more general laws on waste and water management. Although several Directives have an influence on sludge management (such as Directive 1999/31/EEC on the landfill of waste), those having the strongest impact on sludge production, disposal and recycling are Directives 91/271/EEC concerning urban wastewater treatment (EC, 1998) and 86/278/EEC on the use of sludge in agriculture. In particular, the Sewage Sludge Directive 86/278/EEC (EC, 1986), which seeks to encourage the use of sewage sludge in agriculture and to regulate its use in such a way as to prevent

Table 11.6 Summary of the acute and chronic effects of potentially toxic elements (PTEs) (modified after EC, 2001b)

| PTE | Drinking water standards (mg L^{-1}) | | | Acute health effect | Chronic health effects | Carcinogenicity | Notes |
	EU	WHO	USEPA				
Cu	0.1–3	2	1	Irritation of mouth and throat, headaches, dizziness, nausea, diarrhoea, gastric ulcers, jaundice, renal damage, death	Liver and kidney damage, 'pink disease', cirrhosis	No evidence	
Zn	0.1–5		5	Stomach and digestion problems, dehydration, impaired muscular coordination	Immune system damage, interferences with the body's ability to take in and use other essential elements such as copper and iron	No evidence	Taken up by plants, phytotoxic
Cd	0.005	0.003	0.005	Digestive tract irritation, colitis, vomiting, diarrhoea, death	Half-life = 10–40 years. Lung, kidney and hematopoietic system damage	Strong evidence in animals, weak evidence in humans	Taken up by plants, phytotoxic

(continued)

Table 11.6 (*Continued*)

PTE	Drinking water standards (mg L^{-1})			Acute health effect	Chronic health effects	Carcinogenicity	Notes
	EU	WHO	USEPA				
					due to build up, fragile bones, anaemia, nerve or brain damage in animals		
Cr VI	0.05	0.05	0.1	Allergic responses in skin, Cr VI irritates nose, lungs, stomach and intestines, convulsion, death	Damage nose and lungs, increases risks of non-cancer lung diseases, ulcers, kidneys and liver damage. Birth defects and reproductive problems in mice	Evidence in animals and humans	High potential for bioaccumulation in aquatic organisms
Hg	0.001	0.001	0.002	Nausea, vomiting, diarrhoea, increase in blood pressure, skin rashes, eye irritation, renal failure	Brain, lung, kidney and damage to developing foetus, neurological disorders, depression,	Evidence in mice	Binds to dissolved matter not phytotoxic. High potential for bioaccumulation in aquatic organisms

Ni	0.05	0.02	0.04	Allergic reactions, lung damage	vertigo and tremors Chronic bronchitis and reduced lung function, lung disease. Affects blood, liver, kidney, immune system, reproduction and development in mice and rats	Evidence in lung and nasal sinus cancers in humans	Phytotoxic, very mobile in water
Pb	0.05	0.01	0.015	Anaemia, constipation, colic, wrist and foot drop, renal damage. In children, symptoms are irritability, loss of appetite, vomiting and constipation	Non-specific: damage to nervous system, kidneys and immune system. In children, can cause decrease mental ability and reduced growth. In adults, can cause a decrease in reaction time and affect memory, miscarriage, premature births and damage to male reproductive system	Evidence in animals	Not usually phytotoxic, bioconcentration in shellfish

(continued)

Table 11.6 (Continued)

PTE	Drinking water standards (mg L^{-1})			Acute health effect	Chronic health effects	Carcinogenicity	Notes
	EU	WHO	USEPA				
As	0.05	0.01	0.05	Nausea, vomiting, diarrhoea, damage to tissue including nerves, stomach, intestines and skin	Skin keratosis, decrease in the production of blood cells, bone marrow suppression, abnormal heart function, liver/kidney damage, impaired nerve function, damage to foetus in animals	Evidence in humans, of increased risk in liver, bladder, kidney and lung cancer, may inhibit some DNA repair mechanisms	Taken up by plants, which are sensitive to lower concentrations than are animals, bioaccumulation in fish and shellfish, persistent in the environment
Ag	0.01		0.05	Breathing problems, lung and throat irritations, and stomach pains,	Argyria and may affect brain and kidneys	No evidence	

			allergic reactions, necrosis, haemorrhage and pulmonary oedema		
Se	0.01	0.01	Dizziness, fatigue, irritation, collection of fluid in lungs, bronchitis, rashes, swelling	Brittle hair, deformed nails, and loss of feeling and control in arms and legs, reproductive effects in rats and monkeys, and birth defects in birds	Suspected human carcinogen: liver and lung tumours observed in rats and mice

EU: EC drinking water directive (1998); WHO: WHO (2004) and guidelines for drinking water quality vol. 2 (1996); USEPA: Agency for Toxic Substances and Disease Registry (ATSDR) (2000) and Standards for maximum permissible values in sewage sludge/soils (1993). Estimating concern levels for concentration of chemical substances in the environment, Washington DC (1984).

harmful effects on soil, vegetation, animals and man, sets requirements that are a crucial element in the management of sludge currently produced in the Member States. To this end, it prohibits the use of untreated sludge on agricultural land unless it is injected or incorporated into the soil. Treated sludge is defined as having undergone 'biological, chemical or heat treatment, long-term storage or any other appropriate process so as to significantly reduce its fermentability and the health hazards resulting from its use'. To provide protection against potential health risks from residual pathogens, sludge must not be applied to soil in which fruit and vegetable crops are growing or grown, or less than 10 months before fruit and vegetable crops are to be harvested. Grazing animals must not be allowed access to grassland or forage land less than 3 weeks after the application of sludge. The Directive also requires that sludge should be used in such a way that account is taken of the nutrient requirements of plants and that the quality of the soil and of the surface and groundwater is not impaired. The Directive specifies rules for the sampling and analysis of sludges and soils. It sets out requirements for the keeping of detailed records of the quantities of sludge produced, the quantities used in agriculture, the composition and properties of the sludge, the type of treatment and the sites where the sludge is used. Limit values for concentrations of PTE in sewage sludge intended for agricultural use and in sludge-treated soils are in Annexes I A, I B and I C of the Directive.

Even though the EC directive 86/278/EEC tries to regulate these activities, each country implements their own regulations according to technical capabilities and environmental preferences. As a result, recycling rates in the EU can vary from 11% to >50% (Fig. 11.1, EC, 2000a), with landfilling and incineration still widely used as disposal outlets in some Member States despite their environmental drawbacks (Fig. 11.1).

On the other hand, the progressive implementation of the Directives 91/271/EEC and 86/278/EEC in Europe has increased the quantities of sewage sludge requiring disposal. From an annual production of 5.5 million tonnes of dry matter in 1992, the European Community (EC) was heading towards 9 million tonnes in 2005 (Fig. 11.2, Langenkamp and Marmo, 2000).

The Directive's requirements, that have been translated with provisions at national level, present in fact significant differences. In some cases, national regulations introduced on the basis of Directive 86/278/EEC have often introduced provisions which go beyond the requirements of the Directive. In particular, the limit values for concentrations of PTE in sludge are lower than the limit values specified in the Directive in a majority of countries. In five countries (Belgium-Flanders, Denmark, Finland, The Netherlands and Sweden), the limit values for PTE in sludge are even much lower. However, six Member States (Greece, Ireland, Italy, Luxembourg, Portugal and Spain) have implemented limit values, which are identical to those specified in Annex IB of Directive 86/278/EEC. The perspective of the revision of Directive 86/278/EEC, which could lead to the implementation of more stringent limit values for PTE in sludge, could therefore have an impact in the latter countries, at least on the provisions to be set by national regulations (average PTE content in sludge is in most cases well below regulatory requirements). In addition, the regulations on sludge use include limit values for pathogens in

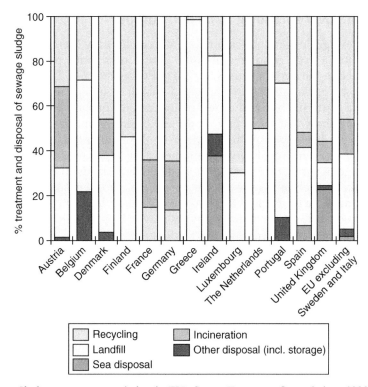

Figure 11.1 Sludge treatment: variation in EU. *Source*: European Commission, 1998, available online: http://reports.eea.eu.int/92–9157–202–0/en/3.7.pdf.

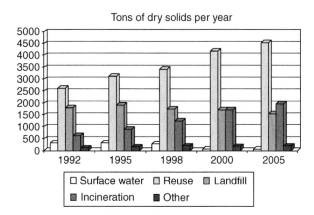

Figure 11.2 Sewage sludge production in the EU from 1992 with projection to 2005 (after Langenkamp and Marmo, 2000). Based on Urban Waste Water Treatment Directive (EC, 1991). http://europa.eu.int/comm/environment/waste/sludge/.

France, Italy and Luxembourg and in a larger number of cases for organic compounds (Austria, Belgium–Flanders, Denmark, France, Germany and Sweden), both of which are not included in the Directive.

3.2. Debate over spreading to land in Europe

The use of sewage sludge in agriculture is important in many Member States, but it is also becoming a very controversial issue. The Community policy in this field (as enunciated in the two Directives 91/271/EEC and 86/278/EEC) is to promote the use of sludge in agriculture, provided that it complies with the applicable requirements in terms of monitoring, treatment and quality. However, past and current events show that the debate on the use of sludge in agriculture has reached different stages depending on the country. In the Netherlands and Flanders, this debate is over, as the regulatory requirements have prevented almost all use of sewage sludge in agriculture since 1991 in The Netherlands and 1999 in Flanders. In Denmark, the debate is now mostly over, since new regulations on the use of sludge in agriculture (Statutory Order no. 49 of January 20, 2000 on the Application of Waste Products for Agricultural Purposes) have played a large part in ending the debate, as they are considered sufficiently strict to reduce risks to an acceptable level. The cases of Germany and Sweden are special. In Sweden, a voluntary agreement was signed in 1994 between the Swedish Environmental Protection Agency (SEPA), the Swedish Federation of Farmers (LRF) and the Swedish Water and Waste Water Association (VAV) concerning quality assurances relating to the use of sludge in agriculture. However, in October 1999, the *LRF* recommended that their members stop using sludge because of concerns about the quality of sludge. In Germany, opinion has recently swung in favour of agricultural land spreading, mainly because this practice is considered economically viable and it is considered that the potential risks are sufficiently reduced by the existing legislation. However, political developments in 2001 have considerably heated the debate, which is quite high at present as some Länder support an increase of regulatory constraints on sludge land spreading. In Austria, France and Walloon, a national (or regional) agreement was under negotiation between the different parties, and hence the debate is heated. The situation is particularly tense in France where the farmers' unions supported, until recently, the development of the agricultural recycling of sewage sludge, on the condition that additional quality controls and an insurance fund system were set up. The situation has now changed, as farmers' unions (the Fédération Nationale des Syndicats d'Exploitants Agricoles (FNSEA) and Centre des Jeunes Agriculteurs (CDJA)) have asked for a ban on the use of sewage sludge, officially because the current methods used are not considered to be sufficient to address the risks related to the agricultural recycling of sludge. In Finland and Luxembourg, the farming community is generally hostile towards the use of sludge for land spreading, mainly because of the pressure to use animal manure for land spreading. For example, the Finnish Union of Agricultural Producers asked for a ban on the use of sewage sludge for land spreading in 1990, and have renewed their stand against the use of sludge in agriculture in 2001. In Ireland and Portugal, farmers support, in some cases, the agricultural use of sludge, both for economic and for agronomic reasons (mainly in terms of organic matter and phosphorus content), although it is difficult to obtain information on this matter. In both countries, the use of sludge seems to be too recent an issue to generate much public debate. In Spain, Italy and Greece, the debate remains limited, as far as can be judged from the available information.

In summary, it seems that the debate is more advanced in Northern Europe but remains limited in Southern Europe. A comparison with the national legal requirements also demonstrates that 'tight' legal constraints (such as very low limit values for pollutants in sludge) do not necessarily imply a greater acceptance of the use of sludge in agriculture. The Swedish example demonstrates this best. Finally, a major trend in the current debate on the use of sludge is clearly the increasing number of agreements regulating the use of sludge. However, whereas voluntary agreements have proven to be successful in the UK (see below), they did not prevent the current crisis in Sweden. In the City of Toulouse (France), our enquiry shows that the national agreement will possibly not allay opponents' fears related to sanitary risks or appease all of the local opposition: the debate largely rests on political and sociological grounds.

3.3. Regulations in the UK

In the UK, the regulating body in these matters is the Environment Agency (EA), which amongst other areas, focuses on 'restored, protected land with healthier soils'. The long-term objective of the EA is to make sure that (i) land and soils in the countryside and towns will be exposed far less to pollutants; (ii) they will support a wide range of uses, including production of healthy, nutritious food and other crops, without damaging wildlife or human health; (iii) contaminated and damaged land will be restored and protected.

The use of sludge as a soil conditioner and fertiliser on agricultural land is regulated in the UK by the Sludge (use in Agriculture) Regulations 1989 (as amended). These regulations specify the tests and the limits on the average annual rates of addition of metals in the sludge HMSO (1989). Furthermore, the receiving soil needs to be tested to check that the limit on the concentration of metals in the soil will not be exceeded by spreading the sludge. There are also prohibitions in spreading on soils with pH of less than 5 and in using sewage sludge and septic tank sludge (i) when fruit, other than fruit trees, or vegetables are growing or are to be harvested in the soil at the time of use, (ii) without taking the nutrient needs of the plants into account and (iii) if it will damage the quality of the soil, surface water or groundwater.

Under the Sludge Regulations, farmers must provide a sewage sludge producer with information of any past sludge use by a different producer. This should include details of where, when and how much sludge was used and who supplied it. On the other hand, sewage sludge producers must

- sample and analyse the sludge according to the Sludge Regulations and give the analysis results to the farmer who receives the sludge
- sample and analyse the soil where sludge is to be used, according to the Sludge Regulations
- prepare and keep a register which contains details of the
  - total quantity of sludge produced and supplied
  - sludge composition and properties
  - quantities of treated sludge supplied and the type of treatment used

    o names and addresses of each farm on which the sludge has been used
    o the quantity of sludge used and anything added to it[2].

There is also a voluntary agreement known as the *Safe Sludge Matrix* which ensures that sludge is only recycled to certain crops and vegetations. This agreement was the result of a heated debate on sludge recycling, until an accord was reached in September 1998 between Water UK, representing the 14 UK water and sewage operators, and the British Retail Consortium (BRC), representing the major retailers. In addition, farmers' associations support the agricultural use of sludge, both for economic and for agronomic reasons. It consists of a table of crop types, together with clear guidance on the minimum acceptable level of treatment for any sewage sludge which may be applied to that crop or rotation.[3] The table (Fig. 11.3) specifies that conventionally treated sludge has been subjected to defined treatment processes and standards that ensure at least 99% of pathogens have been destroyed, whereas enhanced treatment, originally referred to as 'Advanced Treatment', is a term used to describe treatment processes which are capable of virtually eliminating any

The Safe Sludge Matrix

Crop group		Untreated sludges	Conventionally treated sludges	Enhanced treated sludges
Fruit		X	X	✓ ⎤
Salads		X	X (30 month harvest interval applies)	✓
Vegetables		X	X (12 month harvest interval applies)	✓ 10 month harvest interval applies
Horticulture		X	X	✓ ⎦
Combinable and animal feed crops		X	✓	✓
Grass and Forage	- Grazed	X	X (Deep injected or ploughed down only) ✓ (No grazing in season of application) 3 week no grazing and harvest interval applies	✓ 3 week no grazing and harvest interval applies
	- Harvested	X		

Note : ✓ All applications must comply with the sludge (use in agriculture) regulations and DETR code of practice for agricultural use of sewage sludge (to be revised during 2001).

    X Applications not allowed (except where stated conditions apply)

Figure 11.3 Applications of sewage sludge on different types of crops as defined within The Safe Sludge Matrix, available online from www.adas.co.uk.

---

[2] http://www.netregs.gov.uk/netregs/sectors/.
[3] http://www.adas.co.uk/media_files/Publications/SSM.pdf.

pathogens which may be present in the original sludge. Enhanced treated sludge will be free from *Salmonella* and will have been treated so as to ensure that 99.9999% pathogens have been destroyed. The Safe Sludge Matrix cropping categories and treatment processes are regularly reviewed as part of an ongoing process and are subject to possible change and amendment.

In the UK, sewage sludge can also be spread together with other types of non-agricultural wastes (e.g. paper sludge, food wastes, etc.) onto land that is not used for growing commercial food crops or stock-rearing purposes. In this case, the activities are regulated by 'exemption' from waste management licensing (WML) and to qualify for it, the activities must not endanger human health or harm the environment. In particular, the activities must not (i) pose a risk to water, air, soil, plants or animals, (ii) cause a nuisance through noise or odours and (iii) adversely affect the countryside or places of special interest. In general, the activities will only qualify for an exemption for using certain wastes if

- the treatment of agricultural land provides benefit to agriculture or ecological improvement
- the treatment of non-agricultural land provides ecological improvement
- waste and soil analysis are provided, together with a map or plan of the location, a certificate of benefit and a pollution risk assessment
- the treatment of land must comply with certain conditions. These may relate, for instance, to the quantity and location of the spreading. For example, one must not spread more than 250 tonnes of waste per hectare in any 12-month period

A comprehensive Code of Practice has been produced by the Department for Environment, Food and Rural Affairs (DEFRA) and can be consulted online.[4] The Code has been prepared to complement the Sludge Regulations, which enforce the provisions of the EC Directive 86/278/EEC 'on the protection of the environment, and in particular of the soil, when sewage sludge is used in agriculture. Its recommendations are based upon the best available scientific evidence and are in conformity, where relevant, with the requirements of the Regulations'. The Code spells out all the parameters for sludge analysis and methods to apply, including soil sampling procedures, soil analysis parameters, soil limits and permitted rates of application (Table 11.7), application to grassland (which presents particular issues—direct ingestion by livestock), soil pH, sludge concentrations limits and important general guidance on sludge treatment and environmental protection.

3.4. Improved standards and sludge quality in the UK

One of the results of the implementation of EC Directives and Sludge Regulations in the UK is the net improvement of the quality of the sludge over the years. Table 11.8 shows the relative natural abundance, indicative concentrations of some PTEs in soil, sludges and soil pore water, together with typical concentrations after sludge application. The table indicates how concentrations of most metals in sludges

---

[4] http://www.defra.gov.uk/environment/water/quality/sewage/sludge-report.pdf.

Table 11.7 Maximum permissible concentrations of potentially toxic elements (PTEs) in soil after application of sewage sludge and maximum annual rates of addition (modified from DoE, 1996)

PTE	Maximum permissible concentration of PTE in soil (mg kg^{-1} dry solids)				Maximum permissible average annual rate of PTE addition over a 10-year period (kg ha^{-1})[a]
	pH 5.0 < 5.5	pH 5.5 < 6.0	pH 6.0–7.0	pH[b] >7.0	
Zinc[c]	200	200	200	300	15
Copper[c]	80	100	135	200	7.5
Nickel	50	60	75	110	3
For pH 5.0 and above					
Cadmium[c]	3				0.15
Lead[c]	300				15
Mercury	1				0.1
Chromium[d]	400				15
Molybdenum[d,e]	4				0.2
Selenium[d]	3				0.15
Arsenic[d]	50				0.7
Fluoride[d]	500				20

[a] The annual rate of application of PTE to any site shall be determined by averaging over the 10-year period ending with the year of calculation.

[b] The increased permissible PTE concentrations in soils of pH greater than 7.0 apply only to soils containing more than 5% calcium carbonate.

[c] The permitted concentrations of zinc, copper, cadmium and lead are provisional and will be reviewed when current research into their effects on soil fertility and livestock is completed. The pH qualification of limits will also be reviewed with the aim of setting one limit value for copper and one for nickel across pH range 5.0 < 7.0 and therefore ensuring consistency with the approach adopted for zinc in response to the recommendations from the Independent Scientific Committee (MAFF/DOE 1993).

[d] These parameters are not subject to the provisions of Directive 86/278/EEC. (In 1993, the European Commission withdrew its 1988 proposal to set limits for addition of chromium from sewage sludge to agricultural land.)

[e] The accepted safe level of molybdenum in agricultural soils is 4 mg kg^{-1}. However, there are some areas in the UK where, for geological reasons, the natural concentration of this element in the soil exceeds this level. In such cases, there may be no additional problems as a result of applying sludge, but this should not be done except in accordance with expert advice. This advice will take into account the molybdenum content of the sludge, existing soil molybdenum levels and current arrangements to provide copper supplements to livestock.

have decreased in recent years as a result of improved effluent control and waste minimisation. In addition, the proposals for a revision of current EU legislation suggest more restrictive limits to assure sustainable organic waste management up to 2025 (EC, 2000a, c).

Power *et al.* (1999) monitored water in the Thames, UK and found statistically significant reductions in the concentrations of Cd, Cu, Hg, Ni and Zn over the period 1980–1997. For Pb, the initial improvements were reversed by drought in the period 1990–1997 resulting in a slight, though statistically significant rise in

Table 11.8 Potentially toxic element (PTE) concentrations in soils, sludges and soil pore water, including the free ion forms in solution (from Di Bonito, 2005)

Element	Soil (mmol kg⁻¹)	Municipal sludge (mmol kg⁻¹)	UK sludges 1982[a] (mmol kg⁻¹)	UK sludges 1991[a] (mmol kg⁻¹)	UK sludges 1999[b] (mmol kg⁻¹)	EU sludges (limits[c] 2000) (mmol kg⁻¹)	Soil pore water (mmol kg⁻¹) μmol L⁻¹	μmol L⁻¹ at 10% moisture content	'Free ion' fraction x 10⁶
Mn[d]	8.2	7.3	—	1.7	—	—	0.01	0.001	1
Cr	1.0	8	2.4	1.7	—	—(19)	5	0.5	500
B	0.9	6	—	—	—	—	0.08	0.01	10
Zn	0.8	18	18	14	9	38–61 (38)	0.1	0.01	17
Be	0.7	0.07[e]	—	—	—	—	0.17	0.02	730
Ni	0.5	1.7	1.0	0.6	0.4	5.1–6.8 (5.1)	1	0.06	175
Cu	0.3	10	9.8	7.4	6	16–28 (16)	0.08	0.008	63
Co	0.1	0.2	—	—	—	—	0.01	0.0013	17
As[f]	0.1	0.3	—	—	—	—	0.2	0.02	200
Sn	0.08	0.2[e]	—	—	—	—	0.005	0.0005	10
Pb	0.05	1.9	2.0	1.0	0.5	3.6–5.8 (3.6)	0.0004	0.00004	230
Mo[f]	0.02	0.06	—	—	—	—	0.04	0.004	80
Cd	0.001	0.2	0.08	0.03	0.014	0.2–0.4 (0.09)	—	—	—
Sb	0.004	0.1[e]	—	—	—	—			

[a] 50th percentile, DoE (1993).
[b] FWR (1999), from Merrington et al. (2003).
[c] Dry matter related, in brackets are the proposed values in EC (2000b).
[d] Förstner (1991) and references therein.
[e] Eriksson (2001).
[f] www.liv.ac.uk/~rick/BIOL202_Web/Sewage_treatment/metals_in_slu.htm.
Adapted from Wolt (1994) and references therein; dashes (—) refers to 'no data'.

Pb concentrations. However, Pb had fallen prior to this and was still found to have a statistically significant decrease overall in the period 1980–1996 (Table 11.9).

The overall reductions in most metal concentrations in the Thames were greater than 50% over the years 1986–1995, with lower reductions being experienced for Pb and Cd. This complements evidence for sludge quality that overall PTE emissions have markedly improved.

In spite of the reductions achieved by 1995 in the River Thames, PTE concentrations are still high compared to the levels found in the 1980s in the other rivers in the study. It is concluded that while much progress has been made in reducing the anthropogenic sources of PTE pollution discharged into the Thames, improvements are still needed if water is to approximate to background levels. For Cd and Hg, the year 2000 levels were forecast to be about twice estimated background levels (Power et al., 1999).

A comparison of PTE concentrations in sewage sludge applied to farmland in different countries within the EU (Table 11.10) indicates there is some variation apparent in the metal contents of sludges used in agriculture. For example, the reported average concentrations of Cd in German and UK sludges in 1996 were 1.45 and 3.3 mg kg^{-1}, respectively (EC, 2001b). This could indicate differences in

Table 11.9 Achieved concentration reductions of potentially toxic elements (PTEs) in waters in the thames estuary, compared with other European estuaries (from EC, 2001a and references therein)

PTE	Thames 1986/7 (μg L^{-1})	Thames 1995 (μg L^{-1})	Reduction (%)	Elbe 1983 (μg L^{-1})	Rhine 1984 (μg L^{-1})	Humber 1984 (μg L^{-1})	Severn 1988 (μg L^{-1})
Cd	0.43	0.32	24.1	0.1	0.3	0.31	0.25
Cu	31.3	10.7	65.8	2.0	5.5	2.17	5
Hg	0.24	0.09	63	—	—	—	—
Ni	17.3	6.30	63.4	3.25	3.2	1.24	4.5
Pb	16.3	9.90	39.4	1.9	—	—	<0.03
Zn	92	29.1	68.4	—	21.8	—	17.5

Table 11.10 Comparison of potentially toxic element (PTE) concentrations (mg kg^{-1}) in sewage sludge applied to agricultural land in Germany and in the UK in 1996 (EC, 2001a)

Potentially toxic element	Germany (average)	UK (weighed average)	UK (median)
Zn	776	792	559
Cu	305	568	373
Pb	57	221	99
Cr	40	157	24
Ni	24	57	20
Cd	1.45	3.3	1.6
Hg	1.35	2.4	1.5

the amount of Cd discharged to sewer in these countries from industrial, domestic and diffuse inputs or the adoption of different sludge treatment practices and regulatory procedures influencing metal content.

However, these issues are difficult to reconcile given the policies on preventing industrial discharges of Cd followed in both countries and that both states practice extensive sludge stabilisation treatment. A possible explanation may be related to the statistical characteristics of metal concentration data and how data on metal contents in sludge are reported. For example, CEC (2001a) does not state whether arithmetic averages or weighted averages are given for metal contents. The UK figures are weighted according to works size and provide a conservative estimate of sludge metal content because sludge from large treatment works usually have larger metal concentrations compared with smaller works (EA, 1999). For example, the median Cd concentration in sludge used on to farmland from large works (2.9 mg Cd kg^{-1}, pe $>$ 150,000) in the UK was more than twice that from small works (1.3 mg Cd kg^{-1}, pe $<$ 10,000) in 1996/1997. This trend could be interpreted as being the result of higher industrial inputs of metals to the large works, although it may also be explained by the greater interception of atmospheric deposition of metals by paved areas in urban centres that are served by the largest sewage treatment works. The majority of sludge recycled to land in the UK is produced by 55 large works (160,000 t ds y^{-1}) whereas ~840 small works generate sludge (45,000 t ds y^{-1}) for agricultural use. Therefore, an arithmetic mean would indicate a significantly lower concentration was apparent for sludge because all works would have equal weighting. Indeed, the median concentrations recorded for the UK sludge are comparable to the values reported for Germany.

4. REUSE AND DISPOSAL OF SEWAGE SLUDGE IN THE UK

Large-scale application of sewage sludge to agricultural land has been undertaken in the UK for over 100 years (Heaven and Delve, 1997) and is expanding due to the European objective to minimise or reduce disposal at sea or *via* land fill (Fig. 11.4; DETR, 1999; Sonesson *et al.*, 2000). Digested sludges provide a cheap source of nitrogen and phosphorus and improve the soil through their high organic matter content (40% dry weight: DW). The organic content of the sludge can help improve the soil structure and in general, sludge stimulates beneficial biological activity in the soil (DEPA, 1997), even though the usual application rate is below that which would have a significant positive impact on soil structure. For the same reasons, sewage sludge is also a readily available alternative soil-building material and for many years has been used extensively in the UK and elsewhere for successful land restoration or site remediation. Sewage sludge contains major and minor nutrients, and valuable trace elements essential to animals and to plants such as B, Cu and Zn.[5,6] The incorporation of sludge into other materials on derelict land produces a viable growing medium with good properties of soil structure, workability and

---

[5] www.city.toronto.on.ca/water/protecting_quality/biosolids/recyclable.
[6] www.terraecosystems.co.uk.

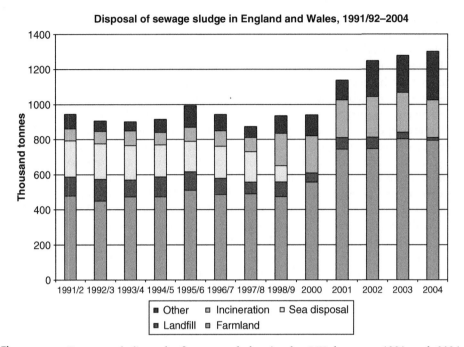

Figure 11.4 Reuse and disposal of sewage sludge in the UK between 1991 and 2004. *Source*: http://www.environment-agency.gov.uk/.

moisture retention, all of which encourage vegetation growth (ADAS, 2002; DEFRA, 2002). Sewage sludge provides a source of slow release nitrogen which is ideal for use in land restoration because one initial application can be used to provide enough nutrients for long-term vegetation growth (FAO, 1992).

The practice of recycling sewage sludge to land is in fact recognised by the EU and the UK Government as often being the Best Practical Environmental Option (EC, 2000b; Scottish Executive, 2002). Historically, about a quarter of sludge was either dumped at sea or discharged to surface waters. This was banned from 1998 under the Directive because it was considered environmentally unacceptable. The changes in UK sludge disposal can be seen in Fig. 11.5.

4.1. Use of sludge for land spreading

Land spreading is a way for recycling the compounds of agricultural value present in sludge to land. All sludge types (liquid, semi–solid, solid or dried sludge) can be spread on land. However, the use of each of them induces practical constraints on storage, transport and spreading itself. The sludge production from a given WWTP is more or less constant throughout the year, but the use on farmland is seasonal. Therefore, storage capacity must be available on the WWTP or on the farm, either separately or in combination with animal slurry when allowed by the national regulations. Average storage duration is about 6 months. Storage on fields may also be practically observed. This however should only be performed shortly before

Sewage sludge disposal outlets 1996 and 1997

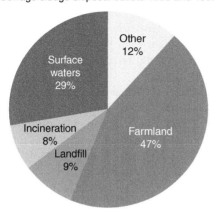

Source: Sewage sludge survey covering
the two years 1996 and 1997

Sewage sludge disposal outlets 1999/00

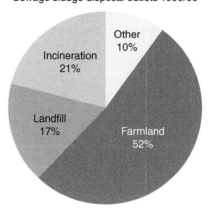

Sources: OFWAT, Scottish executive and DOE
Northern Ireland

Figure 11.5 Sewage sludge disposal outlets before and after the ban of the 1998. *Source*: http://www.environment-agency.gov.uk/.

spreading, and with solid and stabilised sludge in order to reduce risks of leaching. Furthermore, there are some restrictions as to where to store the sludge, in particular no sludge may be stored within (i) 10 m from any watercourse, (ii) 50 m from any spring, well or borehole not used to supply water for domestic or food production purposes, (iii) 250 m from any borehole used for a domestic water supply, (iv) 50 days travel time from a groundwater source used to supply water for domestic or food production purposes (Groundwater Source Protection Zone 1).[7] The Codes for Good Agricultural Practice give guidance on storage.[8]

In the case of de-watered sludge that is not stored in a container or lagoon it must be stored securely and in a way as to prevent pollution. Uncovered stockpiles will be exposed to rainfall and the decision where to locate the storage area should take into account that run-off or leachate may be produced, which could enter a watercourse or groundwater.

Liquid sludge may be stored in concrete tanks (mostly for small WWTP) or lagoons and then being dumped for the transportation. Semi-solid sludge may be stored on a platform, which must be waterproof, or in tanks. Sludge pits may also be found. As in most cases, this type of sludge cannot be pumped, sludge has to be conveyed by using specific hauling equipment such as grabs. Odours may arise when sludge is handled to be conveyed. The structure of solid sludge enables storage on piles. Handling implies the use of a crane or a tractor. Dried sludge does not present any specific constraint. If sludge however is pulverulent, storage must be monitored in order to prevent any explosion and emission of particles to air. Transportation is the most expensive aspect of this route. It is possible to use tankers for liquid sludge

---

[7] http://216.31.193.171/asp/1_introduction.asp.
[8] www.defra.gov.uk/environ/cogap/cogap.htm.

or articulated lorries for other sludge types. Sludge can be applied to the fields by using trailer tank or umbilical delivery system and may be applied by surface spreading (it is however of importance to reduce the formation of aerosols to reduce the risk of odour nuisance) or directly injected into the soil. Dried sludge may be supplied by using the same equipment as for solid mineral fertilisers. The spreading equipment has also to be adapted to the type of sludge.

Land use, soil type, accessibility of the field, meteorological conditions influence land spreading. Mostly, the practice can be performed during two periods of the year: at the end of summer, after harvesting, or in spring, before ploughing and sowing. As already stated, sewage sludge contains compounds of agricultural significance such as nitrogen, phosphorus, potassium, organic matter or calcium, making its use relevant as an organic fertiliser. Moreover, the cost of this route may be cheaper than that of other disposal routes.

The main inputs and outputs of land spreading are shown in Fig. 11.6. Inputs are related to the sludge itself, the additional resources such as petrol for transportation and application and room for storage. The main outputs are the yield improvement and fertiliser substitution. In addition, the emission of pollutants to soil, and indirect emissions to air and water. Other emissions to air are exhaust gases from transportation and application vehicles. Dried sludge may also be emitted to the air during transportation when trucks are not covered.

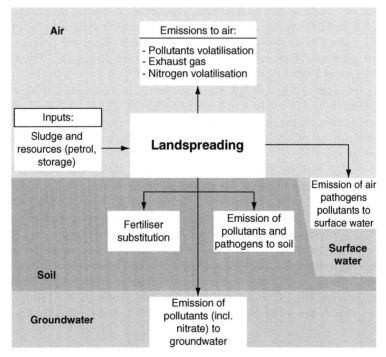

Figure 11.6 Representation of the inputs and outputs for land spreading (from EC 2001c, modified).

Land spreading also involves the application to soil of the pollutants contained in sludge. Those pollutants undergo different transformations and transfer processes, including leaching, runoff, volatilisation, which could enable the transfer of the compounds into the air and water, and their introduction into the food chain. Disamenities may take place because of land spreading operation odour.

4.2. Use of sludge for soil restoration and reclamation

Use of sewage sludge in land reclamation and revegetation aims to restore derelict land or protect the soil from erosion, depending on the previous use of the site. In the case of industrial sites, topsoil may often be absent or if present, damaged by storage or handling. Soil or soil-forming materials on site may be deficient in nutrients and organic matter. Other problems may exist, such as toxicity or adverse pH. All these problems create a hostile environment for the development of vegetation (Hall and Wolstenholme, 1999).

Possible solutions include the use of inorganic fertilisers or imported topsoil, which can be very expensive depending on location and availability. An alternative solution is the use of organic wastes such as sewage sludge, which is already used in the UK as well as Sweden and Finland.

According to the site use, several situations may arise, for which the provision of sludge could be useful (Table 11.11). In the case of industrial sites, the benefits of sludge application also include the improvement of nutrient status, the addition of organic matter and control of acid generation.

Sludge application is performed by using the same machinery as in the case of recycling to agriculture. Some specific machinery for sludge projection may be needed when applying sludge on areas to which access is difficult.

When aiming to increase soil quantity on the site, two techniques are observed in the field: sludge may be either directly applied before mixing with the soil present on the site or mixed with soil before application. The amount of sludge usually applied is much higher than the amount applied in the case of land spreading on agricultural land. As an example, some experiments were made to develop ski runs in La Plagne (France). To reach a soil thickness of 5 cm, it was necessary to use about 100–150

Table 11.11 Objectives of the sludge spreading in land reclamation (modified from EC, 2001c)

	Site restoration	Provision of soil	Erosion reduction	Increase of vegetal covering
Site aftercare (industrial site or landfill)	√	√		√
Forest (before plantation)		√	(√)	
Embankment (railway, motorways, etc.)		√		√
Amenity areas (ski runs, golf courses)		√	√	√

Table 11.12 Summary of typical sludge application rates for different reclamation sites (modified from EC, 2001c)

Type of land	Reason of application	Typical application rates	
		Cake (t DM ha^{-1})	Liquid (m^3 ha^{-1})
Low maintenance amenity	Nutrient status improvement	50	100
Disturbed agricultural soils	Nutrient status improvement	100	—
Sites lacking topsoils	Organic matter addition	100–500	—
Acidic colliery spoil	Acid generation control	>500	—

tonnes ha^{-1} (ADEME 1999). Hall and Wolstenholme (1999) also indicates some typical application rates for different reclamation sites (Table 11.12).

Information is needed to assess the risk of this practice as no sufficient data is available concerning environmental and sanitary impacts. It was assumed that risks are lower than the risks in the case of spreading on agricultural land, when considering that its use is not related to food production. In the case of ski run development, sludge is usually applied in September, ensuring a sufficient time period before the use of the land for grazing. However, no data is available concerning the potential impacts on wild fauna and flora. Moreover, the amount of sludge applied as well as the use of sludge on sloping land for erosion reduction go against actual regulatory prescriptions for the use of sludge in agriculture. A risk may therefore arise because of the amount of pollutants or nitrogen applied to land. Hall and Wolstenholme (1999) recommends that sludge should be sampled and analysed for PTE on a regular basis to ensure that, at the application rates required for land-reclamation, PTE additions are well within acceptable limits.

In any case, sludge used for land reclamation and revegetation in field experiments is always treated sludge, in order to ensure sufficient disinfection and to reduce the presence of odours.

There are already several examples of UK colliery sites where the use of sewage sludge has aided successful site restoration (Scotland: Drumbow, North Lanarkshire, East Ayrshire and Skares). Skares recently won the brownfield category 'Award for Wildlife Champions' from the Scottish Environment Protection Agency in December 2003 for the 'innovative approach to reclaiming a derelict brownfield site, to create a community woodland'). At the site, sewage sludge was used to regenerate poor quality soil, which was then planted with mixed woodland and grass, providing a rich habitat for a wide range of animals and birds.[9] Another example of land reclamation is given by the Scottish Vacant and Derelict Land Survey done in 2002, which identified almost 7800 ha of derelict land within Scotland (Scottish

---

[9] www.sepa.org.uk.

Executive, 2003). This land was so damaged by development or use to render it incapable of being developed for beneficial use without rehabilitation. Importing topsoil in the quantities required for restoration of all the derelict land in Scotland would be neither environmentally sustainable nor feasible and would be very costly. Large volumes of good topsoil are not available for purchase or transportation because of their inherent *in situ* value for other purposes. It is therefore necessary to find a material that contains the required properties and can be used as a substitute and sewage sludge represents a readily available alternative soil-building material.

Recycling sewage sludge to land restores derelict land to a useful purpose. The sludge can be used to stabilise spoil and other waste materials prone to erosion and to re-contour disturbed land, such as former collieries, to blend in with the surrounding landscape. Mixing sewage sludge in with colliery spoil and other materials that lack nutrients creates a growing medium capable of sustaining plant life (SNIFFER, 2007). When applied to acidic colliery spoil, the sewage sludge provides organic matter that helps to balance the soil pH (Water UK, 2006). The risk of PTE being mobilised and moving into groundwater or to watercourses is very low provided that the pH level is controlled.

The incorporation of sludge into other materials on derelict land produces a viable growing medium with good properties of soil structure, workability and moisture retention, all of which encourage vegetation growth (ADAS, 2002; DEFRA, 2002). Sewage sludge provides a source of slow release nitrogen which is ideal for use in land restoration because one initial application can be used to provide enough nutrients for long-term vegetation growth (Pescod, 1992). In fact, the slow release, low cost and long-term nutrient availability of nutrients in sewage sludge makes it more efficient than are mineral fertilisers such as phosphate. Its ready availability also makes it much more sustainable for soil conditioning than are non-renewable resources like phosphate rock, which needs to be mined or quarried and imported.

5. Encouraging the Recycling of Sludge

In order to encourage the recycling of sludge, it should be taken into account that the development of agricultural recycling depends largely on the possibilities to improve the quality of the sludge itself and increase confidence in sludge quality. This implies the prevention of pollution of the wastewater at source by reducing the possibilities for PTE and organic compounds to enter the wastewater sewage system and improving sludge treatment as well as ensuring the monitoring of sludge quality. These technical solutions however require major investment from the water companies or local authorities in charge of treating the wastewater. The possibility to certify the treatment processes involved and the quality of sludge, either through independent 'sewage sludge audits' or by the certification of sludge production and treatment processes, could help to increase confidence in sludge quality. Similarly, the quality standards of sludge recycling practices also need to be guaranteed, especially for agricultural recycling.

One of the main issues with regard to sludge recycling in agriculture is the setting up of guarantee funds or insurance systems in order to cover any loss of profits, damages or other costs related to the use of sludge in agriculture. This would partially address the issue of liability, which is a vital concern for farmers and landowners in the debate over the use of sludge. In addition to economic instruments, legal provisions could be introduced to regulate producer liability. However, according to the City of Düsseldorf officials, the guarantee fund was not considered as a decisive argument leading the City to privilege the use of sludge in agriculture, and has even had negative consequences on the economic conditions of this route.

National regulatory requirements vary greatly from one country to another. In this area, national regulations, based on the same scientific grounds, should be considerably improved by the next Directive on sludge use, in order to provide long-term perspectives for the use of sludge. With regard to increasing confidence in the use of sludge, standardisation initiatives (continuation and completion of CEN TC 308 work on the production and disposal of sewage sludge) have a major role to play. Science is driving the effort to bring some consistency across the borders through valuable international projects,[10] but national regulations for land-applied wastes still restrict only 10–12 metals at most (McBride, 2003). Hence, PTE and organic compounds known to be present in wastes at concentrations much higher than that in soils are being applied to land without regulation. For example, US EPA 503 rule, which regulates loading limits to agricultural soil for eight metals present in sewage sludge, does not impose any limits on Mo and Cr. A recent survey carried out in Canada also revealed variable and sometimes very high levels of unregulated toxic metals, including Tl, Sn, Sb and Ag (Webber and Bedford, 1996; Webber and Nichols, 1995). In the UK, the list of recognised PTE currently includes Zn, Cu, Ni, Cd, Pb, Hg, Cr, Mo, Se, As and F (MAFF, 1998).

On a different level, the evolution of the debate on sludge disposal and recycling in Europe shows that the relationship between farmers and their customers (food industry and retailers) is crucial for the acceptance of the use of sludge in agriculture. Examples at national level show that an agreement at European level between representatives of food industries, retailers, farmers and sludge producers could enhance mutual confidence and information transfer. In this respect, efforts could be made to improve communication between the major stakeholders, for example, by creating 'contact points' similar to the national committees on sludge set up in several Member States.

The current state of the debate on sludge recycling and disposal routes clearly shows that the current uncertainties over possible risks for human health and for the environment play a major part in the resistance against expanding various sludge recycling routes. The areas where scientific results are the most expected by the stakeholders are possible effects of organic pollutants and pathogens in sludge. Progress in the social and political acceptance of sludge recycling could therefore be made by promoting research on these specific aspects, publishing the research results and making them widely available. In particular, there should be better dissemination of the results of current national research programmes on the effects of the agricultural recycling of sludge on health.

---

[10] http://www.ecn.nl/horizontal/.

Previous research has already analysed the impact of sludge application on fertility, bioavailability, plant uptake, fate of pollutants, general soil chemistry and the effect on physical properties. There are, however, still gaps and flaws in the understanding of the chemistry of sludge-amended soils. It is paramount to understand the soil–water media and its interactions with the materials added to it. The biogeochemical availability of nutrients and pollutants accumulating in soil is of concern with respect to potential phytotoxicity, plant uptake and bioaccumulation, transport to ground or surface water and human health (Abrahams, 2002). Prediction of the extent of soil pollution, and restoration of polluted soils, requires an understanding of the processes controlling the fate of pollutants within the soil. Scientists are always interested in the study of their movement through the soil, during which nutrients and polluting substances are subject to complex physical, chemical and biological transformations. In fact, the decision to apply sewage sludge to agricultural land should be governed by a soil's ability to buffer PTEs' input rather than by the general notion of reuse and recycling.

Nevertheless, the nature of these processes is not only influenced by the soil environmental conditions under which the elements or pollutants may be released to soil solution and thereby become environmentally available, but also by the very nature of the waste source. Modern legislation is providing guidelines for an economic and environmentally beneficial waste disposal strategy, including sewage sludges (DoE, 1996; EC, 2000c; MAFF, 1996) where decision support systems (DSS) for optimisation of sewage sludge application are becoming increasingly useful (Horn et al., 2003). The role that regulating bodies, such as the EA in the UK, therefore, is equally important, as they ensure a sustainable and environmental way of managing these expanding practices. This will hopefully address the ecological and human hazard of all unregulated toxic compounds, sometimes assumed to be not significant only because of the lack of evidence indicating otherwise.

In addition to the dissemination of research results, an important effort of communication on sludge use should be carried out. In particular, tools such as codes of practice for the recycling of sludge implemented on a voluntary basis should be considered. Communication should especially aim to promote high-quality sludge (with low levels of contaminants), which could be recognised as fertilisers (or as a component of fertiliser products) at European level. The development of labels at European level would enable users to identify high-quality sludge and to distinguish it from other types of sludge or waste, thus improving the image of sewage sludge itself. Therefore, labels on products could be a useful additional tool to labels on quality assurance, for encouraging the use of sludge in agriculture. The possibilities for providing more training opportunities to specific categories of stakeholders (e.g. farmers) should also be examined.

ACKNOWLEDGMENTS

I would like to thank Mark Cunningham and Mat Davies at the Environment Agency who provided invaluable comments on an earlier draft. Although this chapter is published with the permission of the Environment Agency, the views expressed in the chapter are solely those of the author.

REFERENCES

Abrahams, P. W. (2002). Soils: Their implications to human health. *Sci. Total Environ.* **291**(1–3), 1–32.

ADAS (2002). Beneficial effects of biosolids on soil quality and fertility ADAS research review 2001–2002—Environment ADAS Ltd. www.adas.co.uk.

ADEME (1999). Situation du recyclage agricole des boues d'épuration urbaines en Europe. Agence de l'Environnement et de la Maîtrise de l'Energie (ADEME), Données et références 154 p.—www.ademe.fr.

Alloway, B. J. (1995). "Heavy Metals in Soils," Blackie Academic & Professional, An imprint of Chapman & Hall, London.

Agency for Toxic Substances and Disease Registry (ATSDR) (2000). "Toxicological profile for polychlorinated biphenyls (PCBs)," Public Health Service, U.S. Department of Health and Human Services. November 2000.

Babel, S., and del Mundo Dacera, D. (2006). Heavy metal removal from contaminated sludge for land: A review. *Waste Manage.* **26,** 988–1004.

Brady, N. C., and Weil, R. R. (1999). "The Nature and Properties of Soils," Prentice-Hall, Upper Saddle River, New Jersey.

DEFRA (2002). Growing short rotation coppice—Best practice guidelines rural development programme—England, July 2002. www.defra.gov.uk.

DEPA (1997). Environmental assessment of textiles, p. 220. *In* MoEa Energy (ed.). Danish Environmental Protection Agency.

DETR (1999). Draft waste strategy for England and Wales. 99EP0254, DETR.

Di Bonito, M. (2005). Trace elements in soil pore water: A comparison of sampling methods, Ph.D. thesis University of Nottingham, Nottingham (http://etheses.nottingham.ac.uk/archive/00000123/).

DoE (1993). UK sewage sludge survey. Final report Department of Environment, Consultants in Environmental Sciences Ltd, Gateshead.

DoE (1996). Code of practice for agricultural use of sewage sludge. Department of Environment, Goldthorpe, Rotherham, UK. ISBN 185112005.

EA (1999). Environment agency UK sewage sludge survey—National presentation (1996/7). R&D Technical Report P165, Bristol (1999).

EC (1986). Council Directive 86/278/EEC of 12 June 1986 on the protection of the environment, and in particular of the soil, when sewage sludge is used in agriculture. *Official J. Eur. Commun.* **181,** 6–12.

EC (1991). Council Directive 91/271/EEC of 21 May 1991 concerning urban waste-water treatment. *Official J. Eur. Commun.* **135,** 40–52.

EC (1998). Mis en oeuvre de la directive 91/271/CEE du Conseil du 21 mai 1991 relative au traitement des eaux urbaines résiduaires, modifiée par la directive 98/15/CE de la commission du 27 février 1998. *98/15/CE,* European Commission.

EC (2000a). Report from the Commission to the Council and the European Parliament on the implementation of Community Waste Legislation Directive 75/442/EEC on waste, Directive 91/689/EEC on hazardous waste, Directive 75/439/EEC on waste oils and Directive 86/278/EEC on sewage sludge for the period 1995–1997. European Commission, Bruxelles.

EC (2000b). Working document on sludge, 3rd draft. *ENV.E3/LM.* European Commission, Bruxelles.

EC (2000c). Working document. Biological treatment of biodegradable waste on sludge, first draft. European Commission, Bruxelles.

EC (2001a). Organic contaminants in sewage sludge for agricultural use. *In* H. Langenkamp, and P. Part (eds.), European Commission, Joint Research Centre, Institute for Environment and Sustainability, Soil and Waste Unit.

EC (2001b). "Pollutants in Urban Waste Water and Sewage Sludge." European Commission, Final report prepared by ICON Consultants Ltd., London, United Kingdom, February 2001.

EC (2001c). "Disposal and Recycling Routes for Sewage Sludge." European Commission, Scientific and technical sub-component report, October 2001.

Eriksson, J. (2001). Concentrations of 61 trace elements in sewage sludge, farmyard manure, mineral fertiliser, precipitation and in oil and crops. *5159*, Swedish Environmental Protection Agency, Uppsala, Sweden; available on internet at the URL: www.naturardsverket.se/pdf/620-6246-8.pdf.

European Waste Water Group (EWWG) (1997). European waste water catalogue.

FAO (1992). Wastewater treatment and use in agriculture—FAO irrigation and drainage paper 47 M.B. Pescod, Food and Agriculture Organisation of the United Nations www.fao.org.

Feachem, R. G., Bradley, D. J., Garelick, H., and Mara, D. D. (1983). "Sanitation and Disease: Health Aspects of Excreta and Wastewater Management." John Wiley, Chicester.

Förstner, U. (1991). Soil pollution phenomena—Mobility of heavy metals in contaminated soil. *In* "Interactions at the Soil Colloid-Soil Solution Interface," (G. H. Bolt, M. F. De Boodt, M. H. B. Hayes, and M. B. McBride, eds.). pp. 543–582. Kluwer Academic Publisher, Dordrecht, the Nederlands.

FWR (1999). "UK Sewage Sludge Survey 1996–97." WRc, Medmenham, England.

Gendebien, A., Carlton-Smith, C., Izzo, M., and Hall, J. E. (1999). UK Sewage sludge survey—National presentation. *P. 165.* Environment Agency, England and Wales.

Ginés, F. R. (1997). Barcelona's water flow: A tool for environmental analysis, Paper presented at the Nagoya Conference, November 1997.

Hall, J. E., and Wolstenholme, R. (1999). "Manual of Good Practice for the Use of Sewage Sludge in Land Reclamation." WRc, Medmenhan, UK.

Harrison, E. Z., Oakes, S. R., Hysell, M., and Hay, A. (2006). Organic chemicals in sewage sludges. *Sci. Total Environ.* **367,** 481–497.

Heaven, F. W., and Delve, M. (1997). Detailed investigation of heavy metal and agricultural nutrient content of soils. *352/1,* Land Research associates, Lockington Hall, Lockington, Derby.

Hillman, P. J. (1988). Health aspects of reuse of treated wastewater for irrigation. *In* "Treatment and Use of Sewage Effluent for Irrigation." (M. B. Pescod and A. Arar, eds.). Butterworths, Sevenoaks, Kent.

HMSO (1989). Sludge (Use in Agriculture) Regulations 1989 (SI 1989 No. 1790), amended 1990 (SI 1990 No. 880), available online: http://www.legislation.hmso.gov.uk/si/si1989/Uksi_19891263_en_1.htm.

Horn, A. L., During, R. A., and Gath, S. (2003). Comparison of decision support systems for an optimised application of compost and sewage sludge on agricultural land based on heavy metal accumulation in soil. *Sci. Total Environ.* **311**(1–3), 35–48.

Koch, M., and Rotard, W. (2001). On the contribution of background sources to the heavy metal content of municipal sewage sludge. *Water Sci. Technol.* **43**(2), 67–74.

Koch, M., Knoth, W., and Rotard, W. (2001). Source identification of PCDD/Fs in a sewage treatment plant of a German village. *Chemosphere* **43**(4–7), 737–741.

Langenkamp, H., and Marmo, L. (2000). Problems around sludge, Workshop on Problems around sludge, Stresa (NO) Italy.

MAFF (1996). "Information on the Application of Sewage Sludge to Agricultural Land." *PB2568,* Ministry of Agriculture, Fishieries and Food, London.

MAFF (1998). "Code of Good Agricultural Practice for the Protection of Soil." Ministry of Agriculture, Fisheries and Food, London.

McBride, M. B. (2003). Toxic metals in sewage sludge-amended soils: Has promotion of beneficial use discounted the risks? *Adv. Environ. Res.* **8**(1), 5–19.

Merrington, G., Oliver, I., Smernik, R. J., and McLaughlin, M. J. (2003). The influence of sewage sludge properties on sludge-borne metal availability. *Adv. Environ. Res.* **8,** 21–36.

Pescod, M. B. (1992). Wastewater treatment and use in agriculture. *In* "Irrigation and drainage paper 47," Food and Agriculture Organisation of the United Nations, Rome, Italy—www.fao.org.

Power, M., Attrill, M. J., and Thomas, R. M. (1999). Heavy metal concentration trends in the Thames estuary. *Water Resour.* **33**(7), 1672–1680.

Reilly, C. (1991). "Metal Contamination of Food," Elsevier Applied Science, London and New York.

Scottish Executive (2002). *Safer sludge,* Scottish Executive consultation paper, paper no. 2002/30. www.scotland.gov.uk.

Scottish Executive (2003). Statistical Bulletin—Scottish Vacant and Derelict Land Survey 2002 (ENV/
    2003/1). Scottish Executive, 17th March 2003. www.scotland.gov.uk.
SNIFFER (2007). Human health and environmental impacts of using sewage sludge on forestry
    and for restoration of derelict land, Task 2, Literature review of environmental and ecological
    impacts.Final Report Project UKLQ09, January 2007. SNIFFER, First Floor, Greenside House,
    25 Greenside Place, EDINBURGH EH1 3AA, Scotland, UK -www.sniffer.org.uk.
Sonesson, U., Bjorklund, A., Carlsson, M., and Dalemo, M. (2000). Environmental and economic
    analysis of management systems for biodegradable waste. *Resour. Conserv. Recycl.* **28**(1–2), 29–53.
Sörme, L., and Lagerkvist, R. (2002). Sources of heavy metals in urban wastewater in Stockholm.
    Sci. Total Environ. **298**(1–3), 131–145.
Water UK (2006). Recycling of biosolids to land, Water UK, 1 Queen Anne's Gate, London SW1H
    9BT. www.water.org.uk.
Webber, M. D., and Bedford, J. A. (1996). Organic and metal contaminants in Canadian municipal
    sludges and a sludge compost: Supplemental Report. *1996-RES-3*, WTI, Burlington, Ontario.
Webber, M. D., and Nichols, J. (1995). Organic and metal contaminants in Canadian municipal
    sludges and a sludge compost: Wastewater technology centre. WTI, Burlington, Ontario.
Wolt, J. (1994). "Soil Solution Chemistry: Applications to Environmental Science and Agriculture,"
    John Wiley & Sons, New York.
World Health Organization (1984). "Guidelines for Drinking Water Quality. Vol. 1." WHO, Geneva
    p. 130.
World Health Organization (1996). "Guidelines for Drinking-Water Quality: Volume 2 Health
    Criteria and other Supporting Information," 2nd edn. 1996, addendum to Volume 2 1998,
    WHO, Rome.
World Health Organization (2004). "Guidelines for Drinking-Water Quality. Vol. 143." 3rd edn.
    WHO, Geneva.

CHAPTER TWELVE	

LEAD ISOTOPES AS MONITORS OF ANTHROPOGENIC AND NATURAL SOURCES AFFECTING THE SURFICIAL ENVIRONMENT

Robert A. Ayuso,* Nora K. Foley,* *and* Gail Lipfert[†]

Contents

Abstract

Arsenical pesticides and herbicides were used extensively in New England (USA) during the first half of the twentieth century. The pesticides were used on apple, blueberry, and potato crops. The arsenical pesticides (lead arsenate, calcium arsenate, and sodium arsenate) have similar Pb-isotopic compositions: $^{208}Pb/^{207}Pb = 2.3839-2.4722$ and $^{206}Pb/^{207}Pb = 1.1035-1.2010$. Other arsenical pesticides such as copper acetoarsenite (Paris green), as well as arsanilic acid, are widely variable in isotope composition. Common organoarsenical pesticides and herbicides such as monosodium methyl arsonate (MSMA), methyl arsonic acid, and methane arsonic acid have Pb-isotopic signatures that overlap the historical arsenical pesticides, but extend to significantly less radiogenic ratios of $^{206}Pb/^{207}Pb$ and $^{208}Pb/^{207}Pb$. This Pb-isotopic difference is notable and suggests that these types of pesticides could be distinguished in Pb-bearing minerals formed in the near-surface environment. A full assessment of the environmental impact of the historical use of arsenical pesticides is not available, but initial studies indicate that arsenic and lead concentrations in stream sediments in New England (northeastern USA) are higher in agricultural areas that intensely used arsenical pesticides than in other areas.

* U.S. Geological Survey, MS 954 National Center, Reston, Virginia 20192, USA
† Department of Earth Sciences, University of Maine, Orono, Maine 04469, USA

Environmental Geochemistry
DOI: 10.1016/B978-0-444-53159-9.00012-7

© 2008 Elsevier B.V.
All rights reserved.

The possible lingering effects of arsenical pesticide use were tested in a detailed geochemical and isotopic study of soil profiles from a small watershed containing arsenic-enriched groundwater in coastal Maine. The Pb-isotopic compositions of acid-leach extractions represent lead adsorbed to mineral surfaces or held in soluble minerals (Fe- and Mn-oxyhydroxides, carbonate, and some micaceous minerals) in the soils, whereas residue compositions likely reflect bedrock compositions. Labile Pb-isotopic compositions (acid-leach) show a moderate range in $^{206}Pb/^{207}Pb$ (1.1870–1.2069) and $^{208}Pb/^{207}Pb$ (2.4519–2.4876). Isotope values vary as a function of depth: the lowest Pb-isotopic ratios (e.g., $^{208}Pb/^{206}Pb$) representing labile lead are in the uppermost soil horizons (also containing highest Pb abundances). A multicomponent mixing scheme that includes predominantly lead from the local parent rock (Penobscot Formation) and lead derived from combustion of fossil fuels could account for the observed Pb-isotopic variations in the soil profiles. In general, however, our preliminary data also show that the extensive use of arsenical pesticides and herbicides in agricultural regions can be a notable anthropogenic source of arsenic and lead to stream sediments and soils.

1. INTRODUCTION

The chemical composition of secondary minerals in stream sediments, soils, and other near-surface materials is dominantly contributed by local bedrock, although considerable metal influxes from anthropogenic sources are also found (e.g., Erel and Patterson, 1994; Graney *et al.*, 1995; Hansmann and Koppel, 2000; Nriagu and Pacyna, 1988; Sturges and Barrie, 1987). Man-made sources contribute to near-surface materials because of heavy-metal emissions distributed by atmospheric deposition, as effluents and seepages from industrial sites and plants related to industrialization and urbanization (e.g., burning of gasoline and coal, mining, smelting, manufacturing, sewage treatment, etc.), and as a legacy of historical agricultural activities (e.g., application of pesticides and herbicides). In New England, for example, arsenical pesticides and herbicides extensively used on apple, blueberry, and potato crops during the first half of the twentieth century may have enhanced the contents of arsenic, lead, and other metals in farm soils (Chormann, 1985; D'Angelo *et al.*, 1996; Robinson and Ayuso, 2004; Veneman *et al.*, 1983). The most important historical pesticides and herbicides used in New England were lead arsenate, calcium arsenate, and sodium arsenate, and among the three, lead arsenate was the pesticide most widely employed in apple orchards (Peryea, 1998). Copper acetoarsenite (Paris green) was also commonly used (Peryea, 1998; Veneman *et al.*, 1983). Recently, concern has grown about the role of common organoarsenical herbicides, including monosodium methane arsonate (MSMA), which is extensively used in urban and suburban golf courses and cotton soils (e.g., Solo-Gabriele *et al.*, 2003), and chromated copper arsenate (CCA), which is widely applied as a wood preservative (Stilwell *et al.*, 2003). Recent studies elsewhere have shown that application of such organoarsenical pesticides and herbicides to soils may lead to high rates of soil contamination, arsenic mobility, and possible development of toxic inorganic arsenic that may be transported into groundwater (Cai *et al.*, 2002).

In this chapter, we present for the first time Pb-isotopic data on MSMA, expand our Pb-isotopic database of historical arsenical pesticides used in New England (Ayuso et al., 2004b), and update results of previous studies using Pb-isotopic ratios as tracers of man-made and natural metal sources in the near-surface environment (Ayuso et al., 2006). Several areas in New England have groundwater wells supplied by bedrock aquifers that are anomalously rich in arsenic (Ayotte et al., 2003; Marvinney et al., 1994; Peters et al., 1999). Within one of these areas, we selected a small watershed near Northport, coastal Maine (Fig. 12.1) to test, using the isotopes of lead, whether the effects of arsenical pesticide and MSMA use can be discerned from those of other potential natural and anthropogenic sources. Possible natural sources consist of widely dispersed areas of sulfide mineralization (Pb–Zn and other base-metal mines) as well as areas of disseminated sulfide-rich schists, gneisses, and granitic pegmatites. This small area is also the focus of extensive geological studies on processes leading to arsenic mobilization and groundwater enrichment (Ayuso and Foley, 2002, in press; Foley and Ayuso, in press; Foley et al., 2004a; Horesh, 2001; Lipfert and Reeve, 2004; Lipfert et al., 2006, 2007).

Pb-isotopic compositions are useful tools to investigate the sources and mobility of lead, and by inference, other geochemically similar metals. Biological, physical, and chemical processes in the near-surface environment do not disturb the Pb-isotopic signatures. Among the Pb isotopes, ^{206}Pb, ^{207}Pb, and ^{208}Pb evolve as a result of radioactive decay of ^{238}U, ^{235}U, and ^{232}Th, respectively. ^{204}Pb, in contrast, does not have a radioactive parent. Natural (rock-derived) materials thus contain different amounts of radioactive parent and daughter lead isotopes that evolve as a function of time (half-life) and parent/daughter isotopic ratios (e.g., $^{238}U/^{204}Pb$, $^{232}Th/^{204}Pb$) into distinctive Pb-isotopic signatures that can be used as tracers of geological materials. Anthropogenic metal sources using ore lead during the manufacturing process may also have diagnostic Pb-isotopic ratios and lead contents depending on the source of the ore. Typically, however, environmental samples (e.g., soils) in a given region may range widely in Pb-isotopic ratios and lead contents. It has been postulated that this is due in part to the wide compositional range of lead ore used for industrial applications related to long-lived industrial and urban activities (e.g., burning of fossil fuels, mining and metal alloy industries, smelting, manufacturing, wood preservation, glass and ceramics industries, sewage, etc.), as well as agricultural activities (e.g., pesticides, herbicides, fertilizers, feed additives). Depending on site-specific environmental conditions, nature of geological materials, and flux of anthropogenic sources contributed to near-surface materials, Pb-isotopic ratios can be useful tools with which to evaluate the various contributions from natural and man-made point sources.

1.1. Previous work

Robinson and Ayuso (2004) used spatial statistical tests and lead isotope tracers to measure the influence of arsenical pesticide use on stream sediment chemistry in New England. Factor analysis on metal concentrations in ~1600 stream sediment samples were grouped according to their agricultural-index value (areas inferred to have used arsenical pesticides extensively). Pb and As contents were correlated, and

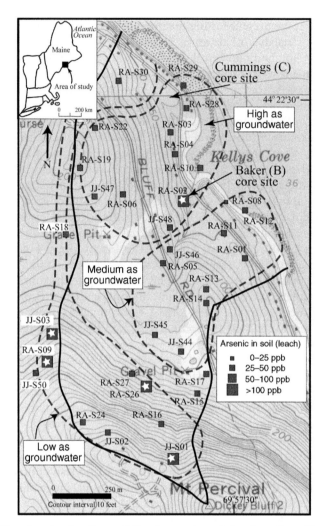

Figure 12.1 Map showing locations of sites (squares) used for the soil profiles and boreholes (C and B), and contents of arsenic in the leach fractions of the top layer of the soil profiles. Squares with stars denote samples used for the Pb isotope study; other squares are locations of profiles used in a geochemical study (Ayuso, unpublished data). Dashed lines enclose areas containing wells characterized by Lipfert and Reeve (2004) and Lipfert *et al.* (2007) as containing high-arsenic groundwater (As > 1.3 μmol L^{-1}), medium arsenic groundwater, and low arsenic groundwater in the Mount Percival recharge area (most wells have <0.13 μmol L^{-1}). Solid line encloses the drainage basin in this study.

the contents of these elements were found to be higher in areas that had higher agricultural-index values. Lead isotope compositions of stream sediments from a range of agricultural-index settings mostly overlapped the isotopic range of bedrock sulfides and their weathering products. A few of the stream sediments representing high agricultural-index settings were more radiogenic than the geologic background values and the radiogenic Pb was mostly attributed to industrial uses (atmospheric deposition). No lead isotope data were available for herbicides and pesticides for comparison, but Robinson and Ayuso (2004) speculated that the more radiogenic sediments collected from high agriculture-index settings may have included Pb and As from the arsenical pesticides.

2. GEOLOGIC SETTING

Rusty-weathering iron sulfide–rich rocks of the Lower Ordovician Penobscot Formation are the predominant rock type in the Northport area (Stewart, 1998). The upper part of the formation underlies the study area and consists of abundant pelitic beds (with sandy siltstone layers) containing andalusite. Limestone and calcareous sandstone are rare. Diorite dikes and muscovite granite pods and dikes are present locally. The formation is thinly covered by a layer of glacial sediments (till) derived from the underlying rocks by the Late Wisconsin Laurentide ice sheet (e.g., Hunter and Smith, 2001). Glaciomarine sediments were deposited as ice withdrew at about 14,000 and 12,200 YBP (Dorion et al., 2001). Till deposits consist of a mixture of clay, silt, sand, cobbles, and boulders.

Several generations of arsenic- and lead-bearing metal sulfides and secondary minerals have been documented in the Penobscot Formation (e.g., Ayuso and Foley, 2002, 2004; Foley et al., 2004a; Horesh, 2001; Lipfert and Reeve, 2004). Also, recently, Lipfert et al. (2007) provided a detailed analysis of the distribution of arsenic-enriched groundwater in fractured, crystalline bedrock in the Northport area. Domestic wells contain groundwater with up to 26.6 μmol L^{-1} and the distribution of As in the groundwater is geographically influenced by the discharge area of the watershed (Lipfert et al., 2006). Such As concentrations in the groundwater exceed the maximum contaminant level value of 0.01 mg L^{-1} for drinking water (US Environmental Protection Agency, 2002). An evaluation of the likely arsenic-bearing primary and secondary source minerals in the Penobscot Formation that could be available for reaction with groundwater was given by Foley and Ayuso (in press). In order of decreasing As content by weight, the primary mineral sources include: löllingite and realgar ($\sim$70%), arsenopyrite, cobaltite, glaucodot, and gersdorffite (in the range of 34–45%), arsenian pyrite (<4%), and pyrrhotite (<0.15%). The most commonly observed highly altered minerals in the Penobscot Formation are pyrrhotite, realgar, niccolite, löllingite, glaucodot, arsenopyrite-cobaltian > arsenopyrite, cobaltite, gersdorffite, fine-grained pyrite, Ni-pyrite > coarse-grained pyrite.

Reactions illustrate that oxidation of Fe–As disulfide group and As-sulfide minerals is the primary release process for As. Liberation of As by carbonation of realgar and orpiment in contact with high pH groundwaters may contribute locally to elevated contents of As in groundwater, especially where As is decoupled from Fe. Released metals are sequestered in secondary minerals by sorption or by incorporation in crystal structures. Secondary minerals acting as intermediate As reservoirs include claudetite (~75%), orpiment (61%), scorodite (~45%), secondary arsenopyrite (~46%), goethite (<4490 ppm), natrojarosite (<42 ppm), rosenite, melanterite, ferrihydrite, and Mn-hydroxide coatings. Some soils also contain Fe–Co–Ni-arsenate, Ca-arsenate, and carbonate minerals. Reductive dissolution of Fe-oxide minerals may govern the ultimate release of iron and arsenic—especially As(V)—to groundwater; however, dissolution of claudetite (arsenic trioxide) may directly contribute As(III).

Foley and Ayuso (2008) suggest that typical processes that could explain the release of arsenic from minerals in bedrock include oxidation of arsenian pyrite or arsenopyrite, or carbonation of As-sulfides, and these in general rely on discrete minerals or on a fairly limited series of minerals. In contrast, in the Penobscot Formation and other metasedimentary rocks of coastal Maine, oxidation of arsenic-bearing iron–cobalt–nickel-sulfide minerals, dissolution (by reduction) of arsenic-bearing secondary arsenic and iron hydroxide and sulfate minerals, carbonation and/or oxidation of As-sulfide minerals, and desorption of arsenic from Fe-hydroxide mineral surfaces are all thought to be implicated. All of these processes contribute to the occurrence of arsenic in groundwaters in coastal Maine, as a result of the variability in composition and overlap in stability of the arsenic source minerals. Also, Lipfert *et al.* (2007) concluded that as sea level rose, environmental conditions favored reduction of bedrock minerals, and that under the current anaerobic conditions in the bedrock, bacteria reduction of the Fe-and Mn-oxyhydroxides are implicated with arsenic releases.

Arsenic contents of soils and groundwater thus reflect the predominant influence and integration of a spectrum of primary mineral reservoirs (not just single or unique mineral reservoirs). Cycling of arsenic through metasedimentary bedrock aquifers may therefore depend on consecutive stages of carbonation, oxidation, and reductive dissolution of primary and secondary arsenic host minerals. Moreover, Lipfert *et al.* (2007) have also shown on the basis of the high concentrations of arsenic and enriched sulfur and oxygen isotopic ratios that the As-enriched Fe-and Mn-oxyhydroxides reflect paleo-aeration and oxidation of the crystalline bedrock.

Ayuso and Foley (2008) concluded that Pb-isotopic signatures establish pathways by which lead, and by inference arsenic, could have been transported from arsenic-bearing minerals, via sulfide oxidation or carbonation reactions into multiple generations of secondary minerals (Fe-hydroxides, Fe-sulfates, and others). Pb-isotopic compositions of the sulfides from the Penobscot Formation and coexisting Fe-hydroxides and Fe-sulfates produced by weathering and alteration overlap, but the secondary minerals extend toward more radiogenic values that broadly indicate the addition of lead from anthropogenic origin. Ayuso and Foley (2008) suggest that the Pb-isotopic compositions are consistent with multiple sources: although predominantly from the Penobscot Formation, the Pb-isotopic signatures include anthropogenic lead from industrial activities and perhaps from agricultural activities. Characterization of the lingering effects of historical arsenical pesticide use and present use of organochemical pesticide (e.g., MSMA) is likely to receive increased interest by the environmental community.

2.1. Soil types in the Northport area

The major protolith of the soils in this portion of coastal Maine is glacial till. Marine and lacustrine sediments can be locally important. The predominant soil type in the watershed (Fig. 12.2) is found along the coast and consists of deep, nearly level to steep soils that range from well drained to poorly drained (Hedstrom and Popp, 1981). The dominant soil associations are characterized by a surface layer of dark brown soils with fine sandy loam at the top ($\sim$10–15 cm thick), overlying a layer of gray to orange brown fine sandy loam ($\sim$4–7 cm). Deeper in the profiles, dark orange brown silty to sandy loam to gray and green brown pebbly to sandy loam ($\sim$25–35 cm) were characteristic. Another soil type in the watershed has moderately well-drained to poorly drained soils that formed in marine and lacustrine sediments. Glaciofluvial sediments, organic material, and alluvium may also constitute significant components in the soils.

The five soil profiles obtained for this study (Ayuso *et al.*, 2006) contain unstratified clay to boulder-sized constituents (generally up to about 10 cm). Soil, till, and

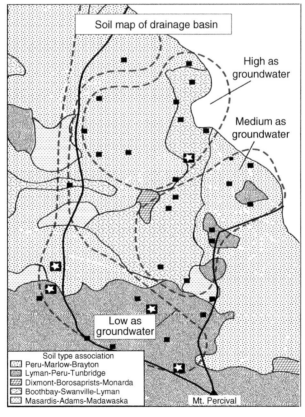

Figure 12.2 Map of the drainage basin showing generalized distribution of the soil type associations (modified from Hedstrom and Popp, 1981). Symbols as in Fig. 12.1.

other near-surface materials make up a relatively thin (<1.5-m thick) layer that overlies bedrock. Glacial till was derived locally. The top of the profiles, to a depth of about 8–12 cm consist of organic-rich and irregularly bedded layers. The bottom of the profiles (~60 cm) consists of intensely weathered, saprolitized, and disaggregated, rusty-weathering sulfidic schistose bedrock.

The soils contain various silicate, carbonate, oxide, and rare sulfide minerals (Foley and Ayuso, 2008; Foley *et al.*, 2004a) and typically show the addition of organic material. Quartz, feldspar, and mica, including sericite and chlorite, are the dominant silciates. Fe-hydroxides and Fe-sulfates are the predominant secondary minerals in the soils and underlying bedrock, particularly below the organic layer (Ayuso and Foley, 2002; Foley and Ayuso, 2008; Foley *et al.*, 2004a). Pyrrhotite, greigite, sphalerite, galena, djurleite, and arsenopyrite are rare; carbonate (calcite and manganoan siderite) and Fe-bearing phosphate minerals also occur.

2.2. Sampling of soil profiles

Soil profiles at five sites were used for the Pb-isotopic study, and these represent a subset of 34 total sites (Fig. 12.1) that have been investigated for an ongoing geochemical study (Ayuso and Foley, 2008; Ayuso *et al.*, 2006). Pits ~30–40 cm in diameter were hand dug in the midsummer to expose profiles, free of forest litter, to depths of at least 60 cm. Five or six samples (about 150–500 g) were obtained, spaced ~10 cm apart, in order to select representatives of the morphological variations in the soils and tills. Samples from the top of the profiles are from the uppermost surface layers of the soils, as characterized by color (e.g., most are dark brown), particle size, and morphology (silt or sandy loam). Most soils in the area exhibited contrasting color and compositional features distinguishing the soils, subsoils, and lower horizons. The surface layer was clearly evident in the watershed (mostly representative of the A1 horizon, and in some cases the Ap soil horizon; Hedstrom and Popp, 1981). This topmost layer is at or near the surface and consists of humified organic matter and minerals, and may include a plowed surface horizon that is part of the underlying, mineral horizon (B). Bedrock was typically found at depths of less than 60 cm.

2.3. Analytical techniques

The pesticides, herbicides, and additives were dissolved in dilute HNO_3 and these solutions were passed through Pb exchange columns using normal elution protocols (Ayuso *et al.*, 2004a). Isotopic compositions were measured using a Finnigan-MAT 262 thermal ionization mass spectrometer (TIMS). Mass fractionation was monitored by frequent analysis of the NIST-SRM 981 standard. Lead blanks during the course of the study were less than 50 ng and thus are not significant compared to the lead abundances in the samples. A mass fractionation correction of about 0.1% per amu was applied to all samples. Maximum analytical uncertainties (± 2 standard error of the mean) are as follows: $^{206}Pb/^{204}Pb$ < 0.005, $^{206}Pb/^{207}Pb$ < 0.00007, and $^{208}Pb/^{207}Pb$ < 0.00006. Table 12.1 summarizes the Pb isotope data for the

Table 12.1 Lead isotopic compositions of pesticides, herbicides, and feed additives

Compound label (purity if known)	Formula	Manufacturer and date (if known)	$^{206}Pb/^{204}Pb$	$^{206}Pb/^{207}Pb$	$^{208}Pb/^{207}Pb$
Historical pesticides and herbicides used in New England					
Calcium arsenate[a]	$Ca_3(AsO_4)_2$	Pfaltz and Bauer Inc., 1981	18.617	1.19345	2.46316
Calcium arsenate[a]	$Ca_3(AsO_4)_2$		18.620	1.19350	2.46320
Calcium arsenate[b]	$Ca_3(AsO_4)_2$	Pfaltz and Bauer Inc., ca. 1981	18.615	1.19343	2.46314
Calcium arsenate[c]	$Ca_3(AsO_4)_2$		18.627	1.19367	2.46326
Calcium arsenate[c]	$Ca_3(AsO_4)_2$		18.630	1.19322	2.46447
Sodium metaarsenite (98.0%)[a]	$NaAsO_2$	Aldrich Chemical Co.	18.759	1.20102	2.47212
Sodium metaarsenite (98.0%)[b]	$NaAsO_2$		18.755	1.20100	2.47209
Sodium metaarsenite (98.0%)[b]	$NaAsO_2$		18.761	1.20104	2.47215
Lead arsenate (acid) (91.6%)[a]	$PbHAsO_4$	Pfaltz and Bauer Inc., 1976	18.395	1.18026	2.46083
Lead arsenate (acid) (91.6%)[b]	$PbHAsO_4$		18.393	1.18024	2.46080
Lead arsenate (acid) (91.6%)[b]	$PbHAsO_4$		18.398	1.18029	2.46086
Acme arsenate of lead (basic)[a]	$Pb_5OH_9(AsO_4)_3$	PBI/Gordon Corp., Acme Quality Paints	17.070	1.10352	2.38394
Acme arsenate of lead (basic)[a]	$Pb_5OH_9(AsO_4)_3$	(from H. Evans, USGS), ca. 1950	17.075	1.10354	2.38397
Acme arsenate of lead (basic)[b]	$Pb_5OH_9(AsO_4)_3$		17.068	1.10348	2.38390
Copper acetoarsenite (90.0%)[a]	$C_4H_6As_6Cu_4O_{16}$	Sherwin–Williams Co., Paris green, 1985	18.153	1.16649	2.44191

Table 12.1 (*Continued*)

Compound label (purity if known)	Formula	Manufacturer and date (if known)	$^{206}Pb/^{204}Pb$	$^{206}Pb/^{207}Pb$	$^{208}Pb/^{207}Pb$
Copper acetoarsenite (90.0%)[b]	$C_4H_6As_6Cu_4O_{16}$		18.149	1.16644	2.44185
Copper acetoarsenite (90.0%)[b]	$C_4H_6As_6Cu_4O_{16}$		18.161	1.16652	2.44197
Other pesticides and herbicides					
Arsenous oxide[a]	As_2O_3	Union Mechling Co. (from J. Ayotte, USGS)	18.153	1.16587	2.43421
Arsenic oxide (99.95%)[a]	As_2O_3	Aldrich Chemical Co.	17.342	1.11826	2.39982
Arsenic oxide (99.95%)[c]	As_2O_3		17.461	1.12251	2.39617
Monosodium methylarsonate (MSMA)					
Methyl arsonic acid (99.9%)[a]	CH_5AsO_3	Ricerca Inc., 1993	18.381	1.17881	2.46001
Methyl arsonic acid (99.9%)[b]	CH_5AsO_3		18.154	1.16534	2.42713
Methane arsonic acid (99.7%)[b]	CH_5AsO_3	Vineland Chemical Co., 1978	18.333	1.17644	2.43727
Methane arsonic acid (99.7%)[b]	CH_5AsO_3	Vineland Chemical Co., 1982	18.042	1.16337	2.44030
	CH_5AsO_3		18.288	1.17139	2.43823

Methane arsonic acid (98.5%)[c]					
Methane arsonic acid (98.5%)[c]	CH_5AsO_3		18.250	1.17063	2.43671
Methane arsonic acid (98.5%)[c]	CH_5AsO_3		18.401	1.17895	2.44452
Methane arsonic acid (98.5%)[c]	CH_5AsO_3		17.103	1.10731	2.37905
Feed additives (poultry and swine growth)					
Arsanilic acid (9.62%)[a]	$C_6H_8AsNO_3$	Fleming Labs Inc., 1990	18.754	1.20321	2.47068
Arsanilic acid (99.86%)[a]	$C_6H_8AsNO_3$	Fleming Labs Inc., 1990	18.411	1.17985	2.46093
Arsanilic acid (99.86%)[c]	$C_6H_8AsNO_3$		18.771	1.20100	2.46899
Arsanilic acid (99.86%)[c]	$C_6H_8AsNO_3$		18.547	1.18833	2.45000

[a] Ayuso et al. (2004b).
[b] Ayuso et al. (2006).
[c] This study.

pesticides, herbicides, and additives determined in this study and from preliminary reports (Ayuso *et al.*, 2004b, 2006).

Soil samples were air dried and prepared for lead isotope analysis using 50–100 mg of the <0.2-mm size fraction. Sulfides, Fe-hydroxides, and other secondary minerals in the Penobscot Formation can be used to monitor the composition of labile Pb and provide the means to discriminate labile (anthropogenic) lead from lead inherited from the parent rocks and sulfides (e.g., Ayuso and Foley, 2008). A cool mild acid leach (1.5N HCl + 3N HNO_3) was used to attack the secondary minerals. This solution likely reflects the labile Pb (e.g., Erel *et al.*, 1997) captured in the Fe-hydroxide, carbonate, or organic materials, or other secondary minerals (clays). These minerals can contain lead, arsenic, and other elements derived from outside of the watershed. Mixed solutions of HF–HNO_3 were used for final dissolution of the residual fractions. Table 12.2 summarizes the Pb-isotopic data for the leach fractions of the soil horizons (together with Pb and As contents); Table 12.3 shows equivalent data for the residues.

Trace element compositions and mineralogy of the soils and Penobscot Formation were obtained by various analytical techniques including XRD, XRF, ICP-MS, and INAA. Table 12.4 is a summary of the trace element data for the leach aliquots obtained on the soils in the Pb isotope study, and for the Penobscot Formation.

3. RESULTS

A preliminary assessment of Pb-isotopic signatures of the arsenical pesticides and herbicides shows that considerable diversity in values exists (Table 12.1 and Fig. 12.3). The pesticides range in values of $^{206}Pb/^{207}Pb$ = 1.1035–1.2010, $^{208}Pb/^{207}Pb$ = 2.3839–2.4722, and $^{206}Pb/^{204}Pb$ = 17.070–18.761. For MSMA, the range in ratios of $^{206}Pb/^{207}Pb$ = 1.1073–1.1790, $^{208}Pb/^{207}Pb$ = 2.3791–2.4600, and $^{206}Pb/^{204}Pb$ = 17.103–18.401 overlaps the historical arsenical pesticides. Feed additives used to promote swine and poultry growth also have isotopic ratios that overlap the arsenical pesticides (Table 12.1). The arsenical pesticides plot along a trend for values of $^{206}Pb/^{207}Pb$ and $^{208}Pb/^{207}Pb$ (Fig. 12.3). Some of the pesticides show large isotope differences even for similar compounds. For example, lead arsenate (acid form) ($^{206}Pb/^{204}Pb$ = 18.395) is significantly more radiogenic than lead arsenate (basic form) ($^{206}Pb/^{204}Pb$ = 17.070), consistent with a different source of lead used in their manufacture.

The three most common historical arsenical pesticides used in New England (lead arsenate-acid form, calcium arsenate, and sodium arsenate) have similar isotopic signatures. As a group, they are somewhat more radiogenic than other pesticides (Fig. 12.3). Arsenious oxide and arsenic oxide are less radiogenic than the more commonly used pesticides in New England (Fig. 12.3). Other compounds such as MSMA (including methyl arsonic acid and methane arsonic acid), copper acetoarsenite (or Paris green) have Pb-isotopic signatures that overlap the historical arsenical pesticides, but extend to significantly less radiogenic ratios of $^{206}Pb/^{207}Pb$ and

Table 12.2 Pb and As contents and Pb isotope compositions of leach samples from soil profiles, Northport area, coastal Maine

Leach sample/ horizon	Pb (ppb) average (range)	As (ppb) average (range)	$^{206}Pb/^{204}Pb$ average (range)	$^{206}Pb/^{207}Pb$ average (range)	$^{208}Pb/^{207}Pb$ average (range)
0–10 cm, all sites	296 (76–606)	372 (35.1–3960)	18.655 (18.601–18.751)	1.19577 (1.19046–1.20111)	2.46014 (2.45188–2.46186)
0–10 cm, isotope study	252 (76–606)	985 (40.3–3960)			
0–10 cm, JJ-ME–02-S01A	86.5	70.5	18.751	1.20111	2.46280
0–10 cm, JJ-ME–02-S03A	606	696	18.601	1.19046	2.45188
0–10 cm, RA-ME–02-S02A	76	40.3	18.621	1.19465	2.45896
0–10 cm, RA-ME–02-S09A	366	159	18.619	1.19274	2.45518
0–10 cm, RA-ME–02-S26A	124	3960	18.685	1.19989	2.46186
10–20 cm, all sites	42.3 (20.4–29.8)	170 (9.4–2030)	18.704 (18.603–18.774)	1.19854 (1.19351–1.20252)	2.46847 (2.46314–2.47236)
10–20 cm, isotope study	42.3 (20.4–29.8)	639 (17.6–2030)			
10–20 cm, JJ-ME–02-S01B	20.4	57.1	18.716	1.19964	2.47119
10–20 cm, JJ-ME–02-S03B	112	1040	18.603	1.19351	2.46314
10–20 cm, RA-ME–02-S02B	22.3	17.6	18.661	1.19564	2.46488
10–20 cm, RA-ME–02-S09B	29.8	47.7	18.774	1.20252	2.47236
10–20 cm, RA-ME–02-S26B	27.1	2030	18.766	1.20138	2.47079
20–30 cm, all sites	26.4 (11.9–47.4)	874 (47.1–2150)	18.734 (18.583–18.836)	1.19418 (1.18703–1.20355)	2.47669 (2.47327–2.48761)
20–30 cm, isotope study	25.9 (11.9–47.4)	702 (14.2–2150)			
20–30 cm, JJ-ME–02-S01C	20.35	48	18.742	1.19957	2.47456
20–30 cm, JJ-ME–02-S03C	47.35	1250	18.583	1.19038	2.47400
20–30 cm, RA-ME–02-S02C	24.1	14.2	18.752	1.19038	2.47400
20–30 cm, RA-ME–02-S09C	26	47.05	18.757	1.20355	2.47327
20–30 cm, RA-ME–02-S26C	11.85	2150	18.836	1.18703	2.48761

(continued)

Table 12.2 (*Continued*)

Leach sample/ horizon	Pb (ppb) average (range)	As (ppb) average (range)	$^{206}Pb/^{204}Pb$ average (range)	$^{206}Pb/^{207}Pb$ average (range)	$^{208}Pb/^{207}Pb$ average (range)
30–40 cm, all sites	29.1 (9.6–65.9)	843 (34.1–1860)			
30–40 cm, isotope study	27.7 (9.60–65.9)	678 (16.0–1860)	18.741 (18.731– 18.822)	1.20048 (1.19147– 1.20685)	2.47232 (2.46895– 2.47630)
30–40 cm, JJ-ME-02-S01D	17.85	34.1	18.730	1.19960	2.47418
30–40 cm, JJ-ME-02-S03D	65.85	1440	18.596	1.19147	2.46588
30–40 cm, RA-ME-02-S02D	22.4	16	18.753	1.20253	2.46895
30–40 cm, RA-ME-02-S09D	22.9	37.95	18.802	1.20197	2.47628
30–40 cm, RA-ME-02-S26D	9.62	1860	18.822	1.20685	2.47630
40–50 cm, all sites	16.9 (12.2–19.7)	716 (48.9–1780)			
40–50 cm, isotope study	19.0 (12.2–27.5)	577 (17.8–1780)	18.902 (18.733– 18.810)	1.20780 (1.19128– 1.23280)	2.47812 (2.46994– 2.49296)
40–50 cm, JJ-ME-02-S01E	17.5	48.9	18.733	1.20106	2.47354
40–50 cm, JJ-ME-02-S03E	18.2	979	18.762	1.19123	2.46994
40–50 cm, RA-ME-02-S02E	27.5	17.8			
40–50 cm, RA-ME-02-S09E	19.7	57.5	19.303	1.23280	2.49296
40–50 cm, RA-ME-02-S26E	12.2	1780	18.810	1.20613	2.47605

Table 12.3 Pb and As contents and Pb isotope compositions of residue (Bulk) samples from soil profiles, Northport area, coastal Maine

Residue (bulk) sample/horizon	Pb (ppm) average (range)	As (ppm) average (range)	$^{206}Pb/^{204}Pb$ average (range)	$^{206}Pb/^{207}Pb$ average (range)	$^{208}Pb/^{207}Pb$ average (range)
0–10 cm, all sites	35.9 (23.7–59.6)	176 (37.9–517)	18.916 (18.831–18.976)	1.20958 (1.20560–1.21375)	2.47517 (2.46654–2.48104)
0–10 cm, isotope study	33.7 (23.7–59.6)	144 (13.9–517)			
0–10 cm, JJ-ME–02-S01A	23.7	102	18.900	1.21075	2.48104
0–10 cm, JJ-ME–02-S03A	59.6	37.9	18.831	1.20597	2.46982
0–10 cm, RA-ME–02-S02A	24.6	13.9	18.965	1.21182	2.47863
0–10 cm, RA-ME–02-S09A	30.0	48.1	18.908	1.20560	2.46654
0–10 cm, RA-ME–02-S26A	30.5	517	18.976	1.21375	2.47984
10–20 cm, all sites	22.8 (15.0–37)	62.7 (9.9–606)	19.075 (18.925–19.303)	1.22054 (1.21182–1.23280)	2.48753 (2.47863–2.49296)
10–20 cm, isotope study	18.7 (15.0–22.1)	172 (13.0–606)			
10–20 cm, JJ-ME–02-S01B	15.0	109	19.054	1.21817	2.48291
10–20 cm, JJ-ME–02-S03B	17.4	71.2	19.303	1.23280	2.49296
10–20 cm, RA-ME–02-S02B	20.2	13	19.076	1.22230	2.49712
10–20 cm, RA-ME–02-S09B	18.7	61.1	19.019	1.21515	2.48477
10–20 cm, RA-ME–02-S26B	22.1	606	18.925	1.21428	2.47989
20–30 cm, all sites	21.8 (16.6–24.3)	221 (62.0–657)	19.031 (18.893–19.267)	1.21812 (1.21042–1.23151)	2.48750 (2.48156–2.50001)
20–30 cm, isotope study	20.9 (16.9–23.2)	183 (32.0–657)			
20–30 cm, JJ-ME–02-S01C	16.6	97.0	18.893	1.21042	2.48182
20–30 cm, JJ-ME–02-S03C	23.0	68.0	19.267	1.23151	2.49179
20–30 cm, RA-ME–02-S02C	16.9	32.0	19.045	1.21853	2.50001
20–30 cm, RA-ME–02-S09C	24.3	62.0	18.953	1.21285	2.48233
20–30 cm, RA-ME–02-S26C	23.5	657.0	18.995	1.21728	2.48156

(continued)

Table 12.3 *(Continued)*

Residue (bulk) sample/horizon	Pb (ppm) average (range)	As (ppm) average (range)	$^{206}Pb/^{204}Pb$ average (range)	$^{206}Pb/^{207}Pb$ average (range)	$^{208}Pb/^{207}Pb$ average (range)
30–40 cm, all sites	21.9 (20.5–23.2)	203 (55.0–616)			
30–40 cm, isotope study	20.9 (16.9–23.2)	170 (34–616)	18.996 (18.937–19.041)	1.21571 (1.21163–1.21884)	2.48674 (2.48024–2.49473)
30–40 cm, JJ-ME–02-S01D	22.1	83	18.961	1.21393	2.48585
30–40 cm, JJ-ME–02-S03D	23.2	55	19.041	1.21884	2.49242
30–40 cm, RA-ME–02-S02D	16.9	34	19.019	1.21659	2.49473
30–40 cm, RA-ME–02-S09D	20.5	61	18.937	1.21163	2.48024
30–40 cm, RA-ME–02-S26D	21.9	616	19.022	1.21755	2.48068
40–50 cm, all sites	19.1 (8.9–47.1)	225 (59.0–640)			
40–50 cm, isotope study	18.9 (8.9–28.9)	192 (59–640)	18.961 (18.905–19.026)	1.21392 (1.20992–1.21774)	2.48641 (2.48008–2.49028)
40–50 cm, JJ-ME–02-S01E	25.1	83	19.026	1.21774	2.49028
40–50 cm, JJ-ME–02-S03E	8.9	118	19.001	1.21648	2.48665
40–50 cm, RA-ME–02-S02E	19.1	60			
40–50 cm, RA-ME–02-S09E	12.8	59	18.914	1.21154	2.48008
40–50 cm, RA-ME–02-S26E	28.9	640	18.905	1.20992	2.48864

Table 12.4 Average of selected trace element compositions of the soil horizons (leach fractions), and the Penobscot Formation (whole rocks), Northport area

	Soils (ppb)					Penobscot formation (ppm)	
	0–10 cm	10–20 cm	20–30 cm	30–40 cm	40–50 cm	Outcrops ($n = 9$)	Cores ($n = 19$)
As	372	170	874	843	716	28.2	82.6
Pb	296	42.3	26.4	29.1	16.9	28.8	17.15
Be	12	17.4	22.7	23.1	20.5	3.44	nd
Sb	10.3	6.98	5.32	5.37	6.27	0.37	0.5
Hg	bdl	1.19	1.08	bdl	bdl	9.32	bdl
Bi	bdl	bdl	0.84	0.75	0.907	0.35	0.4
Se	42.2	65.4	29	24	17	0.58	0.7
Tl	1.6	2.18	1.97	2.17	2.01	1.49	1.3
Sc	nd	nd	138.5	118.8	132	18.3	18.9
V	308	294	270	256	244	179	115
Ni	261	236	63.6	79.5	325	39.9	576.1
Co	463	632	1031	1338	1355	8.75	24.9
Cr	131	167	151	135	88.2	98.8	2404
Ga	18	12.4	10.4	13	11.4	26.9	25
Ge	1.37	1.65	1.22	1.067	1.6	2.27	1.9
Cu	24.5	17.1	24.2	16.1	13.2	29.9	30.8
Zn	934	832	865	734	644	105.5	119.8
Mo	10	6.4	4.16	4.17	4.75	5.43	5.8
Sn	4.88	3.36	2.66	2.98	1.54	3.58	8.7
W	10.4	11.1	12.38	9.3	5.4	1.63	4.5
Cd	18.9	17.6	16.4	16.3	13.5	9.32	nd

bdl, below detection limits; nd, not determined.

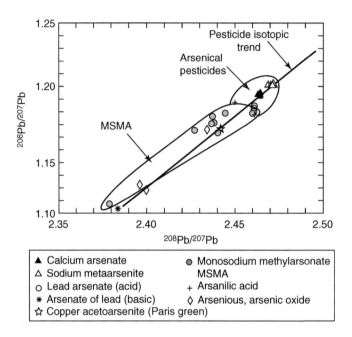

Figure 12.3 $^{206}Pb/^{207}Pb$ versus $^{208}Pb/^{207}Pb$ plot showing the compositions of arsenical pesticides, herbicides, and feed additives.

$^{208}Pb/^{207}Pb$. This isotopic difference is notable and suggests that these types of pesticides could potentially be distinguished in Pb-bearing minerals formed in the near-surface environment.

Soil profiles from the Coastal Maine watershed (Fig. 12.2) were selected to test whether the effects of arsenical pesticide use in the area could be discerned from other potential natural and anthropogenic lead sources. Soil profiles of 48 samples (leachates and residues) show a moderate range in $^{208}Pb/^{207}Pb = 2.4519–2.4876$, and $^{206}Pb/^{204}Pb = 18.583–18.836$ and $^{206}Pb/^{207}Pb = 1.1870–1.2069$ (Table 12.2 and Fig. 12.4). Leach fractions are lowest in $^{208}Pb/^{207}Pb$ for the uppermost soils (surface layers within 10 cm from the surface), in contrast to the values obtained deeper in the profiles (up to about 50 cm from the surface) so that the profiles are stratified isotopically (Fig. 12.4F). Labile Pb isotope compositions are not homogeneous at 20–30 cm. Soil and glacial processes controlling the dissolution and re-precipitation of secondary minerals did not homogenize the lead, similar to results obtained on temperate forest near-surface tills (Emmanuel and Erel, 2002). Compositions of the residues range in $^{206}Pb/^{207}Pb = 1.2056–1.2328$, $^{208}Pb/^{207}Pb = 2.4661–2.5000$, and $^{206}Pb/^{204}Pb = 18.831–19.303$ and are distinctly more radiogenic than the leach fractions (Table 12.2 and Fig. 12.4).

Pb and As contents of the bulk soil samples in the watershed show a wide range (Table 12.3). The content of Pb is highest in the surface layer (average ~36 ppm). As shows no systematic trend with depth (~176 ppm at the top). Variations in the contents of Pb and As in the leach fractions of the depth intervals (Table 12.2) show

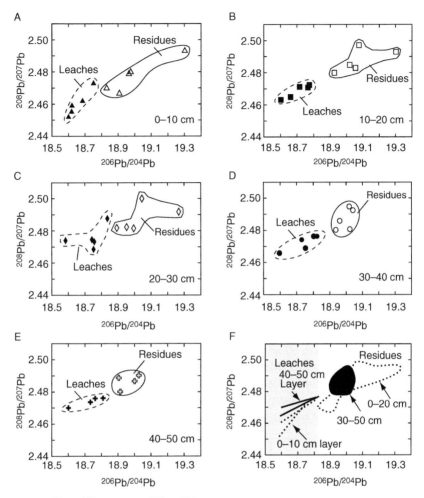

Figure 12.4 ^{208}Pb/^{207}Pb versus ^{206}Pb/^{204}Pb plots showing the acid-leach and the residue fractions at different depth intervals. (A) 0–10 cm. (B) 10–20 cm. (C) 20–30 cm. (D) 30–40 cm. (E) 40–50 cm. (F) Summary plot showing estimated trends of the leach fractions and the fields for the residue fractions. The leach fractions (labile lead) are shown by lines of decreasing slope as a function of depth. The residue fractions (silicate lead) are more homogeneous in the deeper portion (>30 cm) of the profiles.

that Pb contents decrease with depth (average = 252 ppb in the surface soil to 19 ppb in the bottom horizon). Average As contents, however, show no clear trend, but generally decrease from 985 ppb in the uppermost soil surface layer to 577 ppb in the bottom horizon. The leach fractions are also most variable for Pb and As in the soil surface layer (Pb ~76 ppb to ~606 ppb; As ~40 ppb to ~3900 ppb). Average contents of C_{total} (surface soil = 6.17 wt.%, bottom horizon = 1.73 wt.%) and pH (surface soil = 4.55, bottom horizon = 5.11) show consistent differences as a function of depth. The values for S_{total}, however, show no clear trend along the profiles (surface soil = 0.03%, bottom horizon = 0.02%) (Ayuso, unpublished data).

Variations in trace element contents of the sulfidic Penobscot Formation reflect the range from pelite-dominated to sandy silts, and broad variability in contents of sulfide minerals (Table 12.4). Fe_2O_3, Al_2O_3, MgO, and CaO contents, for example, vary widely (Ayuso, unpublished data). Although the outcrop and drill core samples have similar compositional features, arsenic contents can differ widely (Table 12.4). Whole-rock samples from drill core are somewhat depleted in lead (average $\sim$17 ppm) and enriched in arsenic (average $\sim$83 ppm), compared to the outcrop samples (average Pb $\sim$29 ppm; average As $\sim$28 ppm) (Table 12.4).

4. DISCUSSION

Before 1914, the US imported most metallic arsenic from Germany (e.g., Smith, 1945). From 1914 to about 1930, the bulk of the arsenic production in the US (marketed for commercial applications as As_2O_3, arsenic trioxide, or white arsenic) was derived as a metallurgical by-product of the smelting of copper, lead, and gold. Arsenic trioxide was used in the production of fertilizers, herbicides, and insecticides (Kirk-Othmer, 1992; Ullman's Encyclopedia, 1998), or if transformed to arsenic acid, used in the manufacture of CCA, a preservative of wood products (Mineral Commodity Summaries, 2004). Arsenic metal is also used for solders, ammunition, anti-friction additive to bearings, and in the computer and electronics industry for semiconductors (Mineral Commodity Summaries, 2004).

Three major companies accounted for the bulk of US arsenic production in the first half of the twentieth century: American Smelting and Refining Co. (ASARCO; both copper and lead smelting in several domestic plants), Anaconda Copper Co. (copper smelting), and US Smelting Co. (lead smelting), along with several minor producers. Notably, from 1974 to 1985, the domestic supply of arsenic was controlled by ASARCO, and since 1985 by imports primarily from China, Chile, and Mexico (Mineral Commodity Summaries, 2004).

The Pb-isotopic signature of arsenic trioxide (As_2O_3), starting material for the manufacture of the arsenical pesticides, will closely reflect the composition of the sulfide minerals (e.g., arsenian sulfides such as arsenopyrite, and arsenian pyrite) used during the smelting of copper, lead, and gold ore. Production data for the major manufacturers of arsenic trioxide from the 1920s to the 1980s, and isotope data in the literature show that the sulfide ores used in the production of arsenic trioxide had substantial isotopic differences (Fig. 12.5). Also, as a group, the arsenical compounds have distinctive Pb-isotopic signatures that in some cases may be uniquely traced to domestic industrial arsenic trioxide suppliers in the US and perhaps even suggest the dominant sulfide ores from specific mining districts. This information can be useful to constrain the possible provenance of the individual pesticides.

In the case of lead arsenate (acid form), calcium arsenate, and sodium arsenate, the Pb-isotopic compositions closely match those of sulfides from porphyry copper deposits from southeastern Arizona, specifically from the Pima and Silver Bell districts (Bouse *et al.*, 1999). Notably, other major historical producers of arsenic trioxide (e.g., Anaconda Copper Co., US Smelting Co., Jardine Mining Co.) used

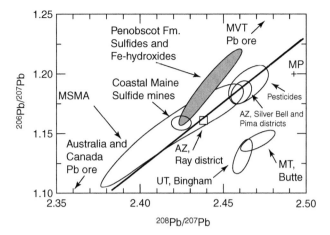

Figure 12.5 $^{206}Pb/^{207}Pb$ versus $^{208}Pb/^{207}Pb$ plot showing the field of arsenical pesticides and herbicides, mining districts from the southern portion of the Southeast Arizona terrane (Arizona, Pima, and Ray), and other districts in Montana (Butte) and Utah (Bingham) likely used in the production of arsenic trioxide. Lead isotope data for porphyry copper deposits in Arizona are from Bouse *et al.* (1999), porphyry copper deposits in Montana are from Murthy and Patterson (1961), and that in Utah are from Stacey *et al.* (1968). Also shown are compositions of Mississippi Valley type (MVT) ore deposits (Doe and Delevaux, 1972), Pb ore from Mexico and Peru (Chow *et al.*, 1975), shown as MP, and Pb ore from Australia and Canada (Cumming and Richards, 1975). The figure also shows the field of galena and other sulfide minerals from massive sulfide deposits from coastal Maine (Foley *et al.*, 2004b), sulfides at various stages of weathering and secondary Fe-hydroxides (representative of the natural background compositions) in the Penobscot Formation (Ayuso and Foley, 2002). Symbols are given in Fig. 12.3.

sulfide ore during their smelting operations that differed greatly from southeastern Arizona because of the different rock ages and metallogenic settings. For example, the porphyry deposits from Montana and Utah (Murthy and Patterson, 1961; Stacey *et al.*, 1968) have higher $^{208}Pb/^{207}Pb$ and lower $^{206}Pb/^{207}Pb$ than ores from southeastern Arizona (Fig. 12.5). Sulfides from the northern domain of southeastern Arizona (e.g., Ray district) also produced ore that is isotopically distinct and is similar to the composition of Paris green (copper acetoarsenite). Other pesticides, such as arsenious oxide and arsenic oxide, arsenate of Pb (basic form), that plot along the pesticide isotope trend of Fig. 12.3 reflect lead sources having much lower values of $^{208}Pb/^{207}Pb$, and $^{206}Pb/^{207}Pb$ than the mines of southeastern Arizona. As a group, currently used MSMA pesticides (Table 12.1) are notably less radiogenic than any of the domestic historical and current lead ore producers (Fig. 12.3). The signatures are low enough in $^{208}Pb/^{207}Pb$ and $^{206}Pb/^{207}Pb$ to suggest foreign sources such as manufacture from lead ore mined from Australia and Canada.

4.1. Pb and As in the soil profiles

A fundamental step toward determining the ultimate source of lead (and by inference other elements) depends on accurately assessing the contributions from all of the possible natural and anthropogenic point sources. We now turn to the question of whether the contributions of the arsenical pesticides can be distinguished from the effects of anthropogenic lead derived from fossil fuel combustion in the Northport watershed area.

Baseline concentrations of arsenic in US near-surface till are $\sim$7.4 mg kg^{-1} (Shacklette *et al.*, 1971), although concentrations can vary broadly because of the heterogeneous distribution of sulfide minerals and redox conditions. Soils that have not been greatly disturbed retain arsenic and lead near the top of the soil horizon, suggesting that metal migration is not significant (Peryea, 1998; Veneman *et al.*, 1983). In the Northport area, the soil profiles show considerable Pb-isotopic variations from top to bottom. Leach fractions are lowest in ^{208}Pb/^{207}Pb in the surface soil layer, 10 cm from the surface, but there is no clear isotope gap that can be used to distinguish a limit for the possible effects of near-surface contaminants such as industrial lead from atmospheric deposition versus lead derived from weathering of bedrock.

The dominant source of anthropogenic lead is industrial emissions involving combustion of alkyl-lead gasoline additives (Chow *et al.*, 1975) distributed by long-range atmospheric transport (e.g., Nriagu and Pacyna, 1988) producing heavily polluted soils enriched in lead, with possible associated enrichments in As, Zn, and Cd (e.g., Steinnes *et al.*, 1989). A direct relation between the Pb contents, Pb-isotopic variations, and As contents remains equivocal. Perhaps this reflects the contrasting solubilities of Pb (low solubility at high soil pH) and As (complexing and solubility contrasts with pH) in the near-surface layers, and the various effects of mechanical disturbance (tillage and digging) near the surface (Peryea, 1998).

Compared to the Pb-isotopic values of lead arsenate, calcium arsenate, and sodium arsenate pesticides, isotope compositions of industrial lead in the 1960s to the 1980s are somewhat shifted toward lower values of ^{208}Pb/^{207}Pb relative to ^{206}Pb/^{207}Pb (Fig. 12.6). Mississippi Valley type (MVT) deposits contain generally radiogenic lead (high values of ^{208}Pb/^{207}Pb and ^{206}Pb/^{207}Pb in Fig. 12.5; Doe and Delevaux, 1972) and provided most of the lead ore used for industrial applications in the US, and for gasoline additives from about 1964 to 1976 (Rosman *et al.*, 1994). From $\sim$1930 to 1970, the period of most intensive application of pesticides in New England, anthropogenic lead throughout the US was primarily derived from combustion of alkyl-lead gasoline additives and input of lead to the atmosphere (^{206}Pb/^{207}Pb $\sim$1.23 and ^{208}Pb/^{207}Pb $\sim$2.46: Chow *et al.*, 1975). Because the supply of lead changed from domestic to foreign sources, ^{206}Pb/^{207}Pb in aerosols and gasoline evolved from $\sim$1.15 in the mid-1960s (Erel *et al.*, 1997) to $\sim$1.22 in the 1970s and early 1980s (Figs. 12.5 and 12.6). Values of ^{206}Pb/^{207}Pb decreased after 1976 (Rosman *et al.*, 1994), partly as a result of decreasing gasoline consumption, lead recycling, and imports from Australia, Canada, Mexico, and Peru (Bollhöfer and Rosman, 2000, 2001; Graney *et al.*, 1995). By 1997–1999 (Fig. 12.6), some eastern US aerosols were less radiogenic than values measured in alkyl-lead in the 1970s (Bollhöfer and Rosman, 2001).

The natural geochemical cycle of lead has also been greatly disturbed by a variety of anthropogenic inputs. These include emissions from industrial plants and ore smelters (Ayuso *et al.*, 2001; De Vivo *et al.*, 2001; Dunlap *et al.*, 1999; Graney *et al.*, 1995; Tarzia *et al.*, 2002), effects of urbanization (Erel *et al.*, 1997; Hansmann and Koppel, 2000; Hopper *et al.*, 1991; Teutsch *et al.*, 2001), and by other human

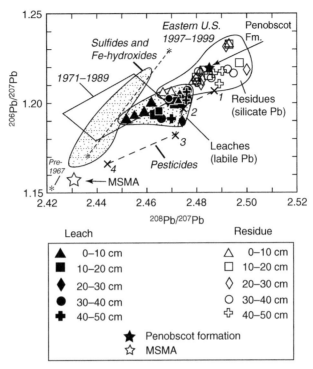

Figure 12.6 $^{206}Pb/^{207}Pb$ versus $^{208}Pb/^{207}Pb$ plot of lead arsenate (#1), sodium metarsenite (#2), calcium arsenate (#3), Cu-acetoarsenite as (#4), and MSMA (open star), compositions of the soil profiles, average composition of the Penobscot Fm., generalized field for sulfides and Fe-hydroxides from the Penboscot Fm. (Foley *et al.*, 2004a), and US aerosols (1971–1989, Rosman *et al.*, 1994) and a dashed line representing the range for atmospheric Pb in the Eastern US for 1997–1999 (Bollhöfer and Rosman, 2001). MP, lead from Mexico and Peru. See text for additional explanation and references.

activities which emit metals (As, Sb, Cd, Se, Hg, etc.) into the atmosphere and surface environment (e.g., Steinnes *et al.*, 1992). However, little arsenic is thought to be contributed to surface and groundwater bodies by precipitation, unless a major source of arsenic exists locally (Smedley and Kinniburgh, 2002). Extensive pesticide use in agricultural lands can significantly increase the amount of anthropogenic lead because of reaction of surface water runoff with contaminated soils (e.g., Erel and Patterson, 1994; Kober *et al.*, 1999). Pb-isotopic compositions of stream sediments inferred to have the highest intensity of pesticide use show considerable overlap in Pb-isotopic compositions with the arsenical pesticides (Ayuso and Foley, 2008; Robinson and Ayuso, 2004, and unpublished data).

The Pb-isotopic trend from top to bottom of the soil profiles indicates that compositions of US aerosols (industrial lead) overlap the leach fractions of the surface soil layers (0–10 cm) (Fig. 12.6). The data suggest that the profiles are influenced by Pb-isotopic contributions from the aerosols perhaps dominated by those in the period of 1997–1999, perhaps generally decreasing in a trend from the top layer to the bottom layer. The Fe-hydroxides also overlap in composition with the leach fractions at the top of the profiles (Fig. 12.6; Ayuso and Foley, 2008). This feature agrees with studies showing that in temperate forest soils, the Fe-hydroxides have a high affinity for lead (Emmanuel and Erel, 2002). Some studies show that the residence time of lead in water (Jaffe and Hites, 1986) and in such temperate forest floors is also thought to be short, probably less than 80 years (e.g., Erel *et al.*, 1997; Miller and Friedland, 1994). Arsenic, as in the case of lead, also has a strong preference for Fe-hydroxides under surface conditions (e.g., Korte and Fernando, 1991).

Leach fractions from deeper in the soil profiles (>10 cm) are somewhat shifted from the field of aerosols and gasoline, and the field of the Fe-hydroxides toward higher values of $^{208}Pb/^{207}Pb$ and toward the field of the arsenical pesticides (Fig. 12.6). Higher values of $^{208}Pb/^{207}Pb$ than the Fe-hydroxides and pesticides in deeper portions of the soil profiles may indicate a more prominent role for imported lead sources (Peruvian and Mexican Pb ores), or a natural source that has not yet been identified (Fig. 12.6).

The lead arsenate-acid form, calcium arsenate, and sodium arsenate pesticides do not have a sufficiently distinct Pb-isotopic range that would allow quantification of their contribution to the anthropogenic lead in the soils. The Pb-isotopic composi-tions of MSMA are too low in $^{208}Pb/^{207}Pb$ and $^{206}Pb/^{207}Pb$ to have affected the soil compositions (Fig. 12.6). Average compositions of the top layers of the profiles, however, plot as a cluster, intermediate to the rock-derived and anthropogenic point sources, and consistent with the idea that labile Pb reflects involvement of several natural and anthropogenic components (Fig. 12.7). This figure shows the average compositions of the top layer of the soils (0–10 cm, A1 and Ap horizons) used in the isotope study, the 40–50 cm layer (near the bottom of the profiles), as well as average isotope compositions of the top layers in areas of the watershed identified as containing high-As groundwater, and the top layers in the area identified as the water recharge zone (Lipfert and Reeve, 2004; Lipfert *et al.*, 2007). The soil field essentially encloses the total isotope variation of the profiles in the watershed. For the purpose of illustrating possible combinations of point sources that may have con-tributed anthropogenic lead to the soils, representative compositions of aerosols and gasoline are shown from pre-1967 to 1999. Qualitative mixing schemes can be devised to explain the observed isotope variations in the soils involving lead from arsenical pesticides, Fe-hydroxides and sulfides, and aerosols (Fig. 12.7). A mixing scheme is illustrated that involves goethite with a high Pb content (Pb ∼492 ppm; As ∼4490 ppm), goethite with a low Pb content (Pb ∼10 ppm; As ∼14 ppm), and lead arsenate. Only the sample representing the average composition of the lowest soil horizon plots outside of the mixing triangle, adjacent to the average composition of the Penobscot Formation. Similar mixing schemes involving sulfides (e.g., pyrite, Pb ∼22–330 ppm; As ∼30–1290 ppm) in the Penobscot Formation, and lead from aerosols would also overlap the data for the soils. The exact role of the pesticides is difficult to evaluate quantitatively due to the inherent isotopic variability of the arsenic-bearing ore used in the production of the pesticides, and a lack of information about the identity, dates, and amounts of pesticide applied in the area.

This study was carried out to test whether Pb isotope compositions of soils and tills can be linked to use of arsenical pesticides, other anthropogenic point sources,

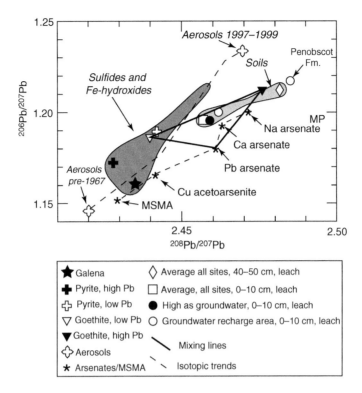

Figure 12.7 $^{206}Pb/^{207}Pb$ versus $^{208}Pb/^{207}Pb$ plot showing the lead isotopic compositions of lead arsenate, sodium metarsenite, calcium arsenate, Cu-acetoarsenite, and MSMA (dashed line connects average compositions), and a field for soils that includes the average compositions of the top layer of the soils (roughly equivalent to the A1 and Ap horizons), the top layer of soils for all sites in the geochemical study, and the average compositions of the deep layer at 40–50 cm (near the bottom of the profiles). Also shown are the isotope compositions of the top layers in areas within the watershed containing wells with groundwater characterized as high-As type, and the top layer in the area identified as the water recharge zone (Lipfert and Reeve, 2004; Lipfert et al., 2007). Pb-isotopic compositions of US aerosols and gasoline produced since the 1960s vary widely and the figure shows a trend of estimated isotopic compositions starting pre-1967 to about 1999 (sources of data include Bollhöfer and Rosman, 2000, 2001; Chow et al., 1975; Erel et al., 1997; Nriagu and Pacyna, 1988; Rosman et al., 1994; Sturges and Barrie, 1987 and references therein).

and likely parental rocks (Penobscot Formation). The Pb-isotopic data indicate that a derivation of Pb in soil and till profiles from the local sulfide-rich country rocks remains the best explanation of the lead isotope compositions, and is the most likely primary source of arsenic in local groundwater. Contribution of Pb from combustion of fossil fuels is likely, especially at the top of the profiles. A major addition of lead from arsenical pesticides is possible but cannot be proven at this time.

   The most important natural source of lead (and by inference arsenic) is likely derived from the breakdown of sulfide-bearing bedrock during glaciation and subsequent weathering of the oxidation zone overlying the groundwater (Ayuso and Foley, 2002, 2004; Foley and Ayuso, 2008; Foley *et al.*, 2004a). Decomposition of the sulfide minerals as a result of fluid migration leached and oxidized bedrock and soils, producing Fe-hydroxides and other secondary minerals. As-bearing sulfides (e.g., pyrite and arsenopyrite), for example, would have been converted to limonite, goethite, and secondary sulfide minerals which captured lead (and by inference arsenic) in the original sulfide minerals. Subsequent reduction of secondary mineral phases is thought to release metals into the groundwater (e.g., As: Matisoff *et al.*, 1982; Stuben *et al.*, 2003).

5. CONCLUSIONS

Although the predominant source of arsenic and metals to most soils and sediments in New England is sulfide-rich rock, the extensive application of arsenical pesticides and herbicides (lead arsenate, calcium arsenate, and sodium arsenate, and others) on apple, blueberry, and potato fields may have been a possible anthropogenic source of arsenic and lead. The main objective of this study was to determine the lead isotopic compositions of commonly used pesticides, such as lead arsenate, sodium metarsenite, and calcium arsenate, in order to assist in future isotopic comparisons and to better characterize this anthropogenic source of Pb. The pesticides plot along a linear trend in isotope diagrams, for example, in values of $^{206}Pb/^{207}Pb$ and $^{208}Pb/^{207}Pb$.

Pb isotope compositions of near-surface soil and till profiles in coastal Maine were measured in an area known to have groundwater with anomalously high arsenic contents to determine the source of the Pb and by inference, to provide constraints on possible sources of As. Labile Pb from the soils and tills shows a moderate range in isotope values, distinctly less radiogenic than the residue fractions. Leach fractions are lowest in values of $^{208}Pb/^{207}Pb$ in the soil surface layers. Acid-leach compositions represent lead that is loosely bound in minerals (Fe- and Mn-hydroxides, carbonate, and micaceous minerals), and likely approximate the composition of Pb available to the regional groundwater. The profiles show that average Pb and As contents are generally highest in the surface soil layer. Pb isotope and other geochemical data show that a contribution of Pb from the local country rocks to the soil and till profiles remains the best explanation for the lead isotope compositions. Contribution of lead from combustion of fossil fuels is likely (especially at the top of the profiles).

ACKNOWLEDGMENTS

We would like to thank John Burns, Anna Colvin, Jeremy Dillingham, John Jackson, Ann Lyon, and Greg Wandless for field and laboratory assistance related to sample collection, mineral separation, and support in the Radiogenic Isotope Laboratory. This chapter is the result of research collaboration and discussion with numerous geologists. In particular, we acknowledge Andrew Reeve (University of Maine), and Robert Marvinney and Marc Loiselle (Geological Survey of Maine) for their contributions related to the understanding of arsenic geochemistry and hydrochemistry in coastal Maine. Thanks to journal reviewers Harvey Belkin and Suzanne Nicholson, whose comments and suggestions improved the chapter.

REFERENCES

Ayotte, J. D., Montgomery, D. L., Flanagan, S. M., and Robinson, K. W. (2003). Arsenic in ground water in eastern New England: Occurrence, controls, and human health implications. *Environ. Sci. Technol.* **37,** 2075–2083.

Ayuso, R. A., and Foley, N. K. (2002). Arsenic in New England: Mineralogical and geochemical studies of sources and enrichment pathways. *United States Geological Survey.* Open-file report 2002–454. [URL http://pubs.usgs.gov/of/2002/of02-454/].

Ayuso, R. A., and Foley, N. K. (2004). Geochemical and isotopic constraints on the sources of arsenic in ground water from Coastal Maine, New England. *In* "Proceedings of the 11th International Symposium on Water Rock Interaction" (R. Wanty and R. Seal, eds.), Vol. 2, pp. 801–805. Balkema, New York.

Ayuso, R. A., and Foley, N. K. (2008). Anthropogenic and natural lead isotopes in Fe-hydroxides and Fe-sulfates in a watershed associated with arsenic-enriched groundwater, Maine, U.S.A. *Geochem. Explor. Environ. Anal.* **8,** 77–89.

Ayuso, R. A., Callender, E., and Van Metre, P. (2001). Pb isotopes of lake sediments and effects of human activities on water quality. *In* "Proceedings of the Tenth International Symposium on Water Rock Interaction" (R. Cidu, ed.), Vol. 2, pp. 1473–1475. Balkema, New York.

Ayuso, R. A., Kelley, K. D., Leach, D. L., Young, L. E., Slack, J. F., Wandless, G., Lyon, A. M., and Dillingham, J. L. (2004a). Origin of the Red Dog Zn-Pb-Ag deposits, Brooks Range, Alaska: Evidence from regional Pb and Sr isotope sources. *Econ. Geol.* **99,** 1533–1554.

Ayuso, R., Foley, N., Robinson, G., Jr., Wandless, G., and Dillingham, J. (2004b). Lead isotopic compositions of common arsenical pesticides used in New England. *United States Geological Survey.* Open-file report 2004–1342, 14. [URL http://pubs.usgs.gov/of/2004/1342/].

Ayuso, R. A., Foley, N. K., Robinson, G. R., Jr., Colvin, A. S., Lipfert, G., and Reeve, A. S. (2006). Tracing lead isotopic compositions of common arsenical pesticides in a coastal Maine watershed containing arsenic-enriched ground water. *In* "Contaminated Soils, Sediments and Water", "Proceedings of the 21st Annual International Conference on Soils Sediments and Water," (P. T. Kostecki, E. J. Calabrese, and J. Dragun, eds.), Vol. 11, University of Massachusetts, Amherst, MA, Compact Disk publication (CD), 25 p.

Bollhöfer, A., and Rosman, K. J. R. (2000). Isotopic source signatures for atmospheric lead: The Southern Hemisphere. *Geochim. Cosmochim. Acta* **64,** 3251–3262.

Bollhöfer, A., and Rosman, K. J. R. (2001). Isotopic source signatures for atmospheric lead: The Northern Hemisphere. *Geochim. Cosmochim. Acta* **65,** 1727–1740.

Bouse, R. M., Ruiz, J., Titley, S. R., Tosdal, R. M., and Wooden, J. L. (1999). Lead isotope compositions of Late Cretaceous and Early Tertiary igneous rocks and sulfide minerals in Arizona: Implications for the sources of plutons and metals in porphyry copper deposits. *Econ. Geol.* **94,** 211–224.

Cai, Y., Cabrera, J. C., Georgiadis, M., and Jayachandran, K (2002). Assessment of arsenic mobility in the soils of some golf courses in South Florida. *Sci. Total Environ.* **291,** 123–134.

Chormann, F. H., Jr. (1985). The occurrence of arsenic in soils and stream sediments, town of Hudson, New Hampshire. *Durham, New Hampshire, University of New Hampshire.* Unpub. M.S. thesis, 155 p.

Chow, T. J., Snyder, C. B., and Earl, J. L. (1975). Isotope ratios of lead as pollutant source indicators. *In* "Symposium on Isotope Ratios as Pollutant Source and Behavior Indicators." Vienna, Austria, Nov. 18–22, 1974. International Atomic Energy Agency, Report: STI/PUB/382, 95–108.

Cumming, G. L., and Richards, J. R. (1975). Ore lead isotope ratios in a continuously changing earth. *Earth Planet. Sci. Lett.* **28,** 155–171.

D'Angelo, D., Norton, S. A., and Loiselle, M. C. (1996). "Historical Uses and Fate of Arsenic in Maine. Water Research Institute Completion Report." University of Maine, Orono, Maine, 24 p.

De Vivo, B., Somma, R., Ayuso, R. A., Calderoni, G., Lima, A., Pagliuca, S., and Sava, A. (2001). Pb isotopes and toxic metals in floodplain and stream sediments from the Volturno basin, Italy. *Environ. Geol.* **41,** 101–112.

Doe, B. R., and Delevaux, M. (1972). Source of lead in southeast Missouri galena ores. *Econ. Geol.* **67,** 409–425.

Dorion, C. G., Balco, G. A., Kaplan, M. R., Kreutz, K. J., Wright, J. D., and Borns, H. W., Jr. (2001). Stratigraphy, paleoceanography, chronology, and environment during deglaciation of eastern Maine. *In* "Deglacial History and Relative Sea-level Changes, Northern New England and Adjacent Canada" (T. K. Weddle and M. J. Retelle, eds.), Geological Society America Special Paper 351, pp. 215–242. Geological Society of America, Boulder, CO.

Dunlap, E. C., Steinnes, E., and Flegal, R. A. (1999). A synthesis of lead isotopes in two millennia of European air. *Earth Planet. Sci. Lett.* **167,** 81–88.

Emmanuel, S., and Erel, Y. (2002). Implications from concentrations and isotopic data for Pb partitioning processes in near-surface tills. *Geochim. Cosmochim. Acta* **66,** 2517–2527.

Erel, Y., and Patterson, C. R. (1994). Leakage of industrial lead into the hydrocycle. *Geochim. Cosmochim. Acta* **58,** 3289–3296.

Erel, Y., Veron, A., and Halicz, L. (1997). Tracing the transport of anthropogenic lead in the atmosphere and in near-surface tills using isotopic ratios. *Geochim. Cosmochim. Acta* **61,** 4495–4505.

Foley, N. K., and Ayuso, R. A. (2008). Mineral sources and pathways of arsenic release in a contaminated coastal watershed. *Geochem. Explor. Environ. Anal.* **8,** 59–75.

Foley, N. K., Ayuso, R. A., West, N., Dillingham, J., and Marvinney, R. G. (2004a). Rock-water pathways for arsenic in weathering metashales: Mineralogical evidence of dominant processes. *In* "Proceedings of the 11th International Symposium on Water Rock Interaction" (R. Wanty and R. Seal, eds.), Vol. 2, pp. 1493–1499. Balkema, New York.

Foley, N., Ayuso, R. A., Culbertson, C., Marvinney, R. G., and Beck, F. (2004b). Mining in downeast Maine: Progress toward a geoenvironmental model assessing consequences of mining metamorphosed massive sulfide deposits on the coast [abs]. *Geol. Soc. Am.* **36,** 82.

Graney, J. R., Halliday, A. N., Keeler, G. J., Nriagu, J. A., Robbins, J. A., and Norton, S. S. (1995). Isotopic record of lead pollution in lake sediments from the northeastern United States. *Geochim. Cosmochim. Acta* **59,** 1715–1728.

Hansmann, W., and Koppel, V. (2000). Lead isotopes as tracers of pollutants in near-surface tills. *Chem. Geol.* **171,** 123–144.

Hedstrom, G. T., and Popp, D. J. (1981). Soil survey of Waldo County, Maine. United States *Department of Agriculture, Soil Conservation Service,* 158 p.

Hopper, J. F., Ross, H. B., Sturges, W. T., and Barrie, L. A. (1991). Regional source discrimination of atmospheric aerosols in Europe using the composition of lead. *Tellus* **43B,** 45–60.

Horesh, M. Y. (2001). Geochemical investigation of a high arsenic cluster, Northport, Maine, U.S.A. M.S. thesis, *University of Maine,* Orono, Maine.

Hunter, L. E., and Smith, G. W. (2001). Morainal banks and the deglaciation of coastal Maine. *In* "Deglacial History and Relative Sea-level Changes, Northern New England and Adjacent Canada" (T. K. Weddle and M. J. Retelle, eds.), Geological Society America Special Paper 351, 151–170. Geological Society of America, Boulder, CO.

Jaffe, R., and Hites, R.A (1986). Fate of hazardous waste derived organic compounds in Lake Ontario. *Environ. Sci. Technol.* **20,** 267–274.

Kirk-Othmer Encyclopedia of Chemical Technology (1992). "Antibiotics to benzene, arsenic and arsenic alloys". 3rd edn., Vol. 1, pp. 624–659. John Wiley & Sons, New York.

Kober, B., Wessels, M., Bollhöfer, A., and Mangini, A. (1999). Pb isotopes in sediments of Lake Constance, Central Europe constrain the heavy metal pathways and the pollution history of the catchment, the lake and the regional atmosphere. *Geochim. Cosmochim. Acta* **63,** 1293–1303.

Korte, N. E., and Fernando, Q. (1991). A review of arsenic (III) in ground water. *Crit. Rev. Environ. Control* **21**(1), 39.

Lipfert, G., and Reeve, A (2004). Fracture-related geochemical controls on As concentrations in ground-water. *In* "Proceedings of the 11th International Symposium on Water Rock Interaction" (R. Wanty and R. Seal, eds.), Vol. 2. pp. 431–434. Balkema, New York.

Lipfert, G., Reeve, A. S., Sidle, W. C., and Marvinney, R. (2006). Geochemical patterns of arsenic-enriched ground water in fractured, crystalline bedrock, Northport, Maine, USA. *Appl. Geochem.* **21,** 528–545.

Lipfert, G., Sidle, W. C., Reeve, A. S., Ayuso, R. A., and Boyce, A. (2007). High arsenic concentrations and enriched sulfur and oxygen isotopes in a fractured-bedrock ground-water system. *Chem. Geol.* **242,** 385–399.

Marvinney, R. G., Loiselle, M. C., Hopeck, J. T., Braley, D., and Krueger, J.A (1994). Arsenic in Maine Ground water: An example from Buxton, Maine. *In* "Focus Conference on Eastern Regional Ground Water Issues Burlington," Vermont, National Ground Water Association. 701–715.

Matisoff, G., Khourey, C. J., Hall, J. F., Varnes, A. W., and Strain, W. H. (1982). The nature and source of arsenic in northeastern Ohio ground water. *Ground Water* **20,** 446–456.

Miller, E. K., and Friedland, A. J. (1994). Lead migration in forest near-surface tills: Response to changing atmospheric inputs. *Environ. Sci. Technol.* **28,** 662–669.

Mineral Commodity Summaries (2004). *Arsenic U.S. Geological Survey,* 197.

Murthy, V. R., and Patterson, C. (1961). Lead isotopes in ores and rocks of Butte, Montana. *Econ. Geol.* **56,** 59–67.

Nriagu, J. O., and Pacyna, J. M. (1988). Quantitative assessment of worldwide contamination of air, water, and soils by trace metals. *Nature* **333,** 134–139.

Peryea, F. J. (1998). Historical use of lead arsenate insecticides, resulting soil contamination and implications for soil remediation. *In* "Proceedings of Symposium 25, 16th World Congress of Soil Science," Reg. No. 274. Montpellier, France. Science, 8.

Peters, S. C., Blum, J. D., Klaue, B., and Karagas, M. R. (1999). Arsenic occurrence in New Hampshire ground water. *Environ. Sci. Technol.* **33,** 1328–1333.

Robinson, G. R., and Ayuso, R. A. (2004). Use of spatial statistics and isotopic tracers to measure the influence of arsenical pesticide use on stream sediment chemistry in New England, USA. *Appl. Geochem.* **19,** 1097–1110.

Rosman, K. J. R., Chisholm, W., Boutron, C. F., Candelone, J. P., and Hong, S. (1994). Isotopic evidence to account for changes in the concentration of lead in Greenland snow between 1960 and 1988. *Geochim. Cosmochim. Acta* **58,** 3265–3269.

Shacklette, H. T., Hamilton, J. C., Boerngen, J. G., and Bowles, J. M. (1971). Elemental composition of surficial materials in the conterminous United States. *U.S. Geological Survey Professional Paper* 574-D, D1-D71.

Smedley, P. L., and Kinniburgh, D. G. (2002). A review of the source, behaviour and distribution of arsenic in natural waters. *Appl. Geochem.* **17,** 517–568.

Smith, W. (1945). Arsenic, Chapter IV. *In* "Handbook of Non-ferrous Metallurgy" (D. M. Liddell, ed.), Vol. II, pp. 94–103. McGraw-Hill Book Company, Inc., New York.

Solo-Gabriele, H., Sakura-Lemessy, D. M., Townsend, T., Dubey, B., and Jembeck, J. (2003). Quantities of arsenic within the State of Florida Report 03–06, State University System of Florida.

Stacey, J. S., Zartman, R. E., and Nkomo, I. T. (1968). A lead isotope study of galenas and selected feldspars from mining districts in Utah. *Econ. Geol.* **63,** 796–814.

Steinnes, E., Solberg, W., Petersen, H. M., and Wren, C. D. (1989). Heavy metal pollution by long range atmospheric transport in natural soils of southern Norway. *Water Air Soil Pollut.* **45,** 207–218.

Steinnes, E., Rambek, J. P., and Hanssen, J. E. (1992). Large-scale multi-element survey of atmospheric deposition using naturally growing moss as a biomonitor. *Chemosphere* **25,** 735–752.

Stewart, D. B. (1998). Geology of the northern Penobscot Bay, Maine. *U.S. Geological Survey, Miscellaneous Investigations Series Map I-2551.*

Stilwell, D., Toner, M., and Sawhney, B. (2003). Dislodgeable copper, chromium, and arsenic from CCA-treated wood surfaces. *Sci. Total Environ.* **312,** 123–131.

Stuben, D., Berner, Z., Chandrasekharam, D., and Karmakar, J. (2003). Arsenic enrichment in ground water of West Bengal, India: Geochemical evidence for remobilization under reducing conditions. *Appl. Geochem.* **18,** 1417–1434.

Sturges, W. T., and Barrie, L. A. (1987). Lead 206/207 isotope ratios in the atmosphere of North America: Tracers of American and Canadian emissions. *Nature* **329,** 144–146.

Tarzia, M., De Vivo, B., Somma, R., Ayuso, R. A., McGill, R. A. R., and Parrish, R. R (2002). Anthropogenic vs. natural pollution: An environmental study of an industrial site under remediation (Naples, Italy). *Geochem. Explor. Environ. Anal.* **2,** 45–56.

Teutsch, N., Erel, Y., Halicz, L., and Banin, A. (2001). Distribution of natural and anthropogenic lead in Mediterranean near-surface tills. *Geochim. Cosmochim. Acta* **65,** 2853–2864.

ULLMAN'S Encyclopedia (1998). "IndustrialInorganic Chemicals and Products, Aluminum Compounds, Inorganic to Carbon Monoxide, Arsenic Compounds." Vol. 1, pp. 343–367. Wiley-VCH, New York.

US Environmental Protection Agency (2002). Arsenic Rule Implementation: Implementation Guidance for the Arsenic Rule-Drinking Water Regulations for Arsenic and Clarifications for Compliance and New Source Contaminant Monitoring. EPA-816-K-02–018 (available on line at URL: http://www.epa.gov/safewater/ars/implement.html).

Veneman, P. L. M., Jr., Murray, J. R., and Baker, J. H. (1983). Spatial distribution of pesticide residues in a former apple orchard. *J. Environ. Qual.* **12,** 101–104.

CHAPTER THIRTEEN

Environmental Impact of the Disposal of Solid By-Products from Municipal Solid Waste Incineration Processes

Francesco Pepe*

Contents

Abstract

Incineration is often regarded as a very efficient technique for municipal solid waste (MSW) management. However, the environmental impacts of MSW incineration need to be carefully taken into account. The most relevant problem with MSW incineration is flue gas treatment. However, another often overlooked issue is the disposal of solid by-products of the incineration process. MSW incinerators essentially produce two types of solid by-products, that is, slag, or bottom ash, and fly ash, often mixed with various other chemicals used for flue gas treatment. Bottom ash and—even more—fly ash are regarded as dangerous wastes mainly due to their potentially toxic elements (PTE) content and their tendency to leach such PTE to the environment.

Different approaches have been proposed to mitigate the impact of fly ash disposal. The most commonly considered technology is stabilization in a cement matrix, with the aim of producing ash–concrete aggregates. Another, more radical, approach consists of using a plasma torch to heat fly ash to the point where it becomes a melt, which is then cooled down into a vitreous, nonleaching material. In light of its characteristics, plasma torch–based vitrification plants appear to constitute a promising answer for the disposal of solid by-products from MSW incineration processes. However, the high costs of plasma-based plants represent a serious drawback, particularly relevant in a context in which energy costs appear set to increase dramatically in the near future.

* Dipartimento di Ingegneria, Università del Sannio, Piazza Roma 21, 82100 Benevento, Italy

Environmental Geochemistry
DOI: 10.1016/B978-0-444-53159-9.00013-9

© 2008 Elsevier B.V.
All rights reserved.

1. INTRODUCTION

Disposal of municipal solid waste (MSW) is one of the most challenging environmental issues modern societies have to deal with. In Italy, the most recent data available is for the year 2004 (APAT-ONR, 2006). These data indicate an MSW production of about 3.1×10^{10} kg/year out of a total waste production of more than 1.3×10^{11} kg/year. This impressive value corresponds to an MSW production of 533 kg/(year·person) (or about 1.5 kg/(day·person)), a value which is similar to the European Union (EU) average of 537 kg/year·person (or 580 kg/year·person, if only the 15 States of the "older" EU are considered). In addition, the data indicate a slow but constant increase in MSW production for the last 10 years. This increase can be estimated at about two percentage points per year, both for Italy and for the EU as a whole. Finally, the same data indicate a linear correlation between MSW production and economic development expressed in terms of Gross National Product (GNP), which exists both on the national scale and, within each state, on a more local scale.

While in the past, landfilling of raw MSW was the preferred option, mainly due to the low direct costs of this disposal technique, this option is currently severely restricted in most developed countries because it is unpopular with the public. The main reason for the hostility encountered by raw MSW landfilling is that, when disposed off in a landfill, the putrescible fraction of the waste undergoes a number of uncontrolled anaerobic decomposition reactions, leading to the production of both leachate and biogas. As reported by Canter *et al.* (1988) and Lee and Jones (1991), leachate can be a heavily polluting liquid stream with high chemical oxygen demand (COD), often in excess of 5×10^4 mg/l and often characterized by the presence of appreciable quantities of potentially toxic elements (PTE), such as Cu ($\sim$5 mg/l), Pb ($\sim$1.0 mg/l), and Cr ($\sim$1.0 mg/l) (Pohland and Harper, 1985). Furthermore, biogas generated from landfills often contains large amounts of greenhouse gases (mainly CH_4), together with considerable amounts of volatile organic compounds (VOCs). Since the degradation processes in landfills may go on for decades (Lee and Jones, 1991), uncontrolled MSW landfills are often viewed as "environmental time bombs," particularly for their potential impacts on groundwater. In fact, even when—after no fewer than 30 years—the degradation processes can be considered as completed and the waste material is mineralized, serious problems often remain with the landfill area and its reuse.

Modern and well-designed landfills tend to limit the adverse environmental impacts cited above. In particular, they are typically insulated from the soil upon which they are constructed by a *multibarrier* insulation (Cossu, 1995), aimed at reducing, or even stopping, leachate percolation. This insulation is usually made of a layer of a low-permeability geological material such as clay, upon which one or more layers of polymeric material such as polyethylene are placed. In addition to multibarrier insulation, modern landfills are equipped with leachate collection systems, which allow interception of the leachate flow prior to its release to the geosphere, and biogas collection systems, which similarly allow biogas to be captured and even used to generate energy. However, despite the relevant progress which has been made in the design and operation of landfills, raw MSW landfilling is prohibited—except in some cases—in many countries, and it is being phased out in the EU. In particular, in Italy, a rather strict regulation (Legislative Decrees no. 22 of 2/2/1997 and no. 36 of 13/1/2003, both incorporating relevant EU Directives into Italian legislation) limits the use of landfills for untreated MSW only to exceptional cases.

To comply with existing laws and to reduce the environmental impact of MSW, a number of possible MSW management options have been considered. Recently, the EU (Directive 2006/12/CE) spelled out a hierarchical approach to the problem of waste management, indicating that priority should be given to the adoption of strategies aimed at reducing both the amount of waste generated and its noxiousness. According to the Directive, these goals should be achieved by (i) adopting "clean" technologies, which minimize the consumption of natural resources; (ii) bringing to the market "environmentally friendly" products, which during the whole course of their life lead to minimal amounts of waste; and (iii) developing appropriate techniques for reducing the quantity of dangerous substances in the waste, with the aim of facilitating its reuse. Since it is not possible to completely eliminate the generation of waste (the so-called zero waste option, which from time to time is mentioned in the public debate), the most realistic approach is the reuse of the waste generated, either in the form of matter (by such techniques as recycling or reusing), or by its conversion to energy.

As for the last option in this hierarchy, no consensus exists concerning the conversion of the waste to energy. In particular, a great deal of attention is currently focused on the choice between "thermal" and "nonthermal" treatments. Thermal treatments aim at converting waste into more stable forms by exposing it to high temperatures (on the order of no less than a few hundreds degrees Celsius), in the process releasing considerable amounts of energy. Nonthermal treatments tend to obtain essentially the same result by using low-temperature treatments (roughly $<100\,^{\circ}C$), which lead to the release of lower (if any) amounts of energy. Nonthermal treatments rely on a combination of mechanical and biological treatments (cumulatively indicated as MBT), and often include size reduction, mechanical sorting and the like, together with aerobic or anaerobic degradation processes. When compared to thermal treatments, MBT have the advantage of producing lower amounts of gaseous by-products and much lower concentrations of potentially harmful species. On the other hand, these processes lead to end products which have a considerably higher mass than those originated by thermal treatments, and therefore require larger landfill areas for final disposal.

2. MSW INCINERATION

MSW incineration is attractive because it greatly reduces both the mass and volume of treated waste. As reported by Arena and Mastellone (1998), the mass of the solid by-products (essentially slag deriving from the incineration process, plus a mixture of fly ash and other solids captured in the flue gas treatment section of the incineration plant) is about 25–30% of the original mass of the waste. Because the solid by-products have a fairly higher density than that of the original waste, their volume is about 10% of the original volume of the waste. Since the ultimate fate of these by-products generally is landfill disposal, and since landfill capacity depends on volume (rather than mass) of the disposed material, waste incineration allows a significant reduction in the need for landfills. In addition, MSW incineration can lead to the release of significant amounts of energy, which can be used for power generation and district heating. Untreated MSW has a higher heating value (HHV[1]) on the order of 7–10 MJ/kg (Niessen, 2002a), which compares rather favorably with coal (HHV of about 30 MJ/kg) and heavy oil (~40 MJ/kg). Since the amount of waste generated is quite high in most industrialized countries, the widespread use of MSW incineration processes has the potential of significantly contributing to the overall power generation requirements even for developed nations.

An important alternative to incineration of raw MSW that is finding an ever-growing interest is the incineration of a fuel produced from selected materials from the raw MSW, usually indicated as *refuse derived fuel*, or RDF (see sketch in Fig. 13.1). RDF is produced by mechanically (and sometimes also biologically) treating raw MSW, with the aim of "concentrating" the waste fractions that have a higher HHV, such as paper, plastic, cardboard, and textiles (provided one or more among these fractions have not previously been removed from the MSW stream to be sent to a material recovery process). On the other hand, the less combustible fractions of the waste (food wastes, yard wastes, glass, metals) are separated and sent for different treatment. RDF, in comparison to raw MSW, typically has a more uniform composition and granulometric distribution as well as a higher HHV (>15 MJ/kg). For these reasons, RDF incineration is characterized by a number of favorable features, including a "smoother" combustion process, lower polluting emissions, lower generation of solid by-products per unit mass of solid fed to the furnace, smaller combustion chamber volume requirements, and reduced ash melting inside the combustion chamber itself. In addition, it is possible to manufacture *high-quality RDF* that can be mixed with "traditional" solid fuels (usually coal) to be used in co-combustion processes inside traditional furnaces. On the other hand, RDF production requires a number of steps, such as grinding, size classification, air classification, and magnetic separation, that use devices generally characterized by high installation costs and even higher mantainance costs. In addition, RDF production leads to the necessity of setting up separate treatment and disposal procedures for the waste fractions which are not sent to incineration, and this unavoidably complicates the overall MSW management strategy and increases its cost.

MSW or RDF incineration is mostly carried out using continuous moving grate furnaces (see sketch in Fig. 13.2). A recent study (Enea and Federambiente, 2006) found that, in Italy, continuous moving grate furnaces account for more than 80% of total waste incineration capacity. Continuous moving grate furnaces generally are rather large installations (the largest furnace operating in Italy can treat more than

---

[1] The higher heating value (HHV) of a combustible substance is defined as the enthalpy which is released when 1 kg (or 1 Nm^3, for gases) of the substance is completely oxidized, and the water (if any) originating from the oxidation process is allowed to condensate as a liquid.

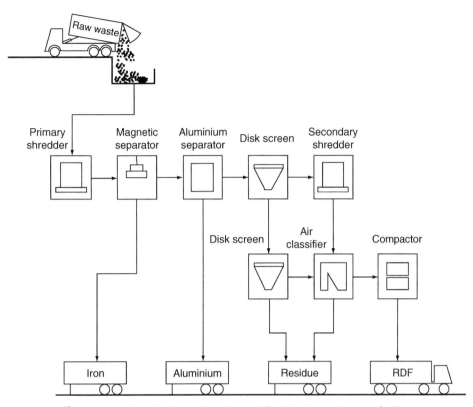

Figure 13.1   Sketch of an refuse derived fuel (RDF) production facility.

8×10^5 kg/day), characterized by a long, inclined refractory-lined combustion chamber, on the pavement of which the fuel is mass burnt. The pavement of the chamber is made of metallic grates through which air is fed to the chamber. The remaining combustion air is fed through the upper parts of the chamber. The air fed from below the grates, besides constituting a large fraction of the combustion air, helps cool the grates upon which the combustion process takes place. However, when RDF is incinerated, due to its higher HHV, water-cooled grates are often used. The metallic grates slowly move, rotate, or vibrate, thus allowing the burning material to travel along the combustion chamber, moving from the entrance section to the discharge section and facilitating mixing between waste and air. The gas originating from the incineration process is fed to a large postcombustion chamber, usually placed above the primary combustion chamber. The postcombustion chamber is maintained at an appropriate temperature ($\sim$900 °C), and retains and mixes the products of combustion, with the aim of reducing the concentration of incomplete combustion by-products, such as soot, CO, and hydrocarbons.

Downstream of the postcombustion chamber, the hot flue gas is fed to the energy recovery section of the plant, where it is cooled down in a sequence of heat exchangers, and its heat is used to produce steam for power generation and/or district heating. Eventually, the "spent" gas is sent to the flue gas treatment section of

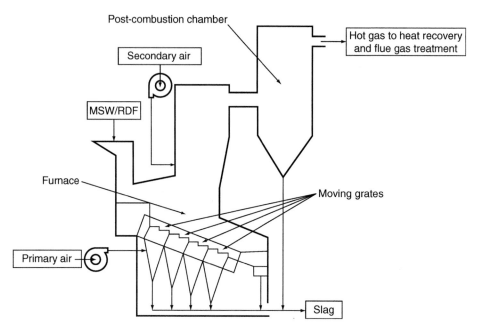

Figure 13.2   Sketch of a large-scale, continuous moving grate furnace.

the plant, where it is usually cooled by water, and then treated by a variety of techniques, such as wet and dry scrubbing, electrostatic precipitation, or bag filtering before being discharged to the atmosphere. The solid by-products are discharged from the combustion chamber in the form of slag and cooled by air or water. The slag is the largest fraction of the solid by-product of the incineration process, the remaining part being constituted by the solids obtained from flue gas treatment. In turn, such solids are a mixture of soot and fly ash captured by the particulate removal devices and other solid by-products of the flue gas treatment devices (e.g., sodium or calcium salts used for acid gas control, activated carbon used for adsorption).

   Other types of furnaces used for waste incineration are fluidized bed combustors and rotary kilns. Fluidized bed combustors (FBC) are currently available in two varieties, known as *bubbling* and *circulating* fluidized bed combustors, and account for about 15% of the overall waste incineration capacity in Italy (Enea and Federambiente, 2006). These devices require a finely divided RDF as fuel, and are characterized by a tall combustion chamber having either a round or a rectangular section (see sketch in Fig. 13.3). Inside the combustion chamber, a mixture between the combusting RDF and a solid, finely ground inert material (usually sand) is kept in suspension by a stream of air. Due to the very intense mixing between air and fuel, FBC have a number of very appealing features in comparison with moving grate furnaces. Their main advantage is that they are characterized by very high combustion rates, so that the volume of the combustion chamber required for the combustion of a given amount of RDF is about one order of magnitude lower than that of a moving grate furnace. In addition, FBC are characterized by a higher combustion efficiency, a lower excess air requirement (leading to a lower flue gas volume), and a lower

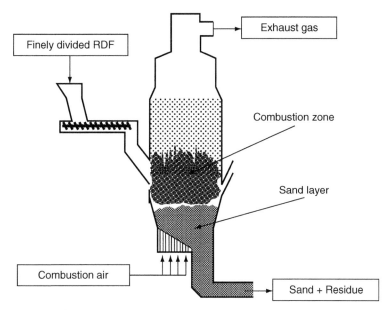

Figure 13.3 Sketch of a bubbling fluidized bed furnace.

combustion temperature (about 900 °C, compared to the 1200–1500 °C for moving grate furnaces), which in turn leads to a reduced NO_x production. On the other hand, due to their installation and operating costs, their greater fragility and the fact that they are still perceived as a relatively novel and not sufficiently proven technology, FBC have not gained that wider application that they could deserve in the field of waste incineration.

The last type of furnace worth mentioning is the rotary kiln furnace, which accounts for less than 5% of the waste incineration capacity in Italy (Enea and Federambiente, 2006). Rotary kilns are characterized by a long, slightly inclined (1–3%), slowly rotating cylindrical combustion chamber, and strongly resemble cement kilns. The combustion chamber is either internally covered by a refractory material or, as in the so called "O'Conner Rotary Combustor," it is lined with a web of water cooled tubes. The waste, either untreated MSW or RDF, is fed to the higher end of the chamber, while the slag is discharged from its lower end. While rotary kilns are often encountered in small-sized applications devoted to toxic waste incinerations, their use in the field of MSW management is severely limited by their poor energy efficiency.

3. TREATMENT OF FLUE GAS FROM MSW INCINERATION

Despite the fact that incineration is a very interesting MSW management technique, there are a number of issues concerning the environmental impact of this process that need to be carefully considered when evaluating the possibility of setting up an incineration plant. The first and foremost of such issues, also provoking the greatest public concern, is flue gas treatment. MSW incineration leads to the significant production of flue gas, resulting from the large volumes of air required for combustion. In the case of a moving grate furnace, more than 6 kg of flue gas (or about 5 Nm^3) are produced by the incineration of 1 kg of waste, and only slightly lower numbers are encountered if an FBC is used. However, the greatest worry posed by MSW incineration flue gas has to do with its composition, rather than with its amount. Untreated incineration flue gas contains a large number of pollutants, which are often divided between "macropollutants" (i.e., those having concentrations in the order of several ppm or more) and "micropollutants" (i.e., those having lower concentrations, but higher toxicity).[2] Among the former are CO, SO$_2$, HCl, NO$_x$, and particulate matter (fly ash, smoke, soot), while among the latter are polycyclic aromatic hydrocarbons (PAH), mercury compounds, and organo-halogenated compounds (e.g., polychlorinated biphenyls, dioxins, and furans).

The necessity of reducing the concentration of polluting species at levels below those set by existing regulations has, in the course of time, led to a growth in the importance of the flue gas treatment section of the incineration plant. In Italy, the latest requirements come from Legislative Decree 133/2005, which enforces EU Directive 2000/76/CE (see Table 13.1). However, it is important to recognize that the first contribution towards lower concentrations of polluting species is given by a close control of combustion conditions both in the primary chamber and in the postcombustion chamber. Indeed, constant combustion temperature and good contact between fuel (MSW or RDF) and oxidant (oxygen from air) leads to a very strong reduction in the concentration of incomplete combustion by-products such as CO, soot, and organic micropollutants.

When considering flue gas treatments itself, the available devices (a combination of which is usually chosen for each incineration plant) can be divided in two categories, depending on whether they are aimed at particulate control or gaseous pollutants control. Particulate control devices can take advantage of different physical processes, and the most common technologies include: cyclones, which remove relatively coarse particles using centrifugal separation; venturi scrubbers, which use liquid (usually water) droplets to capture the solid particles, essentially due to inertial impact, Brownian motion, and condensation effects; and electrostatic precipitators, which rely on electrostatic forces and are capable of removing even very fine particles. Bag filters, which remove particles using a large array of physical effects (sieving, inertial impact, direct interception, Brownian motion, etc.), are the most efficient devices available.

On the other hand, gaseous pollutant control can be carried out by contacting the gaseous stream with a component (either solid, liquid, or gaseous) that is capable of selectively removing one or more pollutants. One of the most common techniques used for the removal of gaseous pollutants is absorption, mainly for removal of

---

[2] In addition to these, some recent publications claim that a third class of pollutants, indicated as "nanopollutants" (essentially made of nm-sized solid particles) should be taken into account. However, scientific evidence available at the moment seems not to be enough to clarify on nanopollutants, their impact on health, and the techniques for their removal from flue gas.

Table 13.1 Emission limits from waste incineration plants according to Italian Legislative Decree 133/2005.

	Pollutant[a,b]	24 h	30 min A[c]	30 min B[c]
1	Particulate matter	10	30	10
2	Total organic carbon	10	20	10
3	HCl	10	60	10
4	HF	1	4	2
5	SO_2	50	200	50
6	NO_x	200	400	200
7	CO	100	50	150
8	Cd + Tl		0.05	
9	Hg		0.05	
10	Metals (Sb + As + Pb + Cr + Co + Cu + Mn + Ni + V)		0.5	
11	PCDD/DF[d] (I-TEQ[e]) (ng/m^3)		0.1	
12	PAH[f]		0.01	

[a] Concentrations expressed in mg/m^3 (except for item no. 11), and normalized to dry flue gas, 101.3 kPa, 273 K, 11% O_2.
[b] Continuous measurement for item nos. 1–7; average over 1 h sampling time for item nos. 8–10; average over 8 h sampling time for item nos. 11 and 12.
[c] Columns (A)/(B): if some of the 30-min averages for item nos. 1–6 do no conform to column (A), then 97% of the 30-min averages has to conform to column (B); if some of the 30-min averages for item no. 7 do no conform to column (A), then 95% of the 10-min averages has to conform to column (B).
[d] PCDD/DF: polychlorinated dibenzodioxins/dibenzofurans.
[e] I-TEQ: International Toxic Equivalent (NATO/CCMS, 1988)
[f] PAH: polycyclic aromatic hydrocarbons.

acidic gases, such as HCl, SO_2, and HF. In the absorption process, the pollutant is selectively absorbed from the gaseous stream using a liquid solution or suspension (often containing basic reactants). The spent suspension is then recycled or, after being concentrated, sent to a separate treatment plant in the form of a sludge. Common variants to the absorption process include *spray-dry* absorption, in which the sensible heat of the gas is used to evaporate most of the solution, with the aim of producing an almost dry solid by-product, rather than a sludge, and *dry* absorption, in which, more than an absorption process, a gas–solid chemical reaction between the gaseous pollutant and a suitable particulate solid (often $Ca(OH)_2$ or $NaHCO_3$) is carried out. Another flue gas–treating technique finding widespread use in connection with MSW incineration is adsorption. In this case, suitable solids, such as activated carbon or sulfur-impregnated activated carbon, having high surface area (Brunauer, Emmet, Teller (BET) in the order of 100 m^2/g: see Brunauer et al., 1938) and good selectivity characteristics, are used to target a number of specific pollutants, such as metallic mercury, mercuric chloride ($HgCl_2$), and organic and organo-halogenated micropollutants. Eventually, another common flue gas–treatment technique, specifically aimed at NO_x control, is selective catalytic conversion (SCR) or thermal (and therefore selective noncatalytic conversion, SNCR) of NO_x to N_2 by reaction with NH_3 or urea ($CO(NH_2)_2$).

As indicated earlier, incineration plants usually rely on a combination of the above mentioned devices. Enea and Federambiente (2006) found that different combinations are used in Italy, even though dry and semi-dry techniques are preferred for acid gas control and SNCR is favored over SCR for NO_x control, with a large use of activated carbon adsorption for removal of micropollutants.

4. SOLID BY-PRODUCTS FROM MSW INCINERATION

In addition to flue gas treatment, another issue posed by MSW incineration is the disposal of solid by-products and their impact on the environment. MSW incinerators essentially produce two kinds of solid wastes. The first is the slag, or bottom ash, which is partly discharged from the outlet of the combustion chamber and partly from the cracks in the grates. Such waste amounts to a sizable fraction of the material charged to the furnace (up to 30% w/w if raw MSW is incinerated, less if RDF is incinerated) and is essentially made of the noncombustible fractions of the waste (glass, metals, and ash deriving from wood and paper), mixed with part of the carbonaceous, combustible fraction which did not get combusted during its residence in the furnace. The second solid by-product is fly ash, the particulate material captured in the particulate removal section of the flue gas treatment plant. Depending on the technology used for the removal of micro- and macro-pollutants, fly ash is often captured in a mixture with various other chemicals (mostly calcium and/or sodium compounds, activated carbon, and various micropollutants), where the overall amount of such mixture is generally in the order of 5% (w/w) of the material originally charged to the furnace.

Bottom and fly ash have widely variable compositions and, in the case of bottom ash, also rather variable granulometric distribution (ranging from large clinkers to fine dusts), color, mechanical strength, and external appearances. The main constituents of bottom and fly ash are silicon, aluminum, iron, and calcium (see Table 13.2); these constituents and their compositions are similar to normal soil, except for iron, which can easily and economically be separated and recovered. Despite the data reported in Table 13.2, the main concerns related to bottom and fly ash originate from the fact that both (and particularly fly ash) have rather high contents of toxic compounds, and particularly of PTE such as lead, nickel, and mercury. Goldin et al. (1992) carried out an extensive analysis of data relative to the composition of MSW/RDF incineration bottom and fly ash (together with the original MSW/RDF) and their results are summarized in Table 13.3 (Niessen, 2002b). Table 13.3 clearly indicates that bottom and particularly fly ash tend to accumulate a number of PTE, posing very serious risks of leaching if exposed to the environment. In addition, some data indicate that polychlorinated biphenyls, dioxins, and furans can also be present in fly ash.

Management of bottom ash is a relatively less severe problem. In the case of a continuous moving grate furnace, bottom ash is discharged at the bottom of the

Table 13.2   Range of major elements in combined MSW incineration bottom and fly ash (from Styron and Gustin, 1992).

Species	Weight (%)
SiO_2	40–50
Al_2O_3	5–15
TiO_2	0.75–1.5
Fe_2O_3	12–25
CaO	8–15
MgO	1–2
K_2O	0.75–1.5
Na_2O	3–6
SO_3	0.5–1.5
P_2O_5	0.5–0.75
CuO	0.06–0.15
PbO	0.04–0.22
ZnO	0.12–0.22
LOI^a	1–3

[a] LOI = loss on ignition at 750 °C.
MSW, municipal solid waste.

combustion chamber directly into a container, onto suitable conveyors, or into water for quenching. After cooling, some iron recovery can be attempted, since the ash usually contains 6–9% iron by weight. Iron recovery, besides having some economic incentive by itself, helps reducing the overall amount of waste to be eventually sent to landfill disposal. After that, the largest particles of the ash can sometimes be used as "clean fill" in roadbeds and in earthworks. The use of bottom ash for these applications is often looked at with suspicion. However, in order to be used, bottom ash must be free of fly ash, must comply with the leaching tests for toxic materials (such as As, Cd, Cr, Pb, Hg, and Se), and must have a low loss on ignition (often below 6%) and a low quantity of putrescible material (often below 2%; van Beurden et al., 1997).

When no use can be found for the ash, then this need to be landfilled as a special waste or sometimes (if it does not comply with the toxicity leaching tests) as a dangerous waste. Clearly, its landfill disposal, particularly if it is labeled as a dangerous waste, poses environmental and economical issues, linked to the risk of leaching and to the sensible increases of the overall cost of MSW management.

Fly ash, on the other hand, even representing roughly just 10% of the overall ash produced by the incineration plant, constitutes a much more serious problem. Fly ash in general has a greater content of PTE (see Table 13.3) and often is mixed with Na/Ca salts and activated carbon from the flue gas pollution control section of the plant. Since fly ash is a finely divided material (mean particle size of about 1 µm, see Gilardoni et al., 2004), it is also characterized by a very high surface area and is therefore highly susceptible to leaching. Due to these characteristics, fly ash disposal is a rather troublesome waste in the overall MSW management strategy.

Table 13.3 Comparison of ranges of metal concentrations (mg/kg) in soil, refuse combustible, bottom ash, and fly ash (Goldin *et al.*, 1992; Niessen, 2002b).

Metal	Common soil	Fuel (MSW/RDF)	Bottom ash	Fly ash
Al	10^4–3×10^5	3×10^3–2.5×10^4	1.8×10^4–1.8×10^5	3.1×10^4–1.8×10^5
As	1–50	0–15	2–2×10^3	3–750
Ca	7×10^3–5×10^5	2.3×10^3–5×10^4	4.1×10^3–9.6×10^4	3.3×10^4–8.6×10^4
Cd	0–1	0–90	0–170	2–7.8×10^4
Cr	1–1000	2–200	10–2000	20–3000
Cu	2–100	20–3.4×10^3	40–1.8×10^4	200–5×10^3
Fe	7×10^3–5.5×10^5	500–4.5×10^4	4×10^3–4.8×10^5	3.1×10^3–3.2×10^5
Hg	0	0–2	0–4	1–100
Ni	5–500	1–90	7–600	10–2.9×10^4
Pb	2–200	30–1.6×10^3	30–4.4×10^4	200–1.4×10^5
Zn	10–300	40–8×10^3	90–1.3×10^5	2×10^3–2.8×10^5

MSW, municipal solid waste; RDF, refuse derived fuel.

Different technologies have been proposed to reduce the impact on the environment of the disposal of MSW incinerator ash. The most commonly used technology is stabilization in a cement matrix. This technology involves grinding of the ash (if bottom ash is used in addition to fly ash), and then preparing an ash–concrete aggregate. These aggregates, while having lower tensile and compression strength compared to traditional aggregates, show a greatly reduced tendency to leaching, and can usually be disposed in landfills for special, nondangerous waste. Many researches carried out in this field have convincingly shown that, if carefully managed, the inertization process can lead to the production of a solid material which—similarly to ground bottom ash—can be used for the preparation of road substrates (Filipponi *et al.*, 2003; Van der Sloot *et al.*, 2001).

5. Plasma Pyrolysis for Waste Treatment

Another, more radical approach to the problems caused by bottom and fly ash consists in the use of a plasma torch to completely inertize such materials. By using a plasma torch, it is possible to use electric power to heat the material treated to temperatures in the excess of 2000 °C. In these conditions, the molecular bonds of the starting materials are broken, and the crystalline structures are destroyed, so that organic compounds are reduced to elemental state, while inorganic compounds are reduced to an amorphous melt. When the by-products are cooled down, the melt solidifies in an amorphous, vitreous form, leading to a solid material which exhibits striking similarities with volcanic obsidian. On the other hand, the gas constituents re-associate to form a synthetic fuel gas (*syngas*), mainly composed of light species, such as H_2 and CO.

It is important to point out that during the syngas cooling, secondary unwanted species, such as dioxins, can be formed (Vogg and Stieglitz, 1986). In order to avoid this risk, special care has to be paid to syngas treatment downstream the reaction vessel. In particular, the first step of such treatment is a prequenching of the gas with water in order to reduce its temperature from 1200 to 1500 °C to about 650 °C. After that, the coarse particles carried by the hot syngas, which are potential catalysts for dioxin synthesis (Schindler and Nelson, 1989) are removed by means of a refractory lined cyclone and are then recycled to the reaction vessel. After the hot cyclone treatment, the flue gas is quenched again with water to about 120 °C, in order to avoid the dioxin formation "window" which exists between 200 and 600 °C (Vogg and Stieglitz, 1986). The cool gas is then passed through a bag filter for further particulate removal, then through a SCR system for NO_x control, then through a wet scrubbing system for acid gas control. Eventually, the gas is stored before being combusted (see sketch in Fig. 13.4).

Plasma torch–based applications for waste treatments include fly ash from incineration processes, asbestos-containing waste, sanitary waste, waste containing organo-halogenated compounds, low-level radioactive waste, and even "traditional" RDF. In particular, plasma applications represent a very interesting technical solution for the treatment of fly ash from MSW/RDF incineration because the solid by-product, with its extremely low tendency to leaching, can be disposed off as a

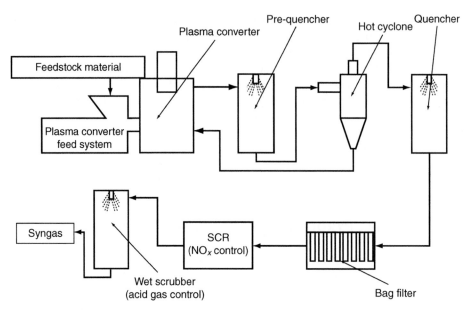

Figure 13.4 Sketch of a plasma torch–based plant with associated syngas treatment.

completely inert material or can even be reused, for example, for the production of bricks for road pavements. Plasma torch–based plants have been in use for some years in Japan, the United States, and many European countries. In particular, a plasma torch–based plant in Cenon (near Bordeaux, France) is used for the vitrification of fly ash coming from a nearby incineration plant, having a capacity of 3.3×10^5 kg/day of waste plus 5.0×10^4 kg/day of sewage sludge. The plasma torch plant has a capacity of 6.5×10^3 kg/day, operates continuously (24 h per day, 7 days per week) and uses a continuously mixed reaction chamber operating at 1500 °C for ash melting and complete destruction of dioxins and furans. One of the most remarkable results of the plasma process, a part from the complete inertization of the ash (the solid "obsidian" produced easily passes any leaching test), is its very high reduction of both mass (25%) and volume (50–60%) of the treated material. Nevertheless, it has to be pointed out that serious concerns exist in relation to the high costs of plasma-based plants. Rough estimates (De Falco, 2006) indicate that a "standard" plant of this kind, with a capacity in the order of 20 tons per day, would require an investment cost of about 8 M€, as well as operating costs of about 0.30 €/kg, and these concerns appear particularly well founded in a context in which energy costs appear set to increase dramatically in the near future.

6. Conclusions

MSW management is a crucial environmental task for all advanced societies, particularly in a context of increasing landfilling restrictions. Incineration can significantly contribute to MSW management. However, MSW (or RDF) incineration has several serious potential environmental impacts that need to be carefully taken into account.

The most relevant problem originating from MSW incineration is flue gas treatment, since untreated incineration flue gas can contain large amounts of macro-pollutants (e.g., CO, SO_2, HCl, NO_x, particulates) and micropollutants (e.g., PAHs, mercury compounds, polychlorinated biphenyls, dioxins, furans). The necessity of reducing polluting emissions to levels compatible with existing regulations dictates the adoption of rather sophisticated- and expensive flue gas treatment sections in incineration plants.

In addition to this, another, often overlooked issue is the disposal of solid by-products of the incineration process. MSW incinerators essentially produce two types of solid by-products. The first is the slag, or bottom ash, which is mostly made of the noncombustible fractions of the waste, plus a small fraction of the combustible fraction. The second is fly ash, the particulate material captured in the particulate removal section of the flue gas treatment plant, which is often mixed with various other chemicals used for flue gas treatment. Bottom ash and fly ash are characterized by very high concentrations of PTE, such as lead, mercury, cadmium, and nickel, which can easily leach into the environment.

Different approaches have been proposed to mitigate the impact of fly ash disposal. The most commonly considered technology is stabilization in a cement matrix, with the aim of producing ash–concrete aggregates, which—while having lower tensile and compression strength compared to traditional aggregates—show a greatly reduced tendency to leach. Another approach consists of using plasma torches to completely stabilize such materials by reducing them to a melt which, when cooled, solidifies into a vitreous, nonleaching form. In light of its character-istics, plasma torch–based vitrification plants appear to constitute a very promising response to the environmental problems posed by the disposal of solid by-products from MSW incineration processes. However, serious concerns exist about the high costs of plasma-based plants, and these concerns appear particularly well founded in an era of rapidly escalating energy costs.

ACKNOWLEDGMENTS

The author is thankful to the reviewers M. Di Bonito (Environmental Agency, Nottingham, UK) and C. Sears (Sears Consulting, LLC, USA), whose comments were useful to improve the chapter.

REFERENCES

APAT-ONR (2006). "Rapporto Rifiuti 2006." APAT (Agenzia per la Protezione dell'Ambiente e i Servizi Tecnici) and ONR (Osservatorio Nazionale sui Rifiuti), Roma (Italy).

Arena, U., and Mastellone, L. (1998). Centrali di termovalorizzazione di rifiuti solidi urbani. La Termotecnica 48(1), 89–97.

Brunauer, S., Emmett, P. H., and Teller, E. (1938). Adsorption of gases in multimolecular layers. J. Am. Chem. Soc. 60, 309–319.

Canter, L. W., Knox, R. C., and Fairchild, D. M. (1988). "Groundwater Quality Protection." Lewis Publishers, Chelsea (MI, USA).

Cossu, R. (1995). The multi-barrier landfill and related engineering problems. In "Proceedings Sardinia 95, Fifth International Landfill Symposium," (T. H. Christensen, R. Cossu, R. Stegmann, Eds.). CISA Publisher, vol. 2. 3–26.

De Falco, A. (2006). "Personal Communication." Eureco European Environmental Company, Napoli (Italy).

Enea and Federambiente (2006). "Rapporto sul Recupero Energetico da Rifiuti in Italia." ENEA (Ente per le Nuove Tecnologie e l'Ambiente), Roma.

Filipponi, P., Polettini, A., Pomi, R., and Sirini, P. (2003). Physical and mechanical properties of cement-based products containing incineration ash. *Waste Manage.* **23**, 145–156.

Gilardoni, S., Fermo, P., Cariati, F., Gianelle, V., Pitea, D., Collina, E., and Lasagni, M. (2004). MSWI fly ash particle analysis by scanning electron microscopy–energy dispersive X-ray spectroscopy. *Environ. Sci. Technol.* **38**, 6669–6675.

Goldin, A., Bigelow, C., and Veneman, P. L. M. (1992). Concentrations of metals in ash from municipal solid waste combustors. *Chemosphere* **24**, 271–280.

Lee, G. F., and Jones, A. R. (1991). Landfills and groundwater quality. *Ground Water* **29**, 482–486.

NATO/CCMS (1988). *In* "International Toxicity Equivalency Factors (I-TEF) Method of Risk Assessment for Complex Mixtures of Dioxin and Related Compounds," NATO, Brussels, Report no.176.

Niessen, W. R. (2002a). "Combustion and Incineration Processes." 3rd edn., p. 106, Marcel Dekker Inc., New York (NY, USA).

Niessen, W. R. (2002b). "Combustion and Incineration Processes." 3rd edn., p. 320, Marcel Dekker Inc., New York (NY, USA).

Pohland, F. G., and Harper, S. R. (1985). "Critical Review and Summary of Leachate and Gas Production from Landfills." EPA–600/2-86-073, US Environmental Protection Agency, Cincinnati (OH, USA).

Schindler, P. J., and Nelson, L. P. (1989). "Municipal Waste Combustion Assessment: Technical Basis for Good Combustion Practice." EPA–600/8–89–063, US Environmental Protection Agency, Cincinnati (OH, USA).

Styron, R. W., and Gustin, F. H. (1992). The production of TAP aggregate from municipal solid waste ash. *In* "Proceedings of 30th Annual Solid Waste Exposition." Tampa (FL, USA), August 3–6 1992.

van Beurden, A., Born, J. G. P., Colnot, E. A., and Keegel, R. H. (1997). High standard upgrading and utilization of MSWI bottom ash: Financial aspects. *In* "Proceedings of Fifth Annual North American Waste–To–Energy Conf. Research Triangle Park (NC, USA)," April. 22–25, 1997.

Van der Sloot, H. A., Kosson, D. S., and Hjelmar, O. (2001). Characteristics, treatment and utilization of residues from municipal waste incineration. *Waste Manage.* **21**, 753–765.

Vogg, H., and Stieglitz, L. (1986). Thermal behavior of PCDD/PCDF in fly ash from municipal incinerators. *Chemosphere* **15**, 1373–1378.

CHAPTER FOURTEEN

INNOVATIVE RESPONSES TO CHALLENGES: REDEVELOPMENT OF COS COB BROWNFIELDS SITE, CONNECTICUT, USA

Cynde Sears*

Contents

* Sears Consulting LLC, 13117 New Parkland Dr., Oak Hill, VA, 20171, USA

Environmental Geochemistry
DOI: 10.1016/B978-0-444-53159-9.00014-0
© 2008 Elsevier Inc.
All rights reserved.

333

Abstract

This case study examines the efforts of the Town of Greenwich, Connecticut, to clean up and redevelop a large, environmentally contaminated "brownfield," a former coal-fired power plant. The contamination posed numerous technical, financial, and regulatory challenges for the Town of Greenwich as it made plans to redevelop the brownfield site; many of these challenges are common to other sites across the country. The Town took several steps to ensure that the site would be turned into a community asset, such as considering multiple options for land use, taking advantage of newly available Federal and state funding for environmental assessment and cleanup, and adopting alternative environmental assessment and cleanup strategies. These decisions have helped this site to transition from a brownfield site to a community asset that will benefit residents of the Town of Greenwich and other citizens of the State of Connecticut.

1. INTRODUCTION

This case study examines the efforts of the Town of Greenwich, Connecticut, to clean up a large, environmentally contaminated brownfield and to return it to beneficial use as a recreational area. In the United States, "brownfields" are defined under Federal statute as "real property, the expansion, development, or reuse of which may be complicated by the presence or potential presence of a hazardous substance, pollutant, or contaminant" (US Congress, 2002). The Cos Cob brownfield is on the site of a former coal-fired power plant. Past practices for the disposal of ash from the power plant, combined with accidental releases of electric transformer fluids, contributed to widespread contamination across the nine-acre (3.6 hectares) site. Multiple environmental investigations revealed that several contaminants present in soils at the site—including polychlorinated biphenyls (PCBs), arsenic, and a range of petroleum hydrocarbons—were high enough to pose a risk to human health and the environment. The contamination posed a series of financial and technical challenges for the Town of Greenwich, which wanted to redevelop the site for beneficial use.

This case study will highlight a variety of technical, financial, and regulatory challenges the Town of Greenwich faced in more than a decade of planning for activities at the brownfield site, many of which are common to other sites across the country. It will also discuss a variety of innovative and creative steps the Town of Greenwich, the State of Connecticut, and the Federal government have taken to support brownfields redevelopment. These steps have helped this site to transition from a brownfield to a community asset that will benefit residents of the Town of Greenwich as well as other Connecticut citizens.

2. INTRODUCTION TO BROWNFIELDS

Communities across the United States are striving to reclaim and redevelop brownfields properties. However, because of the costs and legal burdens associated with environmental cleanup, owners often abandon their brownfields properties.

These properties then revert to local or state government ownership, leaving communities to address existing contamination and find ways to return the properties to beneficial use. However, communities often have limited resources to clean up or redevelop, so the sites remain abandoned and contaminated, contributing to blight and other symptoms of socioeconomic decay.

The dimensions of the brownfields challenge in the United States are staggering. The United States Environmental Protection Agency (US EPA) estimates that between 500,000 and a million brownfield sites exist nationally, and that the cost for cleaning them up will exceed $650 billion [US General Accounting Office (US GAO), 1995]. Cities across the country estimate that they could create more than 550,000 jobs—and recover $2.4 billion in tax revenue—if these properties were redeveloped (US Conference of Mayors, 2000). Despite this promise, many communities have discovered that redevelopment of brownfields is challenging and requires creative approaches and flexibility to succeed.

 ## 3. THE CHALLENGES OF BROWNFIELDS

Communities have historically faced several challenges in their attempts to redevelop brownfields. These have included:

- Extensive and expensive data collection requirements.
- Confusing cleanup standards.
- Technological limitations on ability to clean up sites to protective standards.
- Concerns over long-term liability for environmental contamination.
- Lack of funding for cleanup.

Each of these challenges is discussed in more detail below.

3.1. Extensive and expensive data collection requirements

Detailed site assessments historically required extensive data collection and off-site laboratory analysis of contaminants. These studies are expensive and time consuming. Often, a suite of samples taken during site assessment shows unacceptable levels of contamination, requiring a second or even third round of sampling to determine the dimensions of contaminated areas. These multiple deployments increase both the time and costs associated with site assessment, reducing the resources available for cleanup.

3.2. Confusing cleanup standards

Brownfields can be cleaned up under any number of environmental laws, including the Federal Comprehensive Environmental Response, Compensation and Liability Act (CERLCA, more commonly called Superfund); the Resource Conservation and Recovery Act (RCRA); and a variety of state environmental protection laws that generally parallel the Federal laws. Some sites are regulated under multiple authorities. One example of the confusion of cleanup standards can be seen in the original Superfund regulations. Under Superfund, US EPA and the states determined cleanup standards for each site based on "ARARs," or laws and regulations that were determined to be "applicable, or relevant and appropriate." As a result, nearly identical sites with similar contaminants, geology, and other elements were cleaned up to widely different standards that reflected competing regulatory requirements.

3.3. Technological limitations on ability to clean up sites to protective standards

Before about 1995, many environmental cleanups used a "one-size-fits-all" approach. In order to ensure maximum protection to human health, many regulatory agencies required that contaminated soils be excavated and hauled away for treatment, disposal, or destruction; and contaminated ground water was typically pumped out of the ground and treated, then either reinjected to the subsurface or disposed of. All ground water, whether used as sources of drinking water or not, had to be remediated to drinking water standards. Unfortunately, the costs and disruption of extensive soil excavation for soil cleanup, and the long-term operations (often decades to centuries) required for pump-and-treat ground water systems, discouraged many responsible parties from cleaning up sites.

3.4. Concerns over long-term liability for environmental contamination

Under CERCLA and similar state hazardous waste laws, liability for contamination at a site is "strict, joint and several," as well as retroactive. Anyone who was ever involved with the site—generators of hazardous substances, transporters of materials to or from the site, and any past or present owner—can all be held liable for the costs of cleanup. A fear of perpetual liability kept many private buyers, developers, lenders, and potential future owners of contaminated properties from investing in brownfields sites.

3.5. Lack of funding for cleanup

Cleanup costs for a contaminated site can easily exceed the value of the property, even in areas with high land values. The private sector may be able to recoup its costs through sale or rental of residential projects, commercial space, or industrial facilities. Unfortunately for many public entities, investments in cleanup and redevelopment often burden already limited funding sources, and the types of projects that communities typically undertake (e.g., recreational areas, child care centers, government facilities) do not generate income to offset cleanup and redevelopment costs.

These challenges contributed to situations much like the one that existed at Cos Cob for more than a decade: an environmentally contaminated site, contributing nothing to the community's financial well-being, adding to urban blight, and occupying a highly desirable location. Fortunately, during the same era that the Town of Greenwich was considering new uses for the Cos Cob power plant site, changes were evolving at the state and Federal level and in the private sector that would address these potential barriers.

4. Tools to Respond to Brownfields Challenges

Over the past decade, elected officials, regulatory agencies, and the private sector have started to address many of the brownfields challenges with creative tools, allowing communities to move forward with redevelopment plans; the following describes some of these tools.

4.1. Addressing extensive and expensive data collection requirements

For many years, prospective owners or investors shied away from undertaking the first step—site assessment—needed to determine the actual condition of a brownfields site. In fact, a significant percentage of brownfields sites have only limited amounts of contamination, and much of the contamination present does not pose a threat to human health or safety. Despite this reality, the fear of lengthy and costly site assessment requirements have impeded brownfield redevelopment.

Two tools that have contributed greatly to overcoming the barriers of expensive and extensive data requirements are *Federal funding for site assessment activities*, and a *streamlined approach to project management*. Each of these is described in more detail below.

4.1.1. Federal funding to support site assessment

In 1995, in recognition of the importance of overcoming this barrier to redevelopment, US EPA began to provide funding to states and communities for brownfields site assessments. Communities could receive up to $200,000 over a two-year period, and could use the funding to inventory, characterize, assess, and plan for cleanup and redevelopment, and to perform a range of community involvement activities related to brownfield sites. The incentive worked. From 1995 through mid-2004, more than 6,000 brownfields sites were assessed with these funds, and more than 2,100 properties were made ready for reuse (US Office of Management and Budget, 2006). The program was well-funded in 2005 through 2007, and will likely be extended into the foreseeable future. Many communities have reported that the EPA site assessment grants have "jump started," "catalyzed," or "boosted" their efforts to redevelop brownfields sites.

Support for this initiative has expanded greatly since 1995. In 2001, the US Congress passed the Small Business Liability Relief and Brownfields Revitalization Act (also known as the 2002 Brownfields Law). This act, which updates the Superfund legislation, authorizes US EPA to provide up to $50 million in grants to states and tribes to establish or enhance programs to assess, cleanup, and redevelop brownfields. Each year since 2003, EPA has distributed almost $50 million in brownfields grants to states, territories, and tribal groups.

4.1.2. Streamlining site assessments

Another creative approach is helping to address the extensive sampling and long analysis times that often impede site assessments. US EPA is applying this streamlined scientific approach, called Triad, at several hazardous waste sites in the United States, which enables cleanup project managers to minimize uncertainty, expedite site assessment and cleanup, and reduce project costs. Triad is based on three interrelated elements: Systematic Project Planning, Dynamic Work Strategies, and Real-Time Measurement Tools. Each is described in more detail below.

4.1.2.1. Systematic project planning   This element ensures that the level of detail required for project planning matches the use of the data being collected, which in turn is determined by the intended end use of the site. Such an approach differs from typical site assessment planning, which often approaches all sites as if they are similar, with standardized sampling strategies (e.g., "two soil samples per square meter"), analytical requirements ("test all samples for 36 common contaminants"), and assumed end uses ("all soils will be remediated to residential standards").

The foundation for systematic project planning is the conceptual site model (CSM). Under Triad, a CSM is developed for each site, based on all available historical and current information to estimate where contamination might be located; how much is there and how it varies across the site; possible fate and migration of the contamination; exposure risks associated with the site and contaminants; and means of mitigating exposure risk. Once the project team develops an initial CSM, it identifies data gaps and seeks additional information to resolve those gaps. Throughout the project, the team updates the CSM to reflect new data and identify additional needs for data collection. This iterative process continues until the site project team is confident in its decisions about actions that need to be taken to address the site contamination.

4.1.2.2. Dynamic work strategies   In this element, work planning documents such as the CSM are written and updated in real-time (e.g., while project crews are still in the field). A unique aspect of dynamic work strategies is a quality control process that adapts in real-time, unlike traditional work strategies, which implement quality control on analytical processes that are far removed from the activities of the field.

4.1.2.3. Real-time measurement tools   The use of field analytical instrumentation, rapid sampling platforms, and on-site data management and interpretation make dynamic work strategies possible. Real-time tools include hand-held portable analytical equipment (e.g., immunoassay kits, X-ray fluorescence spectrometers), mobile laboratories with standard analytical equipment (e.g., gas chromatographs, electrochemical detectors), and off-site laboratories with rapid turn-around capabilities. Data from these analyses are incorporated into the CSM as it becomes available, further refining the CSM and allowing the project team to determine the most appropriate means for addressing the contamination, given the intended future use of the property.

4.2. Changing confusing cleanup requirements

In the early 1990s, states and the Federal government began to realize that the ARARs approach was confusing, contributing to skyrocketing cleanup costs, and that sites were often not achieving required cleanup levels. These factors were discouraging responsible parties and potential future investors from undertaking cleanup. Over the next few years, Federal and state regulatory agencies began to implement a range of risk-based cleanup programs that used actual risk to human health as a significant factor in determining site cleanup requirements.

An important component of many of these programs was the concept of "screening," an approach that allowed site managers to eliminate areas, pathways, and/or chemicals of concern and to focus limited resources on addressing those areas that posed unacceptable risks to human health as defined by EPA risk guidelines. Typically, screening levels are based on fate, pathway, and exposure assumptions such as those developed by EPA's Superfund program. These levels are then compared to on-site soil contaminant levels. Generally, areas of a site with contaminant concentrations below the screening levels are eliminated from further assessment. Areas with concentrations above the screening levels generally warrant further evaluation to determine the appropriate next step. Screening levels are often calculated to be protective based on a future use scenario, such as residential or industrial use.

Increasingly, sites are cleaned up to levels that reflect the actual risk posed by the site along with other considerations, such as the technical feasibility of cleaning up the site, costs, and community desires for reuse. This flexibility in approach has allowed many sites to be remediated with a range of noninvasive, lower-cost technologies, and to be redeveloped or reused much more quickly than would have been possible under the previous requirements.

4.3. Overcoming technological limitations on ability to clean up sites to protective standards

Environmental remediation technologies have evolved rapidly in the past decade. In the early days of environmental cleanup, the one-size-fits-all approach reflected the reality that only a limited number of demonstrated, proven alternatives could achieve desired cleanup levels. Since the early 1990s, however, the Federal government and states have invested significant resources in supporting the development, testing, and deployment of innovative technologies. This has resulted in the availability of a wide range of technological options for cleaning up almost any type of contamination in most environments, often more quickly and at lower cost than by use of traditional technologies.

Where it is not technically or economically feasible to undertake large-scale remediation, particularly at low-risk sites, another response to environmental contamination is increasingly popular: combining engineering and institutional controls. *Engineering controls* include permanent caps, typically constructed as horizontal barriers that cover an entire area of contamination to prevent infiltration of water. Caps are usually composed of compacted clay, asphalt, concrete, or other materials, and are considered permanent, although they must be regularly inspected and monitored to ensure that they are maintaining their structural integrity over time. The purpose of an engineered barrier is to limit exposure by eliminating the exposure route (e.g., dermal contact, vapors), and/or to control downward migration of water from the surface, reducing or eliminating the movement of soluble contaminants from the soil matrix into ground water.

Institutional controls are administrative and legal restrictions on the future use of a site to minimize potential exposure to chemicals of concern or to prevent activities that might reduce protectiveness at a site. Institutional controls include deed restrictions or covenants that can prohibit specific types of development or construction on a property, restrict specific activities (e.g., digging, creating dust), or control use of land, surface water, and ground water. These covenants can "attach" to the property, in which case they cannot be changed when the property changes ownership. Other institutional controls include governmental actions, such as local zoning or planning restrictions, in which regulators dictate the types of activities that can occur on residential, commercial, industrial, and other properties.

4.4. Reducing concerns over long-term liability for environmental contamination

To address liability concerns, many states established "voluntary cleanup programs (VCPs)." Connecticut created its voluntary cleanup program in 1995. VCPs typically allow owners or prospective purchasers to investigate and remediate low-risk contaminated sites with limited oversight from regulatory agencies, as long as they work with authorized environmental professionals and undertake all actions necessary to protect human health from the risks posed by the site. US EPA agreed to let states determine whether sites are adequately addressed, and also agreed not to enforce CERCLA requirements for cost recovery and liability if the state program can show that its approach is protective of human health.

In addition, the 2002 Brownfields Law addressed investor and community concerns about future liability for contamination at brownfields sites, even after they have undertaken cleanup and redevelopment. The act relieved prospective purchasers and other "innocent landowners" from future liability for contamination if they take reasonable steps to identify and prevent release of any hazardous substances at the time they purchase or acquire a brownfield property.

4.5. Addressing lack of funding for cleanup

The costs of cleaning up hazardous waste, petroleum, and other contamination can be significant. Superfund sites commonly cost more than $30 million to remediate. The State of New York estimates that it would cost approximately $15–$20 million to excavate and dispose of wastes at a small industrial landfill, or approximately $3 million for the lower-cost alternative of capping [New York Department of Environmental Conservation (NYDEC) 2007]. Although costs are highly variable, it is clear that environmental cleanup is very expensive, and that like site assessments, cleanup will not happen without significant financial support.

4.5.1. Federal and state funding to support cleanup
US EPA's support to brownfields cleanup started in 1995, just as the support to site assessments began. Although US EPA did not initially provide direct cleanup funds, it provided state and local governments with up to $1 million to create a revolving

loan fund that could be used for environmental cleanup and redevelopment. The US Department of Housing and Urban Development (HUD) also began to allow communities to use HUD grants to offset environmental cleanup costs at sites that were planned for redevelopment as community housing. Similarly, several states established cleanup loan programs, both independently and in cooperation with the Federal government, which helped to provide necessary funding for many sites. However, despite these programs, through the 1990s, funding for the high costs of cleanup continued to be a major limitation on the ability of the private sector and communities to redevelop brownfields properties.

The 2002 Brownfields Law greatly expanded the availability of Federal funds for brownfields cleanup. In addition to revolving loan funds, US EPA now offers direct grants to allow communities and nonprofit organizations to clean up brownfields sites. Eligible entities may apply for up to $200,000 per site, and may use the funds to address sites contaminated by petroleum and hazardous substances. The grants require a 20% cost share, which may be in the form of a contribution of money, labor, material, or services, and must be used for eligible and allowable costs.

The role of Federal and other public funding streams for brownfields redevelopment cannot be underestimated. In an evaluation of 107 brownfield projects nationwide, the International Economic Development Council determined that on average, 68% of the funding for brownfield remediation was derived solely from public sources, 22% came from private sector funds, and 10% were remediated using both public and private sector funds. Potentially responsible parties (PRPs) provided limited support for remediation in comparison to public sources (XL International and International Economic Development Council, 2002). It is clear that Federal and state funds have dramatically improved the abilities of communities to get sites evaluated, which has contributed greatly to site remediation and redevelopment.

The remainder of this report will demonstrate the successful application of several of these innovative financial, technical, and policy approaches to a brownfields site in urban Connecticut. This case study will highlight the contribution of these new approaches to the success of this brownfield redevelopment project. Without the changes in the US brownfields program that have evolved over the past decade, this contaminated site would have continued to languish for many more years, continuing to contribute to community degradation and local blight.

5. Introduction to Case Study: Cos Cob Power Plant, Connecticut

The Cos Cob power plant brownfields site is located in the Town of Greenwich, on the banks of the Mianus River in southern Connecticut, ~35 miles (56 km) north of New York City. The Town of Greenwich has a relatively low population density, with approximately 62,000 citizens (in 2005) living in an area of 47 square miles (mi^2) (124 km^2), or 1,320 people per mi^2 (500 people per km^2). Greenwich is one of the most affluent towns in the United States, with an economy based largely on the service and financial industries, several of which have relocated from New York City in recent years.

5.1. History of the site

In 1903, the State of New York passed a law that prohibited all steam locomotives from entering New York City. The State took this action because a series of train accidents had occurred in city railroad tunnels, the result of low visibility from locomotive smoke and steam. Thousands of commuters from the southern communities of Connecticut depended on trains to get to and from their jobs in New York City. Although the local New Haven Railroad operated low-voltage direct current railroad track at this time, the system was inadequate for the operation of the heavy trains that would be required to accommodate the commuters who could no longer travel by steam train. To respond to the increased demands, the New Haven Railroad decided to install high-voltage 11,000-volt overhead wires on its tracks from Woodlawn, New York, into Connecticut. The railroad determined that it would build its own power station to generate high-voltage, single-phase alternating current electricity for its lines, which was a major engineering feat for the time.

In 1905, Westinghouse, Church, Kerr & Company began to construct the coal-powered Cos Cob plant on a 5-acre (2 hectares) site on the banks of the Mianus River. It was the first power plant to be built exclusively for a railroad, and was constructed in the Mission Style, with scalloped gables at both ends and a red tile roof (Fig. 14.1). The plant was a multilevel concrete and metal building, and housed a variety of electrical generation and distribution equipment, including boilers, transformers, circuits, and breakers.

Environmental problems in the area began while the plant was operating. For most of the plant's active life, the boilers used coal to generate steam. The facility's smokestack emitted copious quantities of black soot. However, environmental contamination from the plant was not restricted to its air emissions. For decades, waste products from the power plant, including fly ash, bottom ash and slag, were deposited along the northern and southern ends of the facility. So much ash and slag were deposited that the facility almost doubled in size, from its original 5-acre (2 hectares) lot to an area of 9.4 acres (3.6 hectares). Ash and slag deposits extended to more than 35 ft (10.7 m) below ground surface (bgs) throughout the northern and southern ends of the site.

Figure 14.1 The former Cos Cob power plant (Gryzywacz, 1993).

In 1969, the State of Connecticut sued the owners to force them to reduce the soot output; after years of inaction, in 1982 the owners switched the fuel to oil. Also in that year, the plant building was named on the National Historic Register of Places as an Engineering Landmark in recognition of its unique role as a "pioneering venture" in rail electrification. Despite these upgrades, the plant was in poor condition and required many upgrades to meet environmental and other standards. The railroad company operated the plant until 1986, then closed it and moved to a new generating station nearby. The State Department of Transportation decommissioned the plant in 1987, but left much of the equipment intact, including boilers, transformers, circuits, and circuit breakers.

5.2. Plans for site redevelopment

From the time the plant closed in 1986, the Town of Greenwich was interested in acquiring the site for recreational use. A local elected official reported that the site was desirable for its "vistas, spacious waterfront park space, and value as a critical addition of open space." At the time, the Cos Cob power plant property had an assessed value of $13 million, and when the state announced its intention to sell the property, the Town of Greenwich and several private developers prepared a multi-million dollar bid. However, because the General Assembly was considering selling other properties in the state for $1, the town proposed to buy the Cos Cob property from the state for that amount. To make the offer seem more inviting, a state representative from the area proposed that the town would build low-income housing on at least a portion of the land.

In 1987, Connecticut's Commissioner of Transportation conveyed the Cos Cob power plant and the land on which it was located to the Town of Greenwich in return for payment of $1. The state stipulated that 25% of the land was to be used for senior, low-, and moderate-income housing and 75% for open space that would be accessible to all citizens of the state. This idea was not popular with some of the town's residents, who preferred that all of the land be used for recreation and greenspace, and others who wanted to restore or preserve the historic power plant.

In 1988, the Regional Plan Association, Town of Greenwich, Connecticut Housing Department, and others sponsored a competition for the design of an appropriate affordable housing project to put on the land. The sponsors of the competition selected a relatively dense complex of three apartment buildings, a 25,800-ft^2 commercial structure with stores, restaurants, a health club, and a day-care center (Waite, 1988). However, local residents preferred that the land be used for a large public park. This conflict between the community's desires for a large greenspace and the planning authorities' desire to use at least part of the land for residential development could have slowed down the project for years. However, circumstances would eventually make only one option viable.

5.3. Environmental contamination and its impacts on Cos Cob's redevelopment

In 1988, the Town of Greenwich hired an environmental consulting firm to perform a thorough assessment of the Cos Cob power plant site. This initial investigation included surface soil sampling at nine locations, surface and sub-surface borings at six locations, limited surface water and sediment sampling, and ground water sampling from three monitoring wells (The Selectmen's Cos Cob Power Plant Site Planning Committee, 2003). The investigation revealed the 22–35 ft thick deposits of fly ash and related materials on the northeastern and southern ends of the site; several areas of fuel-oil contamination; heavy metal contamination with arsenic, barium, lead, and silver; and both volatile organic compounds (VOCs) and polycyclic aromatic hydrocarbons (PAHs), primarily in surface soils. The State of Connecticut did not have direct exposure standards for PAHs when these analyses were performed in 1998, so cleanup was not required at the time. (Surface soils at the site did exceed today's PAH exposure standards of 1,000 ppm.) Discovery of this contamination provided a hint that the property would not be appropriate for residential development; however, in 1988, all options were still open for consideration.

For the next decade, the plant sat idle and continued to deteriorate. Reports of historical releases of dielectric fluid and vandalism of transformers generated concerns about the possible release of PCBs. In 1998, the town hired an environmental consultant to perform a site evaluation. Although PCBs were detected in one soil sample from the transformer release area, other areas around the site did not contain them. However, several samples throughout the site—both in building basement soils and outside—contained total petroleum hydrocarbons (TPH) at levels exceeding the state's 1996 Remediation Standards Regulations of 500 ppm. The consultant recommended that the site undergo a comprehensive site assessment to determine the nature and extent of the TPH and other contamination. As an interim action, however, he recommended the removal of soils in the area where the transformer was vandalized. The consultant removed 12–18 in. (0.3–0.45 m) of soil (~15 cubic yards, or 11.5 m³) and disposed it off-site. This appears to have been the first significant environmental remediation at the Cos Cob power plant site, and contributed further to a perception that the site might not be appropriate for residential development.

In 1997, acknowledging that the site might not be appropriate for residential use, the state removed the housing requirement on the site, with the caveat that an equivalent number of units be provided elsewhere in the Town of Greenwich (The Selectmen's Cos Cob Power Plant Site Planning Committee, 2003). This action freed the town to reconsider the preference that its citizens had shown nine years before during the design contest of 1988: a large recreational site.

Another significant environmental issue arose in 1999. The power plant building was found to contain asbestos-containing materials (ACM), primarily transite and galbestos, at levels that constituted a "release." The project environmental consultant concluded that asbestos releases were likely to occur again. Between 1999 and 2000, the ACM were removed and disposed of. To the chagrin of local preservationists, all buildings on the site were demolished. The asbestos removal and disposal and the demolition of the buildings cost almost $6 million, shared equally between the town and the state.

In 2001, the Cos Cob Power Plant Committee, made up of the town's elected representatives, undertook a community survey to which they received more than 1,700 responses. The community clearly indicated its preference for passive recreational use. More than two-thirds of the residents wanted a walking and jogging trail, open meadow, picnic facilities, bicycle path, and flower garden; more than half wanted a rowing facility, playground, wildlife sanctuary, and fishing pier. The community also recognized a need for general sports fields on which residents could play soccer or football (The Selectmen's Cos Cob Power Plant Site Planning Committee, 2003). With these redevelopment goals in mind, the town moved forward to get the Cos Cob power plant site cleaned up.

5.4. Responding to the challenges at Cos Cob

5.4.1. Federal funding for site assessment

Fifteen years after purchasing the Cos Cob power plant site from the state, the Town of Greenwich had undertaken many of the steps necessary to move redevelopment at the site forward. The town had already performed multiple limited environmental assessments and cleaned up some of the most obviously contaminated areas within the site, torn down buildings and other structures contaminated with asbestos and other materials that threaten human health, and achieved a high degree of community consensus that the site should be converted for recreational use. The cleanup and demolition actions had cost the Town of Greenwich close to $2 million, yet the site still needed to be fully assessed, cleaned up, and readied for reuse.

In 2002, the Town of Greenwich applied for and received a $200,000 Targeted Brownfields Assessment grant from US EPA. The primary objective of the assessment was to delineate the nature and extent of surface soil contamination sufficient to allow development of remedial strategies and cost estimates that were consistent with the town's plans to reuse the site as a recreational area. Ground water quality was not investigated because the 1988 investigation found no evidence of ground water contamination, and ground water at the site is not used for drinking water.

An initial plan for sampling at the site was developed. This plan would have required two mobilizations to collect soil samples at 20 locations to confirm the presence or absence of contamination where previous reports had implied potential source areas (US EPA, Office of Solid Waste and Emergency Response, 2004). All samples would be sent off-site for analysis. The total costs for this assessment were estimated at $203,000.

5.4.2. Improving site assessment with triad

Late in 2002, US EPA, working with an organization called the Brownfields Technology Support Center, decided that the Cos Cob site would be appropriate for application of the Triad approach. In February 2003, a Triad-based field investigation was conducted at the Cos Cob site over the course of one week. In a single mobilization, the field team characterized the site for all constituents of concern effectively (US EPA, Office of Solid Waste and Emergency Response, 2004).

As mentioned earlier, an important element in the Triad approach is the conceptual site model (CSM). Early in the project, EPA and the project consultant developed a detailed preliminary CSM based on data from previous investigations. The CSM indicated that potential threats to human health and the environment at this site were limited to direct contact with contaminated surface soil and sediment that contained PCBs, petroleum-related contaminants, and arsenic. The resulting CSM is shown in Fig. 14.2.

Based on these findings, the project team developed a dynamic work strategy calling for random grid sampling and measurement of TPH, PAH, and PCBs. The proposed sampling locations are presented in Fig. 14.3. Soil samples were collected via direct-push methods to a depth of 4 ft (1.2 m) bgs across a 70-ft × 70-ft (21.3 m × 21.3 m) grid across the site. Sample locations within each grid element were selected randomly unless a specific area of potential contamination was identified, in which case a biased sample was collected. This random systematic sampling approach was chosen because it combines the benefits of both random (best employed when historical knowledge of contaminant distribution is limited) and systematic (best for delineating "hot spots") sampling methods (US EPA, Office of Solid Waste and Emergency Response, 2004).

The analysis scheme implemented at the Cos Cob site used three sets of tools: hand-held test kits, an on-site mobile laboratory equipped with gas chromatograph/ electron capture detector (GC/ECD) and X-ray fluorescence (XRF), and an off-site laboratory with rapid turnaround capabilities (<48 h for virtually all analyses). By implementing all of these tools at the same time, the project eliminated the need for multiple sampling events and allowed the team to perform additional real-time sampling, enabling the team to delineate the extent of potential "hot spots" quickly.

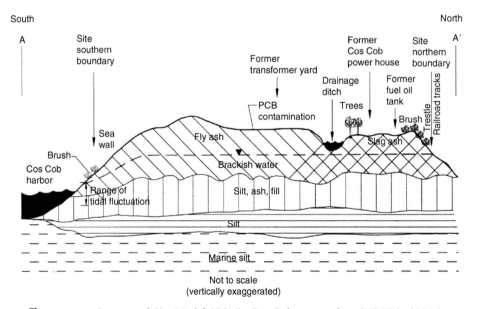

Figure 14.2 Conceptual Site Model (CSM), Cos Cob power plant (US EPA, 2005a).

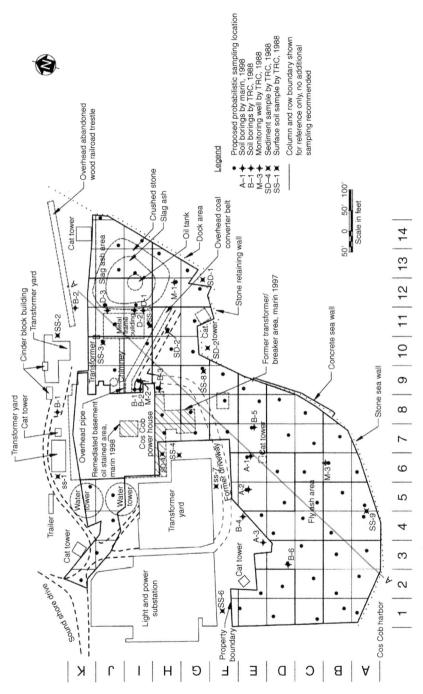

Figure 14.3 Proposed sampling locations, Cos Cob power plant (US EPA, 2005).

At the beginning of the Triad field investigation, only the uppermost two 1-ft intervals at each location were analyzed in the field. Ultraviolet fluorescence test kits, which require ~5 min per sample for collection and analysis, were used for field detection of PAH and TPH at all sampling locations. A percentage of the PAH and TPH samples were sent to an off-site laboratory for independent collaboration of the field results.

PCB analysis was performed at locations where releases were possible based on historical data. US EPA had deployed a mobile laboratory to the site, and all suspected PCB locations were sampled with a gas chromatograph equipped with an electron capture detector. If any PCBs were detected, then subsequent deeper intervals at the location were sampled up to a total depth of 4 ft bgs, as well as at a distance of 10 ft in four directions—north, south, east, and west—from the sampling point. This increased sampling can be seen in Fig. 14.4, near the center of the diagram just east of the power house. The dynamic nature of sampling under the Triad approach allowed the field team to focus on this "hot spot" during a single deployment. Under a more traditional sampling scheme, the team would have taken one sample in this grid and then sent the sample to a laboratory for analysis. Upon learning the results, the team would then have returned to the site to delineate the area of contamination. The Triad approach eliminated the need for multiple deployments and resulted in a well-defined area of concern in one visit.

The EPA mobile lab analyzed soil samples for lead, zinc, copper, and nickel by XRF. Because arsenic had been identified as the element that presents the greatest risk to human health at the Cos Cob site, samples from the 0–1 ft (0–0.3 m) to 1–2 ft (0.3–0.6 m) intervals were sent off-site for analysis by inductively coupled plasma/atomic emission spectrometry. If the off-site analysis showed arsenic concentrations greater than 10 ppm (the Connecticut direct exposure criterion for soil), the field investigators progressively analyzed samples in 1-ft (0.3 m) intervals below and surrounding the sample, delineating "hot spots," until the remaining contamination was lower than 5 ppm.

5.4.3. Results of the investigation

The results of the Triad site assessment showed that site surface soils contain concentrations of arsenic, TPH, and several PAHs at levels that exceed the Connecticut residential direct exposure criteria (RDEC). Although no strong patterns in the distribution of contaminants were observed, concentrations of all contaminants seem to be higher in the areas that contain fly ash (on the southern end of the site). Approximately 30% of the grids tested exceeded the state RDEC for TPHs, and 15% exceeded the RDEC for PAHs. Some PCBs were identified in an area just east of the former transformer yard; a few samples from this area had concentrations 10 times the state standard of 1 ppm for PCBs. Arsenic was found throughout the site, and concentrations generally increased with depth in the fill. Approximately 20% of the grids had concentrations greater than the Connecticut residential direct exposure criterion of 10 ppm.

The costs of the Triad assessment were approximately $132,000, or 30% lower than the initial estimates for the more traditional assessment. Many more data points were collected using Triad than a traditional sampling scheme, and data quality was

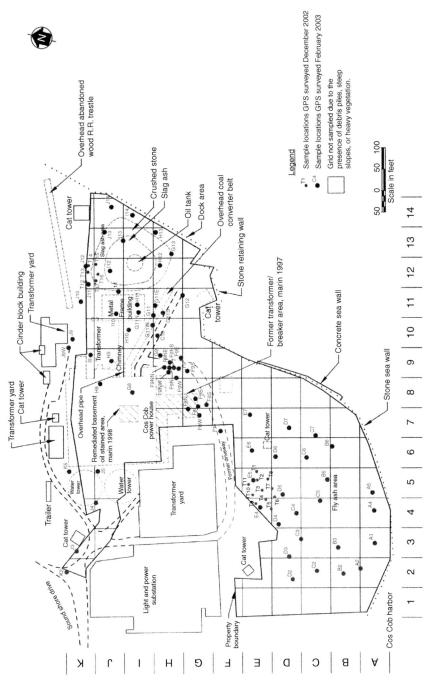

Figure 14.4 Revised sampling locations based on preliminary field results (US EPA, 2005).

determined to be as good as from more traditional site assessment approaches (US EPA, Office of Solid Waste and Emergency Response, 2004). The Triad approach resulted in a far more detailed understanding of the nature and distribution of contaminants at the Cos Cob site than would have been possible if only 20 sampling sites had been tested, as proposed in the original site assessment plan. This information has been useful in determining the next steps for the Cos Cob site.

5.4.4. Selecting a cleanup option

The dynamic investigation at the Cos Cob power plant provided a clear answer to any remaining questions about reuse at the site: risks to human health posed by potential exposure to contaminants in soils at the site must be mitigated. However, to remove all contamination by excavating the large volumes of contaminated fly ash, bottom ash and slag present at the site would be prohibitively expensive. For any reuse scenario, the PCB "hot spots" near the former transformer/breaker area had to be addressed, through excavation, capping, or application of a remedial technology. Preliminary estimates for removal and backfilling of the PCBs were approximately $900,000 (Berlin, 2005), and the costs of applying any other soil remediation technology such as thermal desorption or soil washing would have been at least an order of magnitude greater.

The Town of Greenwich realized that the costs of cleanup were going to be well beyond the capability of the community to pay. In 2004, the town applied for a US EPA brownfields cleanup grant. In its grant application, the town stated that it is committed to redeveloping the site as a public access and recreational park and that it planned to use the grant to excavate PCB-contaminated soil to depth of 4 ft over an area of approximately one-quarter acre. The town committed to backfilling the excavation with clean material, possibly from a local construction project.

Working with its consultants, the town realized that the most appropriate solution to the arsenic contamination, which pervades the site to significant depths, as well as to other contaminants at the site would be an engineering solution, such as an earthen cap, on the entire site. In addition to an engineering solution, institutional controls would provide another level of protection. When the State of Connecticut deeded the property to the Town of Greenwich, it required that the site be used for open space purposes. This institutional control can be made permanent to ensure that residential or commercial construction will not disrupt any cap or other barrier used to protect human health from exposure to contaminated soils on the site. The town decided that the cap and deed restrictions were the best solution for this site.

5.4.5. Financing cleanup

In 2005, the Town of Greenwich applied for, and received, a $200,000 brownfields cleanup grant from US EPA. The town used the funds later that year to excavate the PCBs and areas with high concentrations of TPH and PAHs; the task was completed in March 2006 (Partridge, 2006). As of late 2006, the town was working closely with the state Department of Environmental Protection's Voluntary Cleanup Program and an engineering consulting firm to determine the most appropriate design for the cap; needs for revegetation; and ways to construct such recreational structures as trails, walkways, sports fields, and a boat launch into the Mianus River. The most reasonable approach at the time was determined to be a vegetated earthen cap, constructed to a depth of 4 ft across most of the site, onto which sports fields, trails, walkways, and other relatively impermeable surfaces can be added.

5.4.6. Planning the future for the Cos Cob site

The Town of Greenwich plans to begin construction on the earthen cap (engineering control) and conduct preliminary grading and land contouring activities in 2008. Once these activities are completed, the trails, walkways, and other access structures can be completed. By 2010, the Town of Greenwich will likely be using parts or all of the redeveloped Cos Cob power plant site for recreational purposes.

The costs of redeveloping the Cos Cob power plant site have been significant. As of March 2007, Greenwich has already spent more than $4.7 million to demolish old structures, remove and dispose of asbestos, perform multiple site assessments, undertake remedial actions, and develop plans for the site's future (Vigdor, 2007). The Town of Greenwich's 2007–2008 budget contains $4.8 million for "Cos Cob site remediation," although the total costs will exceed that amount. Current estimates include more than $3 million for grading the land and performing some shoreline stabilization, and another $3 million for landscaping and addition of amenities such as a boat ramp and dock (Vigdor, 2007).

Creative approaches continue to resolve challenges at the Cos Cob site. Cost savings of more than $1 million for the project may be realized if a proposed deal between the state electricity utility—Connecticut Light and Power (CL&P)—and the Town of Greenwich is finalized. CL&P is constructing a pair of underground transmission lines over a distance of almost nine miles. The project will generate thousands of tons of dirt. In early 2007, the town's Planning and Zoning Commission approved the import and storage of more than 30,000 cubic yards of this dirt for use as an earthen cap at the Cos Cob site.

6. Conclusions

The brownfields redevelopment project at the former Cos Cob power plant site in Connecticut has experienced many challenges similar to those faced by sites across the United States. Since the community first expressed an interest in redeveloping the site, it has faced technical, regulatory, and financial challenges that might have resulted in the site remaining abandoned, contaminated, and unused for many years to come. However, recent changes in Federal and state programs have reduced the burdens associated with various aspects of brownfields redevelopment:

- Federal and state funding, first for site assessments, then later for cleanup, greatly accelerated the pace at which communities could start planning their brownfields redevelopment. More than $400,000 in site assessment and cleanup funding helped to move the Cos Cob redevelopment project forward quickly.
- New approaches to site assessment, such as the Triad approach, have reduced the cost and time required for site investigations. This approach provides high-quality

information that can be used to make decisions about future reuse of environmentally contaminated sites. At the Cos Cob site, the Triad approach helped the community to understand the extent and nature—and associated costs of dealing with—the contamination present at the site. Based on that information, the community was able to consider alternatives for the site that might otherwise not have been considered.

- Acceptance of alternatives to active cleanup—such as the construction of engineering barriers and the use of institutional controls—have allowed sites with remaining low levels of contamination to be reused safely. At the Cos Cob site, a more traditional approach might have required complete excavation and backfilling of the entire nine-acre site, an option that was too costly and disruptive for this community. The use of an engineering barrier, and deed restrictions requiring that the land be used for recreation, will protect human health while allowing the community to enjoy a large, open space in an otherwise highly urbanized area.

These benefits will undoubtedly improve the quality of life for the residents of the Town of Greenwich and the citizens of the State of Connecticut.

ACKNOWLEDGMENTS

The author thanks Dr. Benedetto De Vivo of the University of Naples for the invitation to present this paper at the Workshop on Environmental Geochemistry: Site Characterization, Waste Disposal, Data Analysis, Case Histories. Curtis Palmer (USGS) is thanked for his constructive review. She also thanks Harvey Belkin of the U.S. Geological Survey for his review, advice, and technical support.

REFERENCES

Berlin, C. (2005). Cos Cob Power Plant clean-up will cost less with Federal grant Greenwich, CT, : Greenwich Post, Greenwich, CT, May 19.

Gryzywacz, R. W. (1993). (delineator). New York, New Hampshire & Hartford Railroad, Cos Cob Power Plant, Sound Shore Drive, Greenwich, Fairfield County, CT. Historic American Engineering Record, US Library of Congress, Prints and Photograph Division. Card Number CT0568, http://hdl.loc.gov/loc.pnp/hhh.ct0568.

New York Department of Environmental Conservation (NYDEC). (2007). Frequently asked questions about New York's Inactive Hazardous Waste Disposal Site Program. http://www.dec.state.ny.us/website/der/ihws/faqs.html, viewed March 2007.

Partridge, K. (2006). Free fill could save town dough. *Greenwich Post*, Greenwich, CT, October 5.

The Selectmen's Cos Cob Power Plant Site Planning Committee. (2003). Plan for the Cos Cob Power Plant Site, final report to the Board of Selectmen. Town of Greenwich, CT, October 20.

US Conference of Mayors (2000). Recycling America's land: A national report on brownfields redevelopment Vol. 3.

US Congress (2002). Small Business Liability Relief and Brownfields Liability Act, Public Law 107–118 (H.R. 2869). US Office of Management and Budget, January 11 .

US EPA, Office of Solid Waste and Emergency Response. (2004). Triad used for targeted brownfields assessment of former Cos Cob Power Plant. Technology News and Trends, EPA 542-N-04–004.

US EPA, Office of Solid Waste and Emergency Response (2005a). Use of dynamic work strategies under a Triad Approach for Site Assessment and Cleanup – Technology Bulletin EPA 542-F-05–008.

US EPA, Office of Solid Waste and Emergency Response, Office of Superfund Remediation and Technology Innovation (2005b). Road map to understanding innovative technology options for brownfields investigation and cleanup, Fourth Edition. EPA-542-B-05–001.

US General Accounting Office (US GAO). (1995). Community development: Reuse of urban industrial sites. GAO Report #RCED-95–172.

US Office of Management and Budget. (2006). Budget of the United States Government, Fiscal Year 2006.

Vigdor, N. (2007). Panel wants school budgets cut. *Hartford Courant*, Hartford, CT, March 2.

Waite, T. L. (1988). The Cos Cob solution? A winning design for affordability, *New York Times*, New York, NY, December 11.

XL International and International Economic Development Council. (2002). The XL Environmental Land Use Report 2002 Exton, PA and Washington, DC.

CHAPTER FIFTEEN

CHARACTERIZATION AND REMEDIATION OF A BROWNFIELD SITE: THE BAGNOLI CASE IN ITALY

Benedetto De Vivo* *and* Annamaria Lima*

Contents

Abstract

This chapter documents the case history of the Bagnoli brownfield site government remediation project, which is still in progress, being in the remediation phase. The site was the second largest integrated steelworks in Italy and is located in the outskirts of Naples, in an area that is part of the quiescent Campi Flegrei (CF) volcanic caldera. Hundreds of surficial and deep boreholes have been drilled, with the collection of about 3000 samples of soils, scums, slags, and landfill materials. In addition, water samples from underground waters have been collected. The samples have been chemically analyzed for inorganic and organic elements and compounds, as required by Italian

* Dipartimento di Scienze della Terra, Università di Napoli, Federico II, 80134 Napoli, Italy

Environmental Geochemistry
DOI: 10.1016/B978-0-444-53159-9.00015-2
© 2008 Elsevier B.V.
All rights reserved.

Environmental Law DLgs 152/2006. In general, heavy metal enrichments in the cores and water suggest mixing between natural (geogenic) and anthropogenic components. The natural contribution of volcanically related hydrothermal fluids to soil pollution, in addition to the non bioavailability of metal pollutants from industrial materials, indicate that heavy metal remediation of soils in this area would be of little use, because continuous discharge from mineralized hydrothermal solutions would cancel out any remediation effort. The real pollution to be remediated is the occurrence of polycyclic aromatic hydrocarbons (PAH) distributed in different spots across the brownfield site, but mostly in the area sited between two piers along the shoreline that is filled with slag, scum, and landfill material.

1. INTRODUCTION

The Campi Flegrei (CF) volcanic system can be considered a part of the city of Naples. In the CF area and, in particular, in Bagnoli, industrialization and urbanization processes fostered in the last century by the ILVA, Eternit, Cementir, and Federconsorzi industrial factories and plants boosted social and economic development. However, the products and by-products of those processes also altered sensitive natural equilibria and compromised the local environment.

The dismantling of the industrial complexes had a strong social impact on the city of Naples. After all industrial activities ceased, monitoring the area and assessing the requirements for site remediation became a priority. The Italian government funded the remediation plans with two specific Laws (N. 582—18/11/1996 and N. 388—23/12/2000) for the purpose of reusing the areas of ILVA and Eternit for nonindustrial activities. The area of the Federconsorzi has been acquired by the IDIS foundation to build the "City of Science," while the area occupied by Cementir has not been dismantled yet. The work on the brownfield sites concerned both the dismantling of the factories and the environmental remediation of the area, both of which are required before a new future for this site can be planned.

Considering that industries were present in the area for a century, it was reasonable to expect that most of the pollution originated from their activities. The major pollutants would have been expected to be metals derived from the combustion of fossil fuels, industrial waste, dumps, slag, and scum, and similar industrial wastes. However, Bagnoli is located inside an active volcanic field characterized by a strong geothermal activity that generates ascending hydrothermal fluids rich in heavy metals. Thus, we hypothesize that the brownfield site represents an overlap between two contamination components, one natural (originating from the CF hydrothermal activity) and the other anthropogenic (from the industrial activity).

Hydrothermal activity associated with volcanism introduces into the environment high quantities of heavy metals, and in some cases, this activity can even produce ore deposits. Classic examples are porphyry copper and epithermal gold deposits (Bodnar, 1995; Hedenquist and Lowenstern, 1994). For the Bagnoli area, this scenario is confirmed both by research carried out on the waters in front of Bagnoli (Damiani *et al.*, 1987; Sharp and Nardi, 1987) and by recent studies highlighting the existence in the CF, at Vesuvius, and in the Pontine Islands, of hydrothermal fluids similar to those found in porphyry copper systems (Belkin *et al.*, 1996; De Vivo *et al.*, 1995, 2006; Fedele *et al.*, 2006; Tarzia *et al.*, 1999, 2002).

2. ENVIRONMENTAL REMEDIATION OF THE BROWNFIELD SITE

The aim of the remediation plan launched by the Government (CIPE Resolution 20.12.94) was to eliminate the environmental risk due to former industrial activity, and to recover the land to make it usable for a new and different use, in accordance with the new urban development plans of the Naples City Council. The project called for dismantling of plants and structures and subsequent removal of pollutants by means of appropriate actions of environmental recovery.

The reclamation of the industrial area will prepare the Bagnoli area for the building of an urban park (included in the urban development plan for the city's eastern sector), which will represent a tangible sign of the environmental recovery of the area. The park will also preserve some structures as a memento of the industrial history of the area.

The Naples City Council, in agreement with the Sovrintendenza ai Beni Culturali, will recover and preserve 16 structures to represent the former industrial activities (Industrial Archaeological Site), while the original CIPE plan would have preserved only few buildings (up to a volume of 192,000 m^3) to be used for town business. The remediated areas will be the ILVA steel plant (1,945,000 m^2, production stopped in 1991) and the Eternit concrete–asbestos factory (157,000 m^2, production stopped in 1985) (Fig. 15.1).

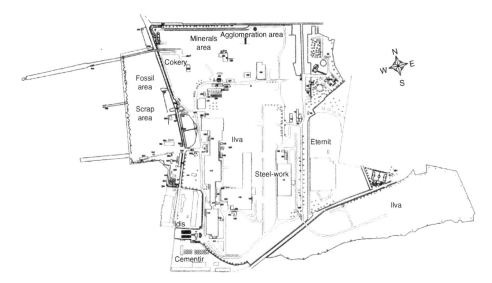

Figure 15.1 Map of the Bagnoli brownfield site.

To carry out the plan, a new company, the Società Bagnoli SpA, was formed on April 1, 1996. In brief, the plan was to: disassemble and dismantle plants and manufacturing structures; demolish buildings, walls, and refractory structures; dispose off raw materials, manufacturing by-products, and decontaminate plants and locations; recycling materials in alternative industrial activities where possible; conduct underground monitoring by means of borehole samples and chemical analyses; perform data elaboration and interpretation using distribution maps; and reclaim the Eternit area.

3. GEOLOGICAL SETTINGS OF THE BAGNOLI–FUORIGROTTA AREA AND STRATIGRAPHY OF THE BROWNFIELD SITE

The Bagnoli–Fuorigrotta Plain is an integral part of the CF, an active quaternary volcanic system, located 10 km W-NW from Naples (Fig. 15.2). On the basis of petrography and geochemistry, the volcanic products can be considered as part of the K-series of the Roman co-magmatic province (Peccerillo, 1985; Washington, 1906) and varies in composition from trachybasalts to phonolitic and peralkaline trachytes (Armienti et al., 1983; Di Girolamo, 1978). According to some authors (Russo et al., 1998 and references therein), the present morphology of the CF is the result of a complex sequence of volcanic and tectonic events, combined with spatial and temporal variations of the relationships between the sea and the ground. In particular, Russo et al. (1998) state that the Bagnoli–Fuorigrotta Plain was formed 12,000 years ago after the Neapolitan Yellow Tuff (NYT) eruption and the collapse that originated the CF caldera. Further activity inside the caldera occurred at 11,000 and 3500 years before present (YBP) in the multivolcanic center of Agnano, and caused the progression of the coastline and the formation of the Bagnoli–Fuorigrotta terrace. However, an environment of marshes and shallow waters was present until the second half of the 1800s, when reclamation and drainage finally established Bagnoli as part of the continental land.

In the central and eastern part of the plain, the substrate is NYT, that outcrops along the margin of the Posillipo ridge and thickens along the Agnano field, whereas the western part is dominated by the Agnano volcanic products. The oldest (11,000–7000 YBP) are intercalated with marine, fossil-rich sediments, whereas the most recent ones (5500–3500 YBP) are intercalated with paleosoils and alluvial volcanic sediments. On top of this sequence are a series of marine fossiliferous, beach, eolian, volcaniclastic, pyroclastic, and anthropogenic sediments.

Shallow stratigraphy: the examination of the surficial borehole core samples shows the presence of a cover made up of waste produced inside the industrial area, in particular, furnace scum and slag, mixed with volcanic ash and tuff, concrete, and brick, all of which overlie the original pyroclastic terrain. The thickness of this cover has been inferred based on core data. In 45% of the cores, the thickness of the cover is between 2 and 4 m; in 30%, it is between 0 and 2 m; in 20%, it is between 4 and 6 m; and in the remaining 5%, it is between 6 and 8 m. The overall volume of the cover waste in the ILVA area is about 5.5 million m^3. Immediately beneath the cover is a deposit of medium-fine sand in an ash matrix, containing pumice from mm to cm in size.

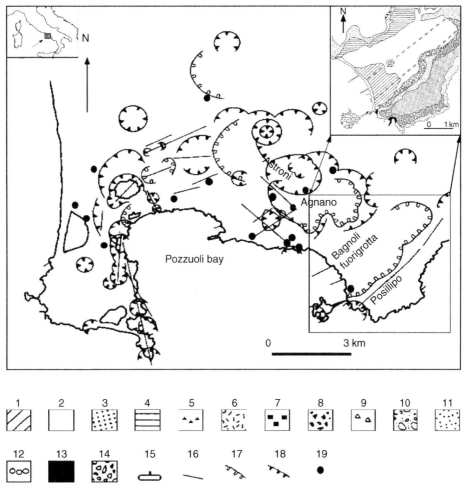

Figure 15.2 Volcanic and tectonic sketch of the Campi Flegrei (CF) and Bagnoli–Fuorigrotta Plain and location of fumaroles and hot springs (after Russo *et al.*, 1998, modif.). (1) Post–Roman age lacustrine and palustrine sediments; (2) volcanics and volcaniclastics (2500 YBP-recent); (3) volcanics and volcanoclastics (5500–3500 YBP); (4) pyroclastics of Agnano volcanic field (4000–3500 YBP); (5) S. Teresa volcanics (5500–3500 YBP); (6) pyroclastics and volcanic breccias of Monte Spina-Agnano eruption (ca. 4400 YBP); (7) pyroclastics of Cella-Monte S. Angelo unit (5500–5000 YBP); (8) Yellow Tuff of Nisida; (9) Yellow Tuff of La Pietra; (10) volcanics of the NYT (12,000 YBP); (11) Neapolitan Yellow Tuff (NYT) (12,000 YBP); (13) stratified Yellow Tuffs of Coroglio-Trentaremi (pre-12,000 YBP); (14) recent and historic volcanic debris; (15) volcano-tectonic lines; (16) faults; (17) post-caldera volcano-tectonic collapse; (18) vents (from Tarzia *et al.*, 2002).

Deep stratigraphy: Six deep boreholes (down to 50 m below the surface) allowed reconstruction of the deep structure of the area (Fig. 15.3). Four horizons (R, A, B, and C) were identified. Horizon R has a thickness that varies from 3 to 11 m, made up of a cover of anthropogenic debris and reworked pyroclastics. Horizon A has a variable thickness ranging from 4 to 10 m, made up of a coarse, ash-rich pyroclastic deposit (with a granulometry of medium- to very fine sand). Horizon B, classified as

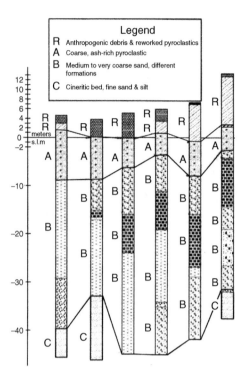

Figure 15.3 Schematic diagram of the stratigraphy of the Bagnoli brownfield site from deep boreholes.

a medium to very coarse sand, has an average thickness of 30 m and comprises different formations. Pumice and lithic lapilli can be found in the matrix, whereas the basal part contains gravel levels with light and dark clasts. Horizon C is a cineritic bed found at 40 m depth, classified as a fine sand-silt.

4. POTENTIAL SOURCES OF ANTHROPOGENIC POLLUTION

Possible pollution sources in the area include dust, ash, scum, slag, carbon coke residues, minerals, heavy oils, hydrocarbons, and combustion residues. The minerals used to produce cast iron and steel were imported mainly from Africa (Liberia and Mauritania), Canada, India, the former USSR, and from the American Continent (L'Industria Mineraria, 1979a).

The coal used as source of energy in smelting furnaces was imported mainly from mines of the eastern USA (Appalachian Basin) (L'industria Mineraria, 1979b). The scum (also known as dross), a by-product of cast iron manufacturing, resulted from melting of limestone and coke ash with the aluminosilicate gangue left over after iron reduction and separation. Slag is a by-product of steel manufacturing that results from oxidation of impurities and compounds generated from inert additives present in the charges. The use of fossil fuels (gasoline) produces many atmospheric pollutants, including Pb, which can be found in atmospheric particulates in form of oxides, carbonates, and sulfides. Most of the Italian production of additives for fuels is a monopoly of the British Associated Octel (AOC) and its Italian subsidiary Società Italiana Additivi Carburanti (SIAC), which use Pb from Broken Hill mines (Australia), and from South Africa, Perù, Mexico, and Italy (Magi *et al.*, 1975; Monna *et al.*, 1999).

5. HYDROGEOLOGICAL CHARACTERISTICS OF THE BAGNOLI–FUORIGROTTA PLAIN

In the Bagnoli–Fuorigrotta area, the water table is found slightly above mean sea level, and can be intercepted at shallow depths, especially south of the railroad (Fig. 15.4). The groundwater of the plain, resupplied directly by rainfall, is part of a wider groundwater body which spans the whole CF area and discharges directly to the sea. Detailed hydrogeological investigation carried out by the "Servizio Urbanistica del Comune di Napoli," in accordance with Italian Law 9/83, showed that groundwater composition of the CF system falls in the Na–Cl and Na-bicarbonate field, while along the coastline, the composition is mainly in the Na–Cl field.

The CF groundwater is affected by anthropogenic pollution from urban and industrial pollution; urban groundwater pollution typically contains nitrates whereas industrial groundwater pollution typically contains heavy metals and hydrocarbons. It is important to recognize the contamination produced by the upwelling

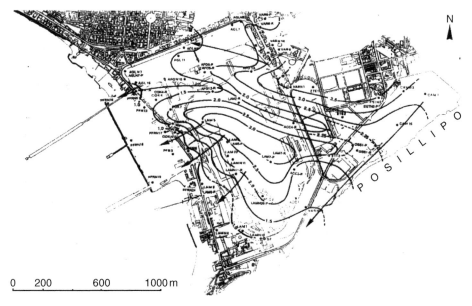

Figure 15.4 Morphology of the piezometric surface obtained from both surface and deep boreholes piezometers.

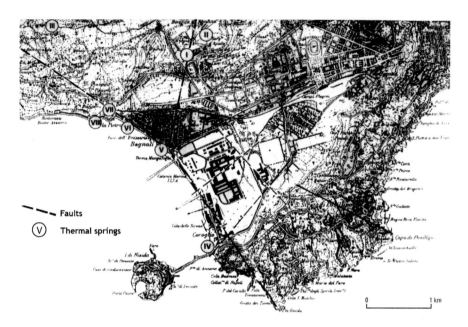

Faults

 Thermal springs

Figure 15.5 Distribution of faults in the Bagnoli–Fuorigrotta Plain and location of the thermal springs (spas).

of geothermal waters, containing heavy and potentially toxic metals such as As, Hg, Cu, Pb, and Cd. Upwelling of this contaminated groundwater occurs mainly along fractures and faults in the Bagnoli brownfield site and in the surroundings (Fig. 15.5).

6. SITE CHARACTERIZATION

In order to properly characterize the brownfield site, before choosing the specific remediation approach, an Expert Committee was nominated by the Government to coordinate and check the remediation activities. This Committee planned the following activities: two monitoring phases, which included waste and soil sampling, groundwater sampling, chemical analyses, map compilation for the pollutant elements, data elaboration and interpretation, asbestos characterization and remediation, and a preliminary operative remediation plan.

6.1. Monitoring: Phases I and II

Before monitoring operations began, the following documentation was gathered:

(a) Cartography and historical photos of the area to evaluate the settlement's evolution starting from 1870.
(b) Description and cartographic representation of industrial activities that occurred in the monitored area.

(c) Maps of the main foundation work.

(d) Maps of the sewer system.

(e) Results of former environmental investigations carried out in the area (e.g., soil, groundwater, air analysis).

(f) Geotechnical and stratigraphic reports made during plant construction.

(g) Report on raw materials and products still present and stored in the area, including information about location, quantity, composition, and their likelihood of dispersal in the environment.

(h) Results of geological and hydrogeological investigation, with particular attention to shallow and deep groundwater.

After obtaining documentation during the preliminary phase, a full-scale investigation was planned in the entire ILVA and Eternit area to locate and define all polluted areas. The investigation was divided in two phases: Phase I was a preliminary general survey and Phase II focused on details from the results of Phase I.

During Phase I, data were collected from shallow boreholes to 5 m, the depth of the local water table, using a 100 m × 100 m grid and sampling at 0.5, 1.5, 3.0, and 5.0 m depths. Six additional deep boreholes were drilled to 50 m, or alternatively to the deep groundwater table, whichever came first. Activities in Phase II were based on the results of the chemical analyses collected on Phase I samples. For Phase II, a 25 m × 25 m sampling grid was used in the polluted areas, and a 50 m × 50 m grid in all the others. The use of the regular grid method in site characterization is dictated in Italy by Law 152/2006; therefore, it was not possible to use a sampling method such as the *random stratified sampling method*, which would have been more appropriate.

To establish the values for natural background to be used as reference for maximum natural concentrations, samples were also taken outside the brownfield site, including 2 deep boreholes in Agnano and Fuorigrotta areas and shallow boreholes in 10 locations in the CF.

During Phase I, a geological survey was carried out from November 1997 to April 1998 at the brownfield site (ILVA and Eternit areas). During the survey, shallow and deep cores were collected; reworked and undisturbed soil was sampled; and groundwater samples were collected. In addition, a detailed geophysical survey was undertaken to establish a terrain lithostratigraphy, to determine the mechanical properties of the terrains, and to map the water table.

Specifically, the following investigations were carried out:

(a) Drilled six, piezometer-equipped deep boreholes, up to 50 m below the surface.

(b) Drilled two deep boreholes, outside the industrial area, up to 50 m below the surface.

(c) Drilled 207 shallow boreholes down to the water table, with an average depth of 3 m. Twenty-four boreholes were equipped with piezometers.

(d) Collected 905 samples (waste and reworked soil), of which 621 were analyzed.

(e) Collected 28 undisturbed soil samples, which were probed for geotechnical properties in laboratory.

(f) Performed 28 standard penetration tests (SPT) during core collection.

(g) Conducted a dipolar geoelectric survey (Eternit area).
(h) Performed a Georadar survey (Eternit area).
(i) Geographically referenced all surveyed sites.

A total of 20,751 chemical analyses for inorganic and organic elements and organic compounds were carried out on collected samples.

Based on the analytical results gathered during Phase I, a second survey was planned and carried out in the ILVA steel brownfield site. During Phase II, additional cores were collected, with shallow boreholes down to the water table, using a 50 m × 50 m and a 25 m × 25 m grid. The wider grid was used in those areas that, based on the results of Phase I, proved to be nonpolluted, whereas the 25 m × 25 m grid was used in the polluted areas.

Phase II started on May 31, 1999, and was completed by October 15, 1999. The following activities were carried out:

(a) Collected 2089 core samples.
(b) Collected 5976 samples (3586 samples to be analyzed for metals and 2390 for organic compounds).

A total of 73,219 analyses were carried out on the collected samples.

6.2. Chemical analysis

The chemical analyses carried out are indicated in Table 15.1. Analytical results produced by the Bagnoli SpA underwent quality controls through use of internationally recognized control standards and duplicated analysis of 5% of the samples at random. Duplicate analysis of 5% of the samples were performed at the British Geological Survey Laboratories.

6.3. Statistical analysis

Table 15.2 shows the univariate statistical parameters for all the elements, metallic and organic. Environmental Law DLgs 152/2006 not only sets the trigger and action limits, but it also states that these limits can (and should) be modified as a function of local background levels. Accordingly, the Expert Committee recommended that sampling be carried out outside the Bagnoli area on sites with the same geolithological characteristics. The Bagnoli SpA collected these samples inside the CF volcanic system. Table 15.3 shows the statistical parameters related to these samples.

Reference background values were established using cumulative frequency distribution curves. Following standard recognized procedures, the background limits were fixed, on a case-by-case basis, on average between the 70th and 90th percentiles.

Using the limits set using the above mentioned procedures, the Bagnoli SpA compiled distribution maps of all inorganic and organic chemical analyses. Only the distributions for some of the chemical parameters, which were found to exceed regulatory limits for a high percentage of the investigated sites, are shown here.

Table 15.1 Analyses carried out at Bagnoli brownfield site

General and anions	Metals	Organics
Conductivity (mS/cm)	As, Ba, Be, Cd, Co, Cr^{VI}, Cu, Hg, Mn, Mo, Ni, Pb, Sn, Th, U, V, Zn	Total hydrocarbons as N-heptane
Sulfides		Aliphatic halogenated solvents (1–2 dichloroethane, 1–1-1 trichloroethane (trichloroethylene)
Fluorides		Nonhalogenated aromatic solvents (benzene; phenols; BTX)
Free cyanides		Aromatic halogenated solvents (monochlorinated benzene; chlorinated phenols)
Complex cyanides		Polycyclic aromatic hydrocarbons (PAH) (benzo(*a*)anthracene, benzo(*a*)pyrene, benzo(*b*)fluorantene, benzo(*j*)fluorantene, benzo(*k*)fluorantene, pyrene, naphthalene, anthracene, fenantrene, fluorantene)
Elemental sulfur		Polychlorinated biphenyls (PCB)
Sulfates		Dioxins
Asbestos		Pesticides and phytopharmaceuticals (DDT)

Figs. 15.6A, 15.6B, 15.7A, and 15.7B show As and polycyclic aromatic hydrocarbon (PAH) distributions detected in Phase II, based on 25 m × 25 m and 50 m × 50 m grids.

In addition to univariate statistical analysis, the data were also examined by means of multivariate statistical techniques. In particular, R-mode factor analysis was used, which is a very effective tool to interpret anomalies and to help identify their sources. Factor analysis allows grouping of anomalies by compatible geochemical associations from a geologic-mineralogical point of view, the presence of mineralizing processes, or processes connected to the surface environment. Based on this analysis, six meaningful chemical associations were identified (Fig. 15.8).

The weight of each single association is quantified for every sampled site using the factor scores distribution. By associating the factor score distribution with lithologies, anthropogenic activities, or other characteristics, it is possible to establish a relationship between a particular association and a possible source. However, it is not useful for defining the trigger and action limits as provided in the guidelines provided by the Ministry of Environment (DLgs 152/2006).

Table 15.2 Statistical parameters of the analytical data from the borehole samples of the phase I monitoring using a 100 m × 100 m network (statistics: all lithologies)

	Parameters	D.M. 471/99 residential use (mg/kg)	Background values (mg/kg)	Analysis number	Detection limit (mg/kg)	Min (mg/kg)	Max (mg/kg)	Mean	Median	Standard deviation
	pH	—		576		5.5	12.8	9.39	9.15	1.38
	EC	—		576		3.2	23,500	835.36	366.50	1836.93
Inorganic	Sulfides	—		576		10	4920	95.30	15.00	390.48
compounds	Sulfates	—		575		5	22,318	655.65	133.00	1567.19
	Fluorides	—		575	1	1	182	11.34	9.75	11.61
	Free cyanides	1		574	1	1	3	1.00	1.00	0.08
	Complex cyanides	—		374	1	1	10.8	1.03	1.00	0.43
	Sulfur	—		525	100	100	612	103.36	100.00	33.62
	Asbestos	100	100	399		Not present				
Heavy metals	As	20	36	365	0.04	1.4	292.2	29.28	19.60	32.32
	Ba	—	—	363	0.07	10	1570	674.07	713.00	342.05
	Be	2	12	365	0.03	0.2	12	4.53	4.80	2.07
	Cd	2	2	365	0.02	0.02	12.6	0.57	0.20	1.20
	Co	20	35	364	0.05	0.3	102	10.10	720	10.40
	Cr total	150	150	365	0.1	2	1380	68.88	25.00	134.97
	Cr VI	2	—	365	5	3	5	5.00	5.00	0.00
	Hg	1	1	364	0.04	0.02	20	0.54	0.20	1.62
	Mo	—	—	365	0.03	0.1	14.4	3.27	3.20	1.40
	Ni	120	120	364	0.1	1.3	904	24.26	12.00	60.43
	Pb	100	110	365	0.5	1	1440	97.79	62.00	154.52
	Cu	120	120	365	0.1	4	644	56.58	33.00	69.19

Group		C1	C2	C3	C4	C5	C6	C7	C8	C9
	Sn	1	15	365	0.04	0.8	149.7	9.46	5.80	12.56
	V	90	100	362	0.05	8.7	2910	123.03	88.00	184.79
	Zn	150	158	364	0.15	2	6159	243.37	116.00	555.52
BTX	Phenols	0.1	0.1	569	0.01	0.1	2.74	0.13	0.10	0.21
	Benzene	0.1	0.1	373	0.005	0.05	9.7	0.07	0.03	0.40
	Toluene	0.5	0.5	575	0.01	0.1	1.25	0.11	0.10	0.07
	Xylene	0.5	0.3	374	0.01	0.1	1.85	0.11	0.10	0.09
	Total hydrocarbon	20	100	376	2	5	68,800	310.30	11.60	3245.33
Halogenated aromatic solvents	Monochlorinated benzene	0.3	0.5	343	0.005	0.005	0.005	0.005	0.005	0.000
	2 Chlorinated phenols	0.3	0.5	343	0.005	0.005	0.035	0.0055	0.005	0.003
	2, 4 Dichlorinated phenols	0.5	0.5	347	0.005	0.005	0.05	0.0056	0.005	0.004
	2, 6 Dichlorinated phenols	n.r.	n.r	547	0.005	0.005	0.05	0.0055	0.005	0.004
	2, 4, 6 Trichlorinated phenols	0.01	0.01	547	0.005	0.005	0.03	0.0053	0.005	0.003
	2, 3, 4, 6- Tetrachlorinated phenols	n.d.	n.d.	347	0.006	0.003	0.06	0.0054	0.003	0.004
	Pentachlorinated phenols	0.01	0.01	547	0.005	0.008	0.05	0.0054	0.005	0.004

Table 15.3 Statistical parameters of the analytical data from the sampling of sites outside the bagnoli brownfield site

Parameters	Mean	Median	Geometric mean	Min	Max	S.D.	Geometric S.D.
As	33.6	23.6	25.1	14.8	217.3	45.0	0.26
Ba	843.7	821.5	784.6	294.0	1545.0	297.2	0.18
Be	7.2	6.6	6.3	2.4	15.4	3.7	0.23
Cd	0.3	0.3	0.2	0.1	0.5	0.1	0.2
Co	44.9	7.15	14.0	4.0	280.0	73.6	0.63
Cr total	27.6	25	21.0	0.8	89.0	19.1	0.41
Hg	0.4	0.1	0.2	0.04	3.8	0.9	0.55
Mo	26.3	4.45	5.5	2.8	400.0	90.6	0.47
Ni	13.8	10	10.9	5.0	76.0	15.3	0.25
Pb	80.5	68.5	75.9	47.0	181.0	31.0	0.14
Cu	30.3	21	23.7	10.0	90.0	24.1	0.3
Sn	10.2	9.42	9.3	5.0	24.0	5.0	0.19
V	75.4	74.9	72.9	46.3	136.0	20.5	0.11
Zn	118.0	111.5	111.3	63.0	202.0	34.3	0.13
Hydrocarbons	45.4	18	24.3	5	204	56.7	0.47

6.4. Monitoring of groundwater

During both Phases I and II, 71 piezometers were installed to monitor groundwater. A total of 221 water samples were collected and 9463 analyses were carried out. Seven field surveys sampled shallow and deep groundwater, analyzing various physicochemical parameters (e.g., pH, Eh, dissolved O_2, temperature, conductivity), and the presence of potentially harmful elements and compounds (e.g., heavy metals, hydrocarbons, PAH). The hydrogeological survey carried out by the Bagnoli SpA, concluded that

(a) The aquifer is made up of different sub-horizontal levels, each with its own lithology and particle size (resulting in different permeabilities). This produces a layered circulation system, where different groundwater bodies are superimposed.

(b) The water table can be divided in subzones, each with unique characteristics. The northwestern zone has a very evident drainage axis and its waters flow toward a small part of the coastline nearby Piazza Bagnoli. The southwestern zone is completely inside the industrial complex and its waters flow directly to the sea along the Via Coroglio coastline. The southeastern zone waters flow toward south and southeast, following the preferential drainage axis located along the base of the northwestern flank of the Posillipo hill.

(c) The theoretical depth of the water table is about 8.5 m in the PFR area, 55 m in the COK area, and 65 m in the AFO area (Fig. 15.1). The morphology of this line is typical, with a slope of about 45 ° and a thickness which increases with distance from the coastline.

(d) Three pumping tests and six Lefranc tests show that permeability values are preferentially low.

A

Geochemical map
As (ppm) in soils (level I)

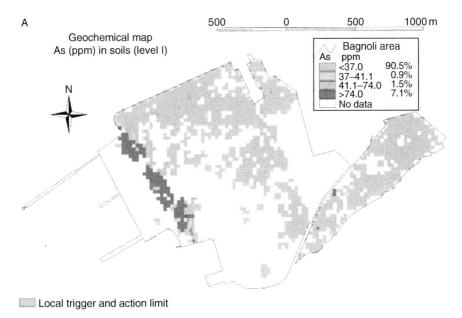

Local trigger and action limit

Figure 15.6A Arsenic distribution in the soil (from 25 m × 25 m network boreholes).

B

Geochemical map
As (ppm) in landfills

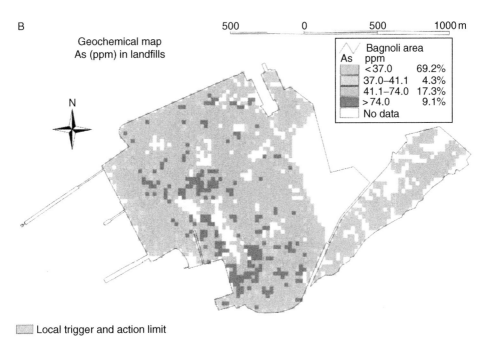

Local trigger and action limit

Figure 15.6B Arsenic distribution in the scum, slag and landfill materials (from 25 m × 25 m network boreholes).

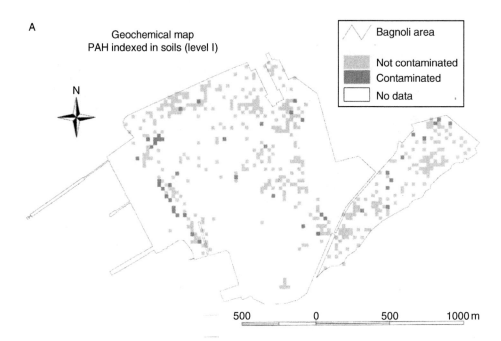

Figure 15.7A Polycyclic aromatic hydrocarbon (PAH) distribution in the soil (from 25 m × 25 m network boreholes).

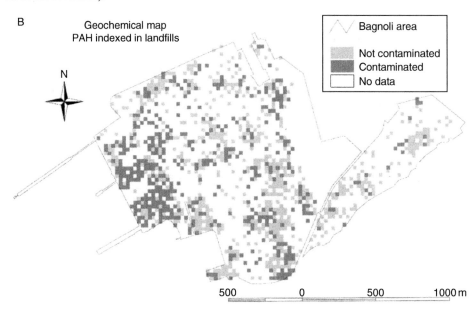

Figure 15.7B Polycyclic aromatic hydrocarbon (PAH) distribution in the scum, slag, and landfill materials (from 25 m × 25 m network boreholes).

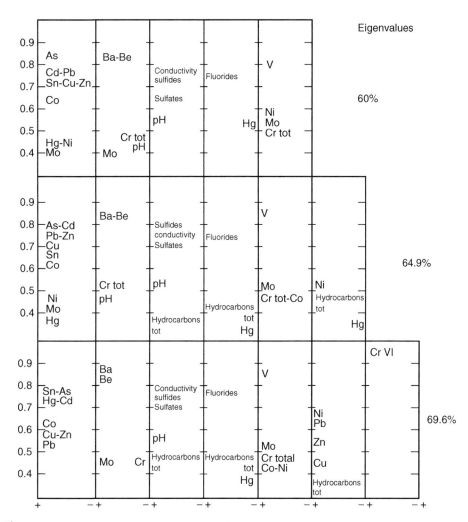

Figure 15.8 R-mode factor analysis models from the 100 m × 100 m network boreholes analytical results.

The highest flow values are found along the north and south drainage axis, along the detritus belt at the base of the Posillipo Hill. Water pH is extremely variable, with the highest basic values (9.7) found near the coastline area between the two piers that are filled with scum and slag waste (*colmata a mare*), and almost neutral values in the northwest sector of the brownfield site (DIR–AGL area). Groundwater temperatures range from 14.8 °C in the hills to 24.3 °C near the coastline. Specific electric conductivity averages 1 mS/cm, with the exception of the *colmata a mare* area where values are at their maximum (16.6 mS/cm) due to the presence of seawater.

Eh positive values are found in the east and north of the brownfield site, whereas negative values are found in the *colmata a mare* area (−167.5 mV); these conditions are favorable for dissolution of metals such as iron and manganese. Dissolved O_2 is generally low (<1 mg/l) and only in the northeast are the values >3 mg/l. The O_2

trend is well correlated with Eh. Water samples reveal high contents of As, Fe, and Mn, all above regulatory intervention limits established in DLgs 152/2006.

The investigations led to the following conclusions:

(a) The high Mn content is not due to leaching of the shallow part of the aquifer by percolating waters. The percolation pathways are too short to explain a Mn enrichment that goes up to 22,500 mg/l. Moreover, there is no correlation between the relatively shallow underground hydrodynamics and Mn contents in the water.

(b) The source of Mn is neither point nor diffuse anthropogenic pollution, since concentrations on the surface are always <50 mg/l. Moreover, no correlation was found between rainfall and variation of Mn concentrations with time.

(c) High Mn content is due to the upwelling of deep fluids characteristic of the Bagnoli–Fuorigrotta Plain substructure. This is based on the following evidence: (1) Sample sites with high Mn are located along four directions, coinciding with faults and fractures found in the tuff bedrock, along the same pathways in which thermal springs and old craters occur. (2) Where sample density is higher, it is possible to detect Mn dilution as a function of the distance from upwelling sites. Electric conductivity is also related to Mn content.

The investigated site is inside an active volcanic area, where geothermal fluids are enriched in As, Cu, Pb, Zn, and Hg. This is of paramount importance while interpreting the geochemical "anomalies" found in Bagnoli. This is also true for Fe concentrations, which are clearly related to Mn contents.

Hydrocarbon concentrations are always above the regulatory intervention limit of 10 mg/l established by DLgs 152/2006, both at the piezometers located inside the site and those located at its margins. External piezometers record high values as well, testifying that these compounds are present in all groundwaters in the Naples area.

PAH distribution patterns are more complex inside the brownfield site. The presence of these compounds in surficial waters is irregular and generally does not reach high values, with few exceptions ("hot spots") localized at the industrial site margins near Via Diocleziano (LAMN4 and PFRN18 areas).

Water surveys and monitoring have revealed the presence of PAH with significant concentrations in two areas: *Colmata a mare* and the LAM area (Laminatoi). A more diffuse contamination has been found in the Acciaieria, Cockeria, and Laminatoi areas.

Suitable barriers have been put in place to minimize contaminant migration from the brownfield site to the surroundings, in particular, toward the sea (see Section 11). One of the pollution sources is the circulation of hydrothermal fluids from thermal springs (spas). Studies conducted on CF thermal waters show high As concentrations in the range 12–5600 ppb. In the Puteolane hot springs (Dazio Bagnoli), located less than 1 km from the brownfield site, As concentrations up to 2600 ppb were found. These concentrations present further evidence of metal enrichment caused by the presence of geothermal fluids. Analytical results also show hydrocarbon contamination uphill from the Bagnoli brownfield site. Inside the brownfield site, the same concentrations were detected as found uphill from the site; therefore, it is likely that this "contamination" is ascribable both to industrial activity and to diffuse contamination from other sources.

7. NATURAL AND ANTHROPOGENIC COMPONENTS FOR THE POLLUTION

As stated in Section 1, the Bagnoli brownfield site contains metallic elements whose source overlaps a natural component (ascribable to the CF hydrothermal activity connected to the quiescent volcanism) and an anthropogenic one (due to the industrial activity). The challenge was demonstrating and separating the contribution of these two components.

Pb isotopes are used to study environmental pollution because of their relative geochemical immobility and the wide use of Pb in industrial processes. Moreover, Pb isotopes are not fractionated by natural or industrial processes; Pb isotopic composition in a material remains constant in time and reflects the nature of the source (Ault *et al.*, 1970).

Pb used in industrial processes is extracted from sulfide ore deposits of different ages and origins. Once released in the environment from industrial activities, this metal is adsorbed by Fe and Mn oxides, whose formation in turn is fostered by atmospheric agents. Determining the isotopic composition and the chemical composition of soils allows us to separate anthropogenic Pb from natural sources, helping us to define the origin and extent of the contribution of anthropogenic sources to pollution.

The use of Pb isotopes in this field dates back to the 1960s (Chow and Johnstone, 1965) and it has been employed in numerous studies in European and Mediterranean regions (Grousset *et al.*, 1995; Hopper *et al.*, 1991; Maring *et al.*, 1987). Fewer studies have been conducted in Italy and they have never been applied to industrial sites (Cochran *et al.*, 1998; Colombo *et al.*, 1988; De Vivo *et al.*, 2001; Facchetti *et al.*, 1982, 1989; Garibaldi *et al.*, 1981; Magi *et al.*, 1975; Monna *et al.*, 1999; Tommasini *et al.*, 2000).

At the Bagnoli brownfield site, Tarzia *et al.* (2002) (as part of his PhD program with University of Naples Federico II) carried out a study aimed to discriminate anthropogenic pollution sources from natural pollution sources. For this study, heavy metals and Pb isotope data from soils, waste materials, scum, and slag samples from the brownfield site were used.

The samples used for isotopic analysis were collected from 20 boreholes during the monitoring activities of Phase I, using a 100 m × 100 m grid. Selected samples were dried in air and sieved to extract the <2 mm fraction. This fraction was then homogenized, quartered, and sieved again to extract the <177 μm (80 mesh) fraction. After processing, the samples were analyzed as follows: (1) inductively coupled plasma–atomic emission spectrometry (ICP-AES) to determine, major, minor, and trace elements; (2) X-ray fluorescence (XRF) to determine mineralogy of scum and slag; and (3) inductively coupled plasma–mass spectrometry (ICP-MS) for Pb isotope analysis. XRF and part of Pb analysis were carried out at the USGS Laboratories (Reston, Virginia), while the remaining Pb isotope analyses were carried out at the British Geological Survey (Nottingham, UK).

ICP-AES analysis determined concentrations for the following elements: Ag, Al, As, Au, B, Ba, Bi, Ca, Cd, Co, Cr, Cu, Fe, Ga, Hg, K, La, Mg, Mn, Mo, Na, Ni, P, Pb, S, Sb, Se, Sr, Te, Th, Ti, Tl, U, V, W, and Zn.

The Pb isotope results showed a linear trend that suggests a mixing between two end members, one natural and related to the CF volcanics, and the other anthropogenic. The large overlap of isotopic data does not allow a precise quantitative discrimination of the contribution of each component, but it is possible to state that the natural component dominates.

Plots of metal concentrations against Pb isotope ratios were extremely useful. For example, Cr values were found to be distributed along two clearly different clusters. One cluster represents the soils of Bagnoli, while the other corresponds to the waste (scum, slag, landfill) material. For decreasing Cr concentrations, the data seem to converge toward soil values. Similar trends are detected for other elements (Fig. 15.9).

The plots confirm the existence of a contamination characterized by isotopic values very similar to those of the soils (i.e., the natural values). Plots of the isotopic ratios ($^{207/204}$Pb vs. $^{208/204}$Pb) show two distribution trends that converge toward values typical of the Neapolitan Yellow Tuff (D'Antonio et al., 1995) (Fig. 15.10).

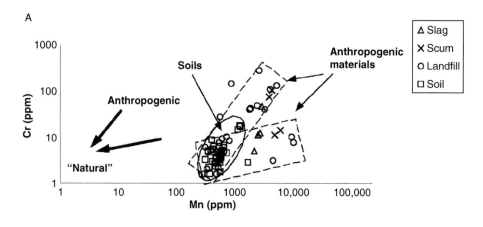

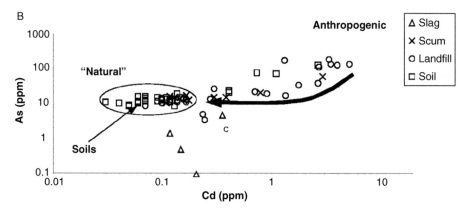

Figure 15.9 Cr versus Mn (A) and As versus Cd (B) concentrations. Such plots show the convergence of data toward natural values (bold arrows) and suggests a relationship between contamination and scum, slag, and landfill materials.

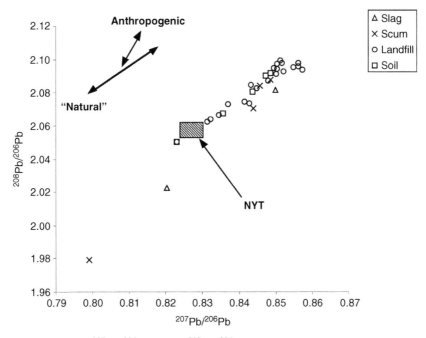

Figure 15.10 Plot of $^{207}Pb/^{206}Pb$ versus $^{208}Pb/^{206}Pb$. The dashed area represents the field of isotopic composition for the NYT (as reported by D'Antonio *et al.*, 1995). The straight arrows underline the general trends of mixing between three potential Pb end members (Tarzia *et al.*, 2002).

Figure 15.11 shows the plot of Pb isotopic ratios for possible end members, such as gasoline, aerosols (Italian and European), and coal, compared to the ratios recorded in Bagnoli. For Italian gasoline, only a few data points (gasoline with additives from Mexico and Peru) are consistent with Bagnoli. The same is true for aerosols, with the only exception being an aerosol from Senegal.

Isotopic data from coal imported from the Appalachian Basin of the eastern United States are instead consistent with Bagnoli, since this coal was used at Bagnoli as an additive in the smelting furnace during the steel manufacturing process. The raw material (i.e., Fe minerals) was imported from Liberia, Canada, India, and other nations; unfortunately, the only isotopic data available in the literature are for Loulo and the Nimba shield Fe formations (Liberia, Eastern Africa), and those are not compatible with Bagnoli data.

Chemical and isotope data clearly show mixing between a major natural component (e.g., reworked subaerial and marine volcanics), and an anthropogenic one. Hydrothermal fluids associated with the CF-active volcanism, an area where fumaroles and hydrothermal springs are quite abundant, provide significant contribution to the metals present at this site. The natural contamination due to the upwelling of geothermal fluids (enriched in heavy and potentially toxic metals such as As, Cu, Pb, and Hg) is confirmed by the high concentrations of heavy metals found in the thermal springs (spas) located at the margins of the brownfield site of Bagnoli (e.g., the Terme di Bagnoli, Dazio, the Terme Puteolane, the Stufe di Nerone) and the nearby Island of Ischia (Daniele, 2000; Lima *et al.*, 2001, 2003). Values at the Stufe di Nerone reach up to 8000 ppb, while Ischia values are >1500 ppb.

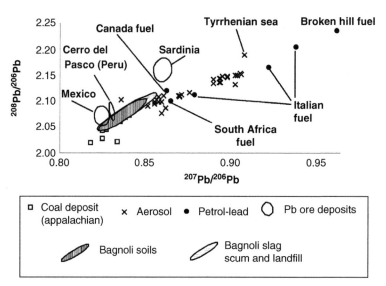

Figure 15.11 $^{208}Pb/^{206}Pb$ versus $^{207}Pb/^{206}Pb$ isotopic composition of samples collected in the study of Tarzia *et al.* (2002) and data reported in literature. The compositional ranges of the natural and anthropogenic Pb sources are reported as dashed and dotted areas, respectively. (Data sources: Elbaz-Poulichet *et al.*, 1984; Grousset *et al.*, 1995; Marcoux and Milesi, 1993; Maring *et al.*, 1987; Monna *et al.*, 1999; Tommasini *et al.*, 2000.).

Isotope values from soils and waste (e.g., scum, slag) are similar, but a large overlap in data does not allow precise discrimination of the natural component from the anthropogenic one. Isotope ratio trends for soils and waste in Bagnoli is partially divergent from the "classic" anthropogenic trend defined by gasoline combustion and aerosols values from other parts of the world. This suggests that for the Bagnoli area, there is a local source of contamination, likely from fossil fuel use and the emission of industrial fumes. Italian gasoline with additives from Mexico and Peru are partially responsible for contamination, but the major role has surely been played by the large amounts of coal used in the industrial process. However, other sources, such as paints, mineral oils, and uncontrolled organic and inorganic sewer waste, for which data are not available, cannot be ruled out as sources for some contaminants.

The presence of a major natural contribution (i.e., hydrothermal fluids) for the metal contamination makes metal remediation of the Bagnoli area to a level below local background futile. This conclusion is further strengthened by the observation that scum and slag proved to be geochemically very stable relative to their metal contents; leachability tests prove that neither scum nor slag releases metals in soils and groundwater at the pH and local conditions existing in the regional area.

8. CHEMICAL–STRUCTURAL CHARACTERIZATION OF WASTE MATERIAL AND LEACHABILITY TESTS

Chemical–structural analysis and leachability tests on scum, slag, and Fe-minerals were carried out to characterize and measure the chemical stability of the metals in waste materials. The investigation was designated to (1) evaluate the environmental issues connected with the industrial waste and (2) define criteria and procedures aimed at recovering the by-products remaining in the brownfield site. The primary objective was to evaluate the rate of pollutant release in the environment and to determine the risk of contamination of groundwater at the site. For this reason, different types of leachability tests and analyses of leached material were carried out on scum, slag, minerals, and steelwork mud.

Microscope analyses and leachability tests showed that the microstructural configuration of materials at the site are stable not only relative to the tests and procedural time frame but also because of the recognition of the absence of isolated microstructures made of heavy elements. Heavy elements are most likely trapped in the lattice of the detected relatively insoluble microstructure phases, and leaching of these elements is highly unlikely. In conclusion, the industrial waste present in the brownfield site of Bagnoli is unlikely to contribute metallic pollution to local groundwater, on a reasonable human time scale. Numerous investigations carried out on industrial steelwork waste in the European Community have reached the same conclusions.

9. ASBESTOS CHARACTERIZATION AND REMEDIATION

The CIPE plan called for a remediation of asbestos-containing materials in the Eternit and in the ILVA steelwork factories. Ninety percent of the buildings, squares, and sites were cleared of asbestos by March 4, 2000. During the remediation activities, and in coordination with the local Health Unities (ASL)), a series of control samples were collected. 915 samples were analyzed to evaluate the presence of aerially dispersed asbestos fibers in nearby locations. No values exceeding WHO limits were detected. 1044 samples and analyses from the Eternit site and 56 from the ILVA site were also collected to monitor fiber dispersion inside the area of the operations.

10. PRELIMINARY OPERATIVE REMEDIATION PLAN

The Bagnoli SpA drafted a preliminary operative remediation plan on the basis of geological, hydrogeological, and chemophysical investigations carried out on the site. It also undertook research to verify the effectiveness and feasibility of some treatments to be applied on underground materials occurring at the Bagnoli site, such as anthropogenic waste covering and natural soils. Effective remediation procedures must be applied to bring concentrations in soils and underground within the limits set by Italian Law DLgs 152/2006. Table 15.1 shows the reported limits set by law, and background levels established during the survey.

The draft reclamation plan adopted the following guidelines:

(a) Minimize the impact of former industrial processes on the natural environment.
(b) Maximize material recycling and reuse.
(c) Contain the costs within acceptable limits.

The project adopts reclamation methods verified by investigations and experimental research, and uses industrial cycles and treatments presently available on an industrial scale. Treatment cycles will depend on contamination levels and presence of organic contaminants.

Regarding the heavy metals, the materials with concentrations of a single metal exceeding the limits of Table 15.4, column B (DLgs 152/2006), and the presence of leachate with concentrations exceeding the limits set for groundwater by Law 152/1999, is considered as highly contaminated ("hot spots").

Table 15.4 Intervention limits for residential and commercial/industrial land use according to the law 152/2006 and background values

Parameters	DLgs 152/2006 residential use (mg/kg)	DLgs 152/2006 industrial and commercial use (mg/kg)	Background (mg/kg)
pH		—	
Electrical conductivity (μS/cm)	—	—	—
Sulfides	—	—	—
Sulfates	—	—	—
Fluorides	—	—	—
Cyanides	1	100	—
Complex cyanides	—	—	—
S	—	—	—
As	20	50	37
Ba	—	—	—
Be	2	10	12
Cd	2	15	2
Co	20	250	130
Cr total	150	800	150
Cr VI	2	15	—
Hg	1	5	1
Mo	—	—	—
Mn	—	—	v
Ni	120	500	120
Pb	100	1000	112
Cu	120	600	120
Sn	1	350	15
V	90	250	110
Zn	150	1500	158
Phenols	1	60	
Benzene	0.1	2	
Toluene	0.5	50	
Xylene	0.5	50	
Total hydrocarbons	50	750	105

Table 15.4 *(Continued)*

Parameters	DLgs 152/2006 residential use (mg/kg)	DLgs 152/2006 industrial and commercial use (mg/kg)	Background (mg/kg)
Monochlorinated benzene	0.5	50	
2-Chlorinated phenols	0.5	25	
2,4-Dichlorinated phenols	0.5	50	
2,4,6-Trichlorinated phenols	0.01	5	
Pentachlorinated phenols	0.01	5	
1,2-Dichlorinated ethane	0.2	5	
1,1,2-Trichlorinated ethane	0.5	15	
Pyrene	5	50	
Benzo(*a*)anthracene	0.5	10	
Chrysene	5	50	
Benzo(*b*)fluoranthene	0.5	10	
Benzo(*k*)fluoranthene	0.5	10	
Benzo(*a*)pyrene	0.1	10	
Dibenzo(*a,h*)anthracene	0.1	10	
Benzo(*g,h,i*)perylene	0.1	10	
Indeno pyrene	0.1		
Dibenzo(*a,i*)pyrene	0.1	10	
PAH total	10	100	
PCB	0.06	5	

PCB, polychlorinated biphenyl; PAH, polycyclic aromatic hydrocarbon.

During this preliminary phase of the remediation plan, based on results obtained at the end of the Phase II of monitoring (25 m × 25 m and 50 m × 50 m sampling grids), it was necessary to accurately determine the amount of waste cover and soil to be treated. The sampling network established during Phase II of the monitoring had an inhomogeneity due to logistic problems, and standard methods used for the volumetric evaluation of the results led to an overestimation of the volumes to be treated. An adjustment has been done using fractal algorithms, which reduced the volumes to be treated by 26%.

Based on these new calculations, 4,063,910 m^3 of the waste cover and 3,190,371 m^3 of soils will be treated down to the water table (Table 15.5). Final volumes to be treated will likely be reduced during operations, since only during preliminary remediation will there be a definitive assessment of the volumes to be considered polluted (based on the 25 m × 25 m grid). To define the volumes to be treated and removed, each 25 m × 25 m cell will be divided into nine subcells. Samples will be collected from the center of the subcells. The eight remaining subcells will be sampled using either new boreholes or trenches, and new chemical analyses will be performed to determine metals, PAH, and hydrocarbons. The new cells will then be classified based on contamination levels (Fig. 15.12).

Table 15.5 Classification criteria of the soil and landfill materials volumes to be remediated for metals, total hydrocarbons + PAH

		Landfills				
		Heavy metals		HC + PAH		
Subarea code	Landfill volumes above the water table (m³)	Landfill volumes with concentration values exceeding limits (m³)	%	Landfill volumes with concentration values exceeding limits (m³)	%
DIR	313,092	2,111,469	67.5	155,029	49.5
DFR	496,705	446,412	89.9	348,759	70.2
AGL	261,988	152,038	58.0	62,499	23.9
OSS	258,125	96,641	37.4	61,305	23.7
CAM	400,678	257,214	64.2	36,778	9.2
ARO-COK	351,778	278,816	79.3	173,660	49.4
TNA	585,611	468,812	80.1	280,419	47.9
LAM-MESTA	416,948	307,347	73.7	158,417	38.0
LAM-MAG	630,608	556,155	88.2	266,577	42.3
ACC	348,377	279,461	80.2	122,129	35.1
Total	4,063,910	3,054,365	75.2	1,665,572	41.0

| | | Soils | | | |
| | | Heavy metals | | HC + PAH | |
Subarea code	Landfill volumes above the water table (m³)	Soil volumes with concentration values exceeding limits (m³)	%	Soil volumes with concentration values exceeding limits (m³)	%
DIR	407,023	80,703	19.8	22,931	5.6
DFR	48,167	46,431	96.4	8659	18.0
AGL	256,437	62,722	24.5	6069	2.4
OSS	291,317	79,001	27.1	18,795	6.5
CAM	1,545,649	666,634	43.1	69,527	4.5
ARO-COK	109,386	29,266	26.8	13,540	12.4
TNA	39,485	13,699	34.7	1974	5.0
LAM-MESTA	352,528	69,395	19.7	7697	2.2
LAM-MAG	28,844	8546	29.6	2747	9.5
ACC	111,536	26,672	23.9	4346	3.9
Total	3,190,372	1,083,079	33.9	156,285	4.9

HC, total hydrocarbons.
PAH, polycyclic aromatic hydrocarbon.

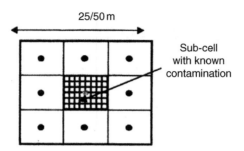

Figure 15.12   Sketch of the subdivision of the 25 m × 50 m networks in subnetwork of lower order.

Reclamation will be carried on different materials down to the water table, with the following goals:

(a) Full removal of the organic component exceeding Italian regulatory limits for the materials occurring in the site
(b) Removal of the inorganic component with leachate exceeding Italian regulatory limits set by Law 152/1999
(c) Surficial restoration and rebuilding of soil cover in the Parco and Parco–Sport areas. In the area designated as beach, the actual surface will be lowered about 2.2 m

To reach these goals, many treatment cycles will be required, with each cycle consisting of

(a) treating of materials contaminated by organic waste
(b) treating of materials contaminated by heavy metals
(c) surface restoration of the area

The project will also secure the groundwater system to reduce the impact of operations on the hydrologic system. Wherever groundwater contamination is detected, stations will be installed locally to pump and send polluted water to treatment plants.

All reclaimed areas where demolition and digging has been conducted will be leveled out to prepare the sites for the soil layer reconstitution and the subsequent construction of the Parco and Parco–Sport structure (which will cover 1,200,000 m²).

11. Securing the Site

During site characterization, contamination of groundwater by organic compounds (total hydrocarbons and PAH) has been detected. PAH contamination is diffuse, with concentrations of about 1 mg/l, but also localized, with hot spots in the

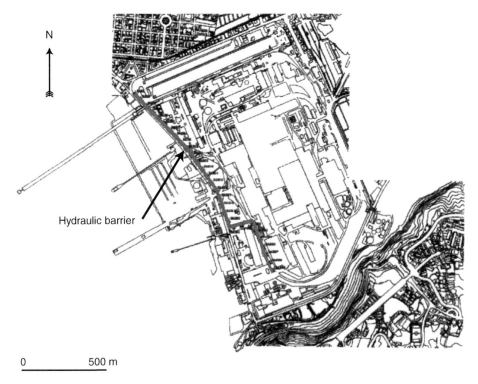

0 _____ 500 m

Figure 15.13   Hydraulic barrier along the coastline in the Bagnoli brownfield site.

northern (VAR6) and southern sectors (LAM N4), where levels are >1 mg/l. To secure the site, three approaches were considered:

(a) Form a hydrologic barrier using pumping stations (i.e., drawing wells).
(b) Construct a plastic diaphragm (concrete-bentonite).
(c) Create a jet grouting barrier.

Solution (a) has been selected, and pumping stations connected to a water treatment plant will be built to stop organic pollutants from flowing off-site. The barrier will be made up of 29 wells spanning a total length of 1500 m (Fig. 15.13). This structure will allow complete blockage of the pollution front and optimize pumping, minimizing water table depression and seawater ingression. The water treatment plant has been designed to meet the requirements of the pumping stations and to meet the needs of the reclamation operations.

ACKNOWLEDGMENTS

We thank C. Sears and H. E. Belkin for their constructive reviews and we also thank S. Albanese for his help in figure graphics.

REFERENCES

Armienti, P., Barberi, F., Bizouard, H., Clocchiatti, R., Innocenti, F., Metrich, N., Rossi, M., and Sbrana, A. (1983). The Phlegraean Fields: Magma evolution within a shallow magma chamber. *J. Volcanol. Geotherm. Res.* **17**, 289–311.

Ault, W. A., Senechal, R. G., and Erlebach, W. E. (1970). Isotopic composition as a natural tracer of lead in the environment. *Environ. Sci. Technol.* **4**, 305–313.

Belkin, H. E., De Vivo, B., Lima, A., and Torok, K. (1996). Magmatic silicate/saline/sulphur rich/ CO_2 immiscibility and zirconium and REE enrichment from alkaline magma chamber margins evidence from Ponza island, Pontine archipelago, Italy. *Eur. J. Miner.* **8**, 1401–1420.

Bodnar, R. J. (1995). Fluid inclusion evidence for magmatic source of metals in porphyry copper deposits. *Miner. Assoc. Canada, Short Course Series* **23**, 139–152.

Chow, T. J., and Johnstone, M. S. (1965). Lead isotopes in gasoline and aerosols of Los Angeles basin, California. *Science* **147**, 502–503.

Cochran, J. K., Frignani, M., Salamanca, M., Bellocci, L. G., and Guerzoni, S. (1998). Lead-210 as a tracer of atmospheric input of heavy metals in the northern Venice lagoon. *Mar. Chem.* **62**, 15–29.

Colombo, A., Facchetti, S., Gaglione, P., Geiss, F., Leyendecker, W., Rodari, R., Trincherini, P. R., Versino, B., and Garibaldi, G. (1988). The isotopic lead experiment: Impact of petrol lead on human blood and air. Final Report Ispra, Commission of the European Communities, EUR 12002.

D'Antonio, M., Tilton, G. R., and Civetta, L. (1995). Petrogenesis of Italian alkaline lavas deduced from Pb-Sr-Nd isotope relationships. *In* "Isotopic Studies of Crust-Mantle Evolution" (A. Basu and S.R Hart, eds.), pp. 253–267. Amer. Geoph. Union Monograph, Washington, USA.

Damiani, V., Baudo, R., De Rosa, S., De Simone, R., Ferretti, O., Izzo, G., and Serena, F. (1987). A case study: Bay of Pozzuoli (Gulf of Naples, Italy). *Hydrobiologia* **149**, 210–211.

Daniele, L. (2000). Geochimica degli elementi metallici nelle acque sotterranee dell'isola d'Ischia: Esempio di prospezione nel settore occidentale : Tesi di Laurea, Università degli Studi di Napoli Federico II. A.A. 1999/2000.

De Vivo, B., Torok, K., Ayuso, R. A., Lima, A., and Lirer, L. (1995). Fluid inclusion evidence for magmatic silicate/saline/CO_2 immiscibility and geochemistry of alkaline xenoliths from Ventotene island (Italy). *Geochim. Cosmochim. Acta* **59**, 2941–2953.

De Vivo, B., Somma, R., Ayuso, R. A., Calderoni, G., Lima, A., Pagliuca, S., and Sava, A. (2001). Pb isotopes and toxic metals in floodplain and stream sediments from the Volturno river basin, Italy. *Environ. Geol.* **41**, 101–112.

De Vivo, B., Lima, A., Kamenetsky, V. S., and Danyushevsky, L. V. (2006). Fluid and melt inclusions in the sub-volcanic environments from volcanic systems: Examples from the Neapolitan area and Pontine islands (Italy). *In* "Melt Inclusions in Plutonic Rocks" (J. D. Webster, ed.), pp. 211–237. Mineralogical Association of Canada Short Course 36, Montreal, Quebec.

Di Girolamo, P. (1978). Geotectonic setting of Miocene-Quaternary volcanism in and around the eastern Thyrrenian sea border (Italy) as deduced from major element geochemistry. *Bull. Volcanol.* **41**, 229–250.

Elbaz-Poulichet, F., Holliger, P., and Huang, W. W. (1984). Lead cycling in estuaries, illustrated by the Gyronde esuary, France. *Nature* **308**, 409–414.

Facchetti, S. (1989). Lead in petrol. *Acc. Chem. Res.* **22**, 370–374.

Facchetti, S., Geiss, F., Gaglione, P., Colombo, A., Garibaldi, G., Spallanzani, G., and Gilli, G. (1982). Isotopic lead experiment: Status report Luxembourg, Commission of European Communities, EUR 8352 EN.

Fedele, L., Tarzia, M., Belkin, H. E., De Vivo, B., Lima, A., and Lowenstern, J. B. (2006). Magmatic-hydrothermal fluid interaction and mineralization in alkali-syenite nodules from the Breccia Museo pyroclastic deposit, Naples, Italy. *In* "Volcanism in the Campania Plain: Vesuvius, Campi Flegrei and Ignimbrites" (B. De Vivo, ed.). Developments in Volcanology, 9, pp. 125–161. Elsevier, Amsterdam, The Netherlands.

Garibaldi, P., Vanini, G., and Gilli, G. (1981). Isotopic tracing of lead into children from automobile exhaust. In "Environmental Lead" (D. Lynam, L. Piantanida, and J. Cole, eds.), pp. 9–21. Academic Press, Amsterdam, The Netherlands.

Grousset, F. E., Quetel, C. R., Thomas, B., Lambert, C. E., Guillard, F., Donard, O. F. X., and Monaco, A. (1995). Anthropogenic vs lithologic origins of trace elements (As, Cd, Pb, Sn, Zn) in water column particles: Northwestern Mediterranean Sea. Mar. Chem. 48, 291–310.

Hedenquist, J. W., and Lowenstern, J. B. (1994). The role of magmas in the formation of hydrothermal ore deposits. Nature 370, 519–527.

Hopper, J. F., Ross, H. B., Sturges, W. T., and Barrie, L. A. (1991). Regional source discrimination of atmospheric aerosols in Europe using the isotopic composition of lead. Tellus 43, 45–60.

L'Industria Mineraria (1979a). Italsider—l'approvvigionamento dei minerali di ferro. Gennaio-Febbraio 1, 1–9.

L'industria Mineraria (1979b). Carbon fossile: La realtà Italsider. Marzo-Aprile 2, 76–88.

Lima, A., Daniele, L., De Vivo, B., and Sava, A. (2001). Minor and trace elements investigation on thermal groundwaters of Ischia Island (Southern Italy). In "Proceedings of Water-Rock Interaction-10" (R. Cidu, ed.), Vol. 2, pp. 981–984. Balkema, Rotterdam, The Netherlands.

Lima, A., Cicchella, D., and Di Francia, S. (2003). Natural contribution of harmful elements in thermal groundwaters of Ischia Island (Southern Italy). Environ. Geol. 43, 930–940.

Magi, F., Facchetti, S., and Garibaldi, P. (1975). Essences additionnees de plomb isotopiquement differencie. In "Proceedings of United Nations FAO and International Atomic Energy Association Symposium," 191, pp. 109–119. Vienna, Austria (IAEA-SM).

Marcoux, E., and Milesi, J. P. (1993). Lead isotope signature of early Proterozoic ore deposits in western Africa: Comparison with gold deposits in French Guyana. Econ. Geol. 88, 1862–1879.

Maring, H., Settle, D. M., Buat-Meanard, P., Dulac, F., and Patterson, C. C. (1987). Stable lead isotope tracers of air mass trajectories in the Mediterranean region. Nature 330, 154–156.

Monna, F., Aiuppa, A., Barrica, D., and Dongarrà, G. (1999). Pb isotope composition in lichens and aerosols from Eastern Sicily: Insights into the regional impact of volcanoes on the environment. Environ. Sci. Technol. 33, 2517–2523.

Peccerillo, A. (1985). Roman comagmatic province (Central Italy): Evidence for subduction-related magma genesis. Geology 13, 103–106.

Russo, F., Calderoni, G., and Lombardo, M. (1998). Evoluzione geomorfologica della depressione Bagnoli-Fuorigrotta: Periferia urbana della città di Napoli. Boll. Soc. Geol. It. 116, 21–38.

Sharp, W. E., and Nardi, G. (1987). A study of the heavy metal pollution in the bottom sediments at Porto di Bagnoli (Naples), Italy. J. Geochem. Explor. 29, 49–73.

Tarzia, M., Lima, A., De Vivo, B., and Belkin, H. E. (1999). Uranium, zirconium and rare earth element enrichment in alkali syenite nodules from the Breccia Museo deposit, Naples, Italy. Geol. Soc. Amer. Annual Meeeting, Abstracts with Programs, Vol. 31, n. 7, p. A-69.

Tarzia, M., De Vivo, B., Somma, R., Ayuso, R. A., McGill, R. A. R., and Parrish, R. R. (2002). Anthropogenic vs. natural pollution: An environmental study of an industrial site under remediation (Naples, Italy). Geochem. Explor. Environ. Anal. 2, 45–56.

Tommasini, S., Davies, G., and Elliott, T. (2000). Lead isotope composition of tree rings as biogeochemical tracers of heavy metal pollution: A reconnaissance study from Firenze, Italy. Appl. Geochem. 15, 891–900.

Washington, H. S. (1906). "The Roman Comagmatic Region." Carnegie Inst. of Washington. Vol. 57, p. 199.

CHAPTER SIXTEEN

Relationships Between Heavy Metal Distribution and Cancer Mortality Rates in the Campania Region, Italy

Stefano Albanese,* Maria Luisa De Luca,* Benedetto De Vivo,* Annamaria Lima,* *and* Giuseppe Grezzi*

Contents

Abstract

We report geochemical and epidemiological data as maps representing the patterns of toxic metal concentrations and some, potentially, related pathologies in the Campania region of Italy. The comparison of a particular element distribution with specific pathologies, at regional scale, has been carried out taking into account previous epidemiological research, that demonstrated the existence of relationships between anomalous concentrations of some metals and incidence of some pathologies. This study shows that some types of cancer are found in Campania, in areas characterized by relatively high concentration of heavy metals, though, in epidemiological study, correlation does not automatically imply causation. For instance, Zn–Cd-rich areas overlap with areas of high prostate-cancer mortality; bladder and pancreatic cancer are correlated with Pb–Sb-rich areas, whereas, bronchial–tracheal–lung cancer is correlated with As-, Cd- and Pb-rich areas.

1. Introduction

An important factor to relate epidemiology with the presence of potentially toxic metals in the shallow environment, hence potentially bioavailable, is the accessibility of maps reporting territorial distribution of both toxic metals and organics and significant

* Dipartimento di Scienze della Terra, Università di Napoli "Federico II", 80134 Napoli, Italy

Environmental Geochemistry
DOI: 10.1016/B978-0-444-53159-9.00016-4

© 2008 Elsevier B.V.
All rights reserved.

pathologies. Their representation constitutes a valid instrument to establish a correct comparison and to find interactions between element distribution in rocks, soils, and water and health of humans and other living organisms (Berger, 2003).

Illness history reflects the history of modifications which have occurred in the environment surrounding humans and living organisms. Environmental pollution has arisen mostly with the development of modern technologies and consequently, the diseases related to pollution have increased, but this linkage is often difficult to demonstrate on a cause–effect base. In fact, a purely geochemical approach to epidemiology study problem has the limitation in that correlation does not imply causation. Therefore, the fact that a correlation is observed between relatively high concentrations of some toxic metals (e.g., As, Pb, Cd, and others) in soils or sediments and a particular cancer type does not mean that these high metallic concentrations have any role in causing the cancer, or even that there is any increased probability that a specific toxic metal is a cause of the cancer.

In recent years, geo-medicine, a subsidiary of environmental medicine (Möller, 2000) which is born from the synergy between medicine and geology (Bølviken, 1998), studied the influence of environmental factors on the geographic distribution of the human and animal pathologies. Geo-medicine is, hence, important for public health and preventive medicine. Trace elements, as micronutrients—metals and nonmetals—are very important for human well-being, but the same elements have negative consequences on human health if they are ingested in anomalous amount (either in excess or in deficiency).

The objective of this work, in particular, is to provide comprehensive geographic and geologic data to help understand possible interactions between the occurrence of anomalous amounts of toxic metal concentrations and pathologies in Campania region. The latter is, with Sardinia (De Vivo *et al.*, 1997, 2001, 2006c), the only Italian region covered by a systematic sampling of soils and stream sediments with the distribution of toxic metals.

At this stage of the study, no cause–effect relationships between toxic metal distribution and cancer rates can be established, as a joint study with medical and epidemiological professionals is needed to accurately interpret the epidemiological data available for the Campania region.

2. Geology, Geochemical Data, and Cancer Mortality Data of Campania Region

The lithologies of Campania region (Fig. 16.1) can be grouped into three domains:

1. Mountainous sector represented by the Campanian Apennine Mountains, made up mostly by limestones and classified as a Neogenic Nappe edifice
2. Plain sector made up by graben structures forming the Campanian Plain and other structures, where there is occurrence of pre-, syn-, and postorogenic sedimentation (mostly, fine-grained sediments)
3. Volcanic sector made up by volcanics of the Neapolitan potassic province (Somma-Vesuvio, Campi-Flegrei, Ischia, and Roccamonfina) (Peccerillo, 2005).

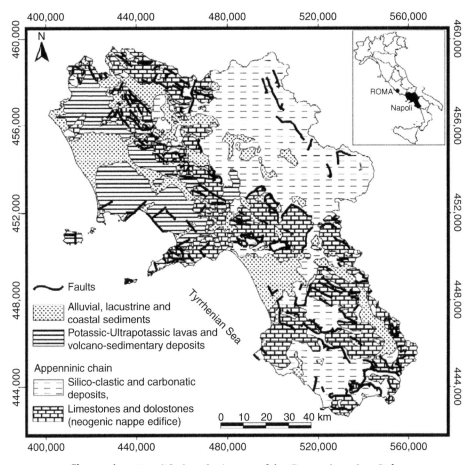

Figure 16.1 Simplified geologic map of the Campania region, Italy.

Geochemical data used for this study are chemical analyses of soil and stream sediment samples covering the entire Campania region (Albanese *et al.*, 2007; De Vivo *et al.*, 2006a, b). The Geochemical Atlases of Campania region (De Vivo *et al.*, 2006a) and of urban and provincial areas of Naples (De Vivo *et al.*, 2006b) include geochemical maps generated from chemical data from 2389 stream sediment and 982 soil samples. Samples were analyzed for 37 elements: Ag, Al, As, Au, B, Ba, Bi, Ca, Cd, Co, Cr, Cu, Fe, Ga, Hg, K, La, Mg, Mn, Mo, Na, Ni, P, Pb, S, Sb, Sc, Se, Sr, Te, Th, Ti, Tl, U, V, W, and Zn by ICP-MS and ICP-AES (inductively coupled plasma mass spectrometry and atomic emission spectrometry) at ACME Analytical Laboratories (Vancouver, Canada).

Specifically, for each sample, a 15-g split of pulp was digested in 45 ml of Aqua Regia (equal quantities of HCl, HNO_3, and distilled water) at 90 °C for 1 h. The solution was taken to a final volume of 300 ml with 5% HCl. Aliquots of sample solution were aspirated into a Jarrel Ash Atomcomp 975 ICP-AES and a Perkin Elmer Elan 6000 ICP-MS instruments.

Precision of the analysis was calculated using three in-house replicates, and two blind duplicates submitted by the authors. Accuracy was determined using ACME's in-house reference material, DS2 (HMTRI, 1997) (Table 16.1).

Table 16.1 Detection limits, accuracy, and precision

Elements	Unit	Detection limit (DL)	Accuracy (%)	Precision (%RPD)
Al	%	0.01	0	1.8
Ca	%	0.01	3.9	2.2
Fe	%	0.01	0.7	1.3
K	%	0.01	6.3	5.3
Mg	%	0.01	0	1.5
Na	%	0.001	3.6	2.9
P	%	0.001	0	3.6
S	%	0.02	30	11.9
Ti	%	0.001	0	5.7
As	mg/kg	0.1	0.3	3
B	mg/kg	1	0	11
Ba	mg/kg	0.5	0.3	1.5
Bi	mg/kg	0.02	1.8	3.2
Cd	mg/kg	0.01	1.4	5.6
Co	mg/kg	0.1	0	2.7
Cr	mg/kg	0.5	1.5	3.2
Cu	mg/kg	0.01	1.6	3.7
Ga	mg/kg	0.1	3.2	2.2
La	mg/kg	0.5	3.5	3.4
Mn	mg/kg	1	0.5	1.9
Mo	mg/kg	0.01	1.2	3.1
Ni	mg/kg	0.1	0.6	1.7
Pb	mg/kg	0.01	0.6	3.5
Sb	mg/kg	0.02	1.2	3.1
Sc	mg/kg	0.1	0	4.4
Se	mg/kg	0.1	0	28
Sr	mg/kg	0.5	5.3	2.4
Te	mg/kg	0.02	0.9	8.4
Th	mg/kg	0.1	5.1	3.6
Tl	mg/kg	0.02	1	3.6
U	mg/kg	0.1	1.6	3.7
V	mg/kg	2	1.3	2.4
W	mg/kg	0.2	2.7	4.4
Zn	mg/kg	0.1	0.5	2.6
Ag	mg/kg	2	0.4	7.9
Au	μg/kg	0.2	4.8	28.9
Hg	μg/kg	5	0	8

RPD, relative percent difference.

For each sampled site, radioactivity was measured by a portable scintillometer (Lima *et al.*, 2005). The data set was used to produce various types of geochemical maps, including dot maps, baseline maps, factor analysis association maps, risk, partial and total radioactivity maps.

Mortality data for four groups of cancer, grouped per ASL (ASL are local health units), have been used for this study and they have been extracted from the Atlas of Cancer Mortality in Campania region in the period 1989–1992 (Montella *et al.*, 1996). Since the area of influence of each local health unit is established on the basis of population density, Campania region territory is divided into 13 ASL and 5 of them belong to the Neapolitan province. For each ASL, mortality data have been expressed as Regional Standard Mortality Ratio (SMR-REG). The SMR-REG, expressed as a percentage, is the ratio between the number of observed deaths from a defined cause for each considered ASL and the number of expected deaths (expected deaths REG) computed on the basis of the population sorted per age group in each ASL and the regional age-specific mortality rates (Table 16.2).

To calculate the number of expected deaths for each ASL, the following formula has been used:

$$\text{Expected deaths REG} = \sum_i \text{Tr}_{i*}p_i$$

where Tr_i = regional age-specific mortality rates calculated for 100,000 in population for the i-esim age group referred to the average four calendar years considered (1998–2001) and p_i= ASL population in the i-esim age group.

To take into account the uncertainties of mortality data, a 95% confidence interval (CI) has been calculated for each SMR-REG using the following formula

$$\text{CI} = \text{SMR} \pm 1.96\text{SE}$$

where SE = standard error of SMR = square root of $[\text{SMR}(1 - \text{SMR})]/P$ and P = total ASL population.

SMR values included within the lower and the upper limit of the CI have been assumed as equal. As consequence, if the value of 100 is included in the CI a nonsignificant difference occurs between the number of observed deaths and the expected death in a considered ASL. Otherwise, if the lower limit or the higher limit of the CI are respectively above or below 100, there is a 95% probability that observed deaths are significantly less or more than the expected ones.

3. METHODS

To compare the epidemiological data with the distribution of toxic metals in the study area, geochemical data for 13 harmful elements (although some are micronutrients) (As, Cd, Co, Cr, Cu, Hg, Ni, Pb, Se, Sb, Tl, V, and Zn) from the complete geochemical database of De Vivo *et al.* (2006a, b) have been grouped and elemental average concentration values have been calculated for each ASL, as well (Fig. 16.2).

Table 16.2 For each ASL and considered cancer type, for males and females, observed deaths, expected number of deaths, SMR-REG, and SE are reported

	Males															
	Trachea, bronchus, and lung				Prostate				Bladder				Pancreas			
ASL	Observed deaths	Reg. expected deaths	SMR-REG	SE	Observed deaths	Reg. expected deaths	SMR-REG	SE	Observed deaths	Reg. expected deaths	SMR-REG	SE	Observed deaths	Reg. expected deaths	SMR-REG	SE
NA 1	1947	1505	129.4	2.9	598	499	119.8	4.9	391	310	126.1	6.4	156	129	120.9	9.7
NA 2	530	427	124.1	5.4	138	148	93.2	7.9	116	84	138.1	12.9	33	37	89.2	15.6
NA 3	434	348	124.7	6.0	100	104	96.2	9.6	89	65	136.9	14.4	37	30	123.3	20.5
NA 4	536	551	97.3	4.2	147	171	86.0	6.9	140	109	128.4	10.9	42	47	89.4	13.7
NA 5	969	874	110.9	3.6	287	288	99.7	5.9	201	177	113.6	8.0	73	75	97.3	11.4
CE 1	432	479	90.2	4.3	223	200	111.5	7.4	87	103	84.5	9.1	34	41	82.9	14.1
CE 2	644	605	106.4	4.2	187	111	168.5	7.6	120	122	98.4	9.0	56	52	107.7	14.4
SA 1	394	405	97.3	4.9	159	165	96.4	7.6	68	82	82.9	10.0	28	35	80.0	15.2
SA 2	671	750	89.5	3.5	227	256	88.7	5.9	126	160	78.8	7.0	67	65	103.1	12.6
SA 3	368	551	66.8	3.5	195	230	84.8	6.1	81	132	61.4	6.8	41	49	83.7	13.1
AV 1	208	373	55.8	3.9	153	165	92.7	7.5	56	90	62.2	8.3	30	33	90.9	16.6
AV 2	329	423	77.8	4.3	134	168	79.8	6.9	66	95	69.5	8.5	25	37	67.6	13.5
BN	391	562	69.6	3.5	261	234	111.5	6.9	119	130	91.5	8.4	57	49	116.3	15.3
Campania	7853	7853	100.0	—	2809	2739	102.6	—	1660	1660	100.0	—	679	679	100.0	—

| | Females | | | | | | | | | | | |
| | Trachea, bronchus, and lung | | | | Bladder | | | | Pancreas | | | |
ASL	Observed deaths	Reg. expected deaths	SMR-REG	SE	Observed deaths	Reg. expected deaths	SMR-REG	SE	Observed deaths	Reg. expected deaths	SMR-REG	SE
NA 1	314	226	138.9	7.9	88	63	139.7	14.8	158	124	127.4	10.1
NA 2	70	57	122.8	14.6	14	15	93.3	24.9	39	30	130.0	20.8
NA 3	44	48	91.7	13.9	7	12	58.3	22.5	27	24	112.5	21.4
NA 4	63	74	85.1	10.8	20	19	105.3	23.3	38	39	97.4	15.9
NA 5	128	119	107.6	9.5	40	33	121.2	19.4	63	64	98.4	12.4
CE 1	62	63	98.4	12.4	15	18	83.3	21.7	27	35	77.1	14.9
CE 2	91	82	111.0	11.6	22	22	100.0	21.5	42	44	95.5	14.8
SA 1	50	55	90.9	13.0	13	15	86.7	24.5	32	29	110.3	19.3
SA 2	86	98	87.8	9.5	24	28	85.7	17.7	54	54	100.0	13.7
SA 3	48	69	69.6	10.1	13	21	61.9	16.8	32	39	82.1	14.3
AV 1	24	48	50.0	10.3	10	15	66.7	21.2	19	28	67.9	15.8
AV 2	53	58	91.4	12.6	10	17	58.8	18.4	22	32	68.8	14.4
BN	36	74	48.6	8.1	24	23	104.3	21.7	32	42	76.2	13.4
Campania	1069	1069	100.0	—	300	300	100.0	—	585	585	100.0	—

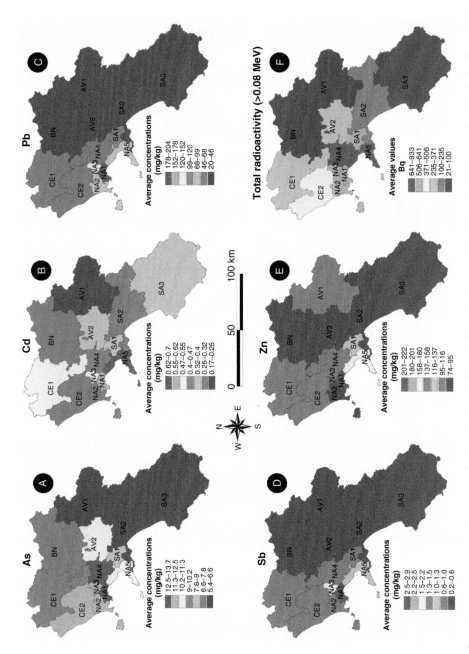

Figure 16.2 Average concentration maps of As (A), Cd (B), Pb (C), Sb (D), Zn (E), and total radioactivity (MeV > 0.08) (E) distribution in ASL territories.

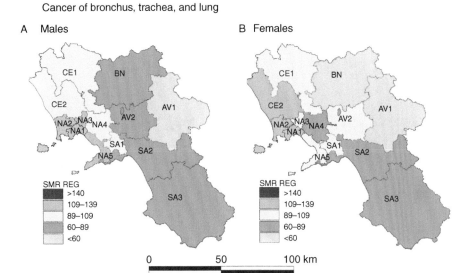

Figure 16.3 Maps representing the Regional Standardized Mortality ratio (SMR-REG) distribution for each ASL referred to bronchus, trachea, and lung neoplasm cancer, for male (A) and female (B) in Campania region. NA, AV, BN, CE, and SA prefix in the labels indicates the pertinence of an ASL to a provincial territory (NA = Naples; AV = Avellino; BN = Benevento; CE = Caserta; SA = Salerno).

Average concentrations of considered elements and SMR-REG data for various cancer types data have been mapped, by means of vector maps, handled with Geographical Information System (GIS) software, representing ASL territories with polygons (Figs. 16.3–16.5). Basically, for each element, the geochemical map has been produced classifying polygons on the basis of the respective average concentrations whereas, for each considered cancer type, mortality maps have been produced classifying polygons on the basis of the respective SMR-REG.

In addition, based on the same mapping criteria, the maps for partial (K^{40}, Th^{232}, U^{238}) and total radioactivity have been compiled.

Since the average concentrations for the 5 ASL of the Naples province have been calculated exclusively on soil sample data whereas stream sediments samples have been used for the rest of the ASL, geochemical data from different media were not mixed in statistical analysis.

4. DISCUSSION OF RESULTS

4.1. Cancer of trachea, bronchus, and lung

Cancers of the trachea, bronchus, and lung cause many deaths in both men and women in the Campania region. It is the leading cause of death for men and the third leading cause of death for women, after breast cancer. The map which report

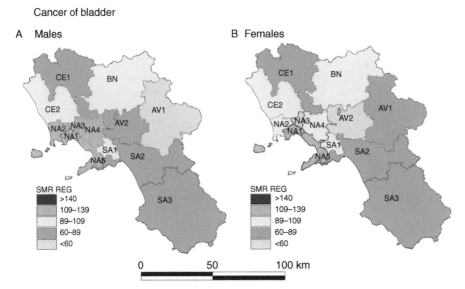

Figure 16.4 Map representing the SMR–REG distribution for each ASL referred to bladder cancer, for male (A) and female (B) in the Campania region. For explanations of labels, see Fig. 16.3.

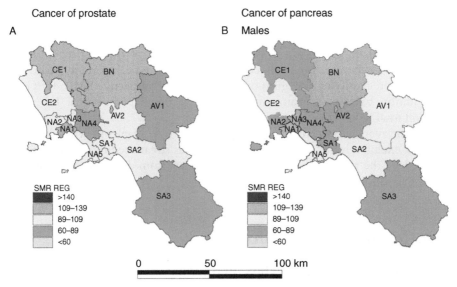

Figure 16.5 Map representing the SMR–REG distribution for each ASL referred to prostate cancer (A) and pancreatic cancer, for female (B), in Campania region. For labels explanations, see Fig. 16.3.

SMR–REG values for males (Fig. 16.3A) together with the observation of the respective SE show that mortality in 4 ASL of Naples province is, generally, significantly higher than regional average, whereas in Avellino, Benevento Salerno, and Caserta ASL deaths are equal or fewer than expected.

SMR–REG map generated for females (Fig. 16.3B) show an increase of mortality for the sole ASL NA1 (corresponding to the city of Naples) and SMR–REG values significantly below the regional average for ASL AV1, BN, and SA3.

The maps of average concentrations for As and Cd (Fig. 16.2A and B) show the highest values in correspondence of ASL territories of Naples and Caserta provinces. Lead shows high values only in the Naples province, especially in correspondence with ASL NA1 territory where the highest values for the SMR–REG for the cancer of trachea, bronchus, and lung are registered, as well. The Pb present in the urban area of Naples has been related principally with pollution from vehicular traffic (Cicchella et al., 2005; De Vivo et al., 2006b).

The correspondence between high SMR–REG and high anomalous values for As, Cd, and Pb mostly in Naples province represents a probability that a cause–effect relationship might exist between tracheal, bronchial, and lung cancer and toxic metal environmental pollution (Boyd et al., 1970; Goyer, 1993). This hypothesis should be tested by means of biomonitoring data on blood and/or urine samples in residents of the Campania region.

Furthermore, in Campania, a roughly spatial correspondence has been found between the highest values of SMR–REG for lung cancer (in Naples and Caserta provinces) and the strong amount of gamma radiation emitted by the alkaline volcanics of the Neapolitan province (Fig. 16.2F). Since Rn gas, widely recognized as a cause of lung cancer (Field et al., 2000), is a direct product of U^{238} decay process, in Naples and Caserta provinces, natural radioactivity can be considered a potential cause of the increased mortality for this cancer, though the spatial correspondence between SMR–REG and gamma radiation is not prima facie evidence in terms of a causative factor for the cancer.

4.2. Prostate cancer

The prostate is a gland present only in men, and produces prostatic liquid. The factors that influence the onset of prostate cancer are age, hormones, sexual activity, virus, genetic factors, and chronic exposure to Zn and Cd (Dinse et al., 1999; Plant and Davis, 2003; Smith, 1999). Mortality in ASL NA1 (corresponding to the city of Naples), CE1, and BN (corresponding to the whole provincial territory of Benevento) are significantly higher than the regional average (Fig. 16.5A). Zinc is an element strongly correlated to prostate function because it is normally present in the prostatic gland to maintain its vitality; it can determine the weakening of immunity system and an antagonism with Se and Cu, increasing cancer risk (Bertholf, 1981). The Zn concentration map (Fig. 16.2E) shows the highest concentration values in the Naples urban area, where the highest SMR–REG values are found as well. The Cd concentration map (Fig. 16.2B), like Zn, presents higher values in Naples province (especially ASL NA5 and NA1), and subordinately in ASL CE1.

4.3. Bladder and pancreatic cancer

The bladder is an organ that accumulates urine via the ureters from the kidneys. For bladder cancer, there are some factors of risk, such as smoking, occupational exposure to toxic agents (i.e., toxic metal concentrations), use of particular medicines, and bacterial infections (Zeegers *et al.*, 2004). The pancreas is a gland that produces insulin and others enzymes necessary for digestion. Factors of risk for pancreatic cancer are smoking, some occupational exposures to industrial and agricultural solvents, and exposure to petroleum derivatives (Lowenfels and Maisonneuve, 2006). Both these cancers may be connected to Pb and Sb anomalous concentrations. The highest significant SMR–REG values for both cancer types (Figs. 16.4A and B, and 16.5B) are localized mostly in the Naples urban (ASL NA1) and provincial territory. The SMR–REG map generated for pancreatic cancer (Fig. 16.5B) shows an increase of mortality in correspondence with the Benevento province (ASL BN) where, locally, some elevated Zn values occur related to malfunctioning water and dust purification apparatus of local industries that use the metal to galvanize steel. Anthropogenic Pb and Sb values (Fig. 16.2C and D) are specially concentrated in the Naples urban territory (ASL NA1).

5. CONCLUSIONS

Maps of geochemical and epidemiological data in Campania region indicate some correlation between concentrations of some toxic metals and specific cancers, whose SMR–REG is above the regional standard value. The most evident correlations occur in the Naples urban and provincial areas, where the highest mortality values for all types of cancer occur, and where there are also high concentrations of heavy metals from anthropogenic pollution. The latter factor is confirmed by the observation that specific cancers are mostly localized in the highly urbanized Naples areas (e.g., trachea, bronchus, and lung cancer). In less urbanized areas, mortality is equal to or much lower than the regional average. For the Campania region therefore, it is possible to state, that, in some areas, a good spatial correspondence occurs between high SMR–REG values and certain types of cancer, namely, between cancer of the bronchus, trachea, and lung and anomalous concentrations of As, Cd, and Pb; prostate cancer and anomalous Zn–Cd concentrations; and bladder and pancreatic cancer and anomalous Pb and Sb concentrations.

This preliminary study, with the limitations of the epidemiological method of not being able to establish and demonstrate a direct cause–effect relationship between pollution and pathologies, nevertheless suggests strong evidence that a compromised environment such as the urban and provincial areas of Naples has a very negative effect for human health. The results in this study could be the base for a joint environmental geochemistry and epidemiological study whose aim would be to provide scientifically sound information to local and national health and environmental authorities with the hope of improving the quality of the environment and human health in the highly urbanized areas of the Campania region.

ACKNOWLEDGMENTS

The authors are grateful to David Smith (U.S. Geol. Survey), Harvey E. Belkin (U.S. Geol. Survey), and to an anonymous reviewer, whose very constructive comments helped to improve significantly the final manuscript.

REFERENCES

Albanese, S., De Vivo, B., Lima, A., and Cicchella, D. (2007). Geochemical background and baseline values of toxic elements in stream sediments of Campania region (Italy). *J. Geochem. Explor.* **93,** 21–34.

Berger, A. R. (2003). Linking health to geology. *In* "Geology and Health. Closing the Gap" (H. C. Skinner and A. R. Berger, eds.), pp. 5–11. Oxford University Press, New York.

Bertholf, L. A. (1981). "Zinc. Handbook on Toxicity of Inorganic Compounds." Marcel Dekker, New York, pp. 787–800.

Bølviken, B. (1998). Geomedisin (Geomedicine—in Norwegian with an English summary). *In* "Geographisk Epidemiologi" (A. Aase, ed.), Norsk J. Epidemiol. 8/1, 29–39. Norskforening for epidemiologi, Oslo.

Boyd, J. T., Doll, R., Foulds, J. S., and Leiper, J. (1970). Cancer of lung in iron ore (haematite) miners. *Br. J. Ind. Med.* **27,** 97–103.

Cicchella, D., De Vivo, B., and Lima, A. (2005). Background and baseline concentration values of elements harmful to human health in the volcanic soils of the metropolitan and provincial areas of Napoli (Italy). *Geochem. Explor. Environ. Anal.* **5,** 29–40.

De Vivo, B., Boni, M., Marcello, A., Di Bonito, M., and Russo, A. (1997). Baseline geochemical mapping of Sardinia (Italy). *J. Geochem. Explor.* **60,** 77–90.

De Vivo, B., Boni, M., and Costabile, S. (2001). Cartografia geochimica ambientale della Sardegna. Carte di intervento per l'uso del territorio. *In* "Monografia Memorie Descrittive della Carta Geologica d'Italia LV II" (B. De Vivo and M. Boni, eds.), pp. 7–32. Servizio Geologico Nazionale, Roma, Italy.

De Vivo, B., Lima, A., Albanese, S., and Cicchella, D. (2006a). "Atlante geochimico-ambientale della Regione Campania." Aracne Editrice, Roma. ISBN 88-548-0819-9 pp. 216.

De Vivo, B., Cicchella, D., Lima, A., and Albanese, S. (2006b). "Atlante Geochimico-Ambientale dei suoli dell'area urbana e della provincia di Napoli." Aracne Editrice, Roma. ISBN 88-548-0563-7 pp. 324.

De Vivo, B., Boni, M., Lima, A., Marcello, A., Pretti, S., Costabile, S., Gasparrini, M., Iachetta, A., and Tarzia, M. (2006c). Cartografia geochimica ambientale e carte di intervento per l'uso del territorio del Foglio Cagliari, Sardegna Meridionale. *In* "Monografia Memorie Descrittive della Carta Geologica d'Italia LXIX" (B. De Vivo, ed.), pp. 5–40. APAT-Servizio Geologico Nazionale, Roma, Italy.

Dinse, G. E., Umbach, D. M., Sasco, A. J., Hoel, D. G., and Davis, D. L. (1999). Unexplained increased in cancer incidence in the United States from 1975 to 1994: Possible sentient health indicators? *Annu. Rev. Public Health* **20,** 173–209.

Field, R. W., Steck, D. J., Smith, B. J., Brus, C. P., Fisher, E. L., Neuberger, J. S., Platz, C. E., Robinson, R. A., Woolson, R. F., and Lynch, C. F. (2000). Residential radon gas exposure and lung cancer. The Iowa radon lung cancer study. *Am. J. Epidemiol.* **151,** 1091–1102.

Goyer, R. A. (1993). Lead toxicity: Current concerns. *Environ. Health Perspect.* **100,** 177–187.

HMTRI (Hazardous Materials Training and Research Institute) (1997). "Site Characterization: Sampling and Analysis." Van Nostrand Reinhold, New York, USA. pp. 336.

Lima, A., Albanese, S., and Cicchella, D. (2005). Geochemical baselines for the radioelements K, U, and Th in the Campania region, Italy: A comparison of stream-sediment geochemistry and gamma-ray surveys. *Appl. Geochem.* **20,** 611–625.

Lowenfels, A. B., and Maisonneuve, P. (2006). Epidemiology and risk factors for pancreatic cancer. *Best Pract. Res. Clin. Gastroenterol.* **20**(2), 197–209.

Möller L. (ed.), (2000). Environmental Medicine. Joint industrial Safety Council, Product 33, Sweden.

Montella, M., Bidoli, E., De Marco, M. R., Redivo, A., and Francesci, S. (1996). "Atlante della mortalità per tumori nella Regione campania, 1998–92." Lega Italiana per la Lotta contro i Tumori, Istituto nazionale tumori, Napoli, pp. 136.

Peccerillo, A. (2005). "Plio-Quaternary Volcanism in Italy." Springer-Verlag, Berlin, pp. 361.

Plant, J. A., and Davis, D. L. (2003). Breast and prostate cancer: sources and pathways of endocrine-disrupting chemicals (ECS). *In* "Geology and Health. Closing the gap" (H. C. Skinner and A. R. Berger, eds.), pp. 95–98. Oxford University Press, New York.

Smith, S. K. (1999). Cadmium. *In* "Encyclopedia of Geochemistry" (C. P. Marshall and R. W. Fairbridge, eds.), Kluwer Academic Publishers, Dordrecht, The Netherlands, pp. 656.

Zeegers, M. P. A., Kellen, E., Buntinx, F., and Brandt, P. A. (2004). The association between smoking, beverage consumption, diet and bladder cancer: A systematic literature review. *World J. Urol.* **21**(6), 392–401.

CHAPTER SEVENTEEN

CHRONIC ARSENIC POISONING FROM DOMESTIC COMBUSTION OF COAL IN RURAL CHINA: A CASE STUDY OF THE RELATIONSHIP BETWEEN EARTH MATERIALS AND HUMAN HEALTH

Harvey E. Belkin,[*] Baoshan Zheng,[†] Daixing Zhou,[‡]
and Robert B. Finkelman[§]

Contents

Abstract

The use of locally mined, high-arsenic (>100 ppm) coals for domestic heating and cooking has caused arsenic poisoning in several villages in southwestern Guizhou Province, China. Extensive epidemiological and geochemical studies have revealed that the prime poisoning pathway is ingestion of arsenic-contaminated foodstuffs, especially chili peppers. A collaborative program between Chinese and American medical- and earth-science researchers has addressed this specific occurrence of arsenosis from

[*] 956 National Center, U.S. Geological Survey, Reston, Virginia 20192, USA
[†] State Key Laboratory of Environmental Geochemistry, Institute of Geochemistry, 550002 Guiyang, Guizhou, PR China
[‡] Sanitation and Anti-Epidemic Station of Qianxinan Autonomous Prefecture, 562400 Xingyi, Guizhou, PR China
[§] Department of Geosciences, University of Texas at Dallas, Richardson, Texas 75083, USA

Environmental Geochemistry
DOI: 10.1016/B978-0-444-53159-9.00017-6

© 2008 Elsevier B.V.
All rights reserved.

domestic coal combustion. Samples of the Longtan coal (late Permian age) used in this region have been studied to determine the concentrations, distributions, and form(s) of arsenic. The coal contains various As^{3+}- and As^{5+}-bearing phases and some coal has As contents as high as 35,000 ppm, on a whole-coal, as-determined basis. Knowledge of the mode of occurrence of arsenic in the coal, the disease etiology, and the geology of the area has led to substantial mitigation of chronic arsenic poisoning. The relationship and information exchange between earth scientists and local medical and public health officials was essential in this accomplishment.

1. Introduction

Determining the relationship between release of hazardous substances into the environment and effects on human health is a challenging responsibility. Such investigations are often hindered by lack of knowledge regarding the nature and origin of the dangerous material and the pathways and timing of exposure. We have been engaged in a collaborative program between Chinese and American medical- and earth-science researchers to address the specific occurrence of chronic arsenic poisoning (arsenosis) in southwest Guizhou Province, People's Republic of China (Figs. 17.1 and 17.2; Finkelman *et al.*, 1999, 2003). China has a large, vigorous urban population that is rapidly embracing new millennium technologies and a larger rural population which often is hindered by poverty, circumstance, and geography (Fig. 17.3A). Domestic combustion of coal for heating and cooking is very widespread in rural China (also India, Indonesia, and Africa) and may involve ~200–300 million households (Finkelman *et al.*, 1999). Such domestic coal combustion, commonly in nonvented dwellings using open stoves (Fig. 17.3B and C), may significantly affect health and has been implicated in cases of arsenic and fluorine poisoning (Finkelman *et al.*, 1999, 2003). Domestic combustion of coal for residence heating and food preparation is pervasive in the mountainous regions of Guizhou Province. Approximately 10 million people in this area are impacted deleteriously by domestic combustion of coal. Arsenic and fluorine are the principal causes of the health problems attributed to domestic coal combustion, although there may be some evidence for mercury poisoning also (Belkin *et al.*, 1997). However, the environmental and geological factors are somewhat different. In this report, we address the case of chronic arsenic poisoning; additional publications will address adverse human health effects related to fluorine and mercury. Our goal here is to describe our past and ongoing research that relates human health issues to properties of earth materials with emphasis on the geochemistry and mineralogy of the coal as a source of arsenic.

2. Previous Studies

The use of locally mined, high-arsenic (>100 ppm) coals has caused arsenic poisoning in several villages in Guizhou Province, China (Fig. 17.2; Zhou *et al.*, 1993). Although coal has been produced and used in this area for 100 years, only with the increase in population and subsequent deforestation have some villages and

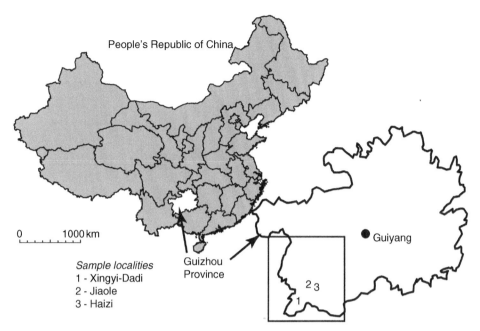

Figure 17.1   Map showing the position of Guizhou Province in the People's Republic of China and the location of the three sampling areas. Box shows the outline of the geologic index map (Fig. 17.2).

areas used coal. The first description of arsenosis (chronic arsenic poisoning) was reported in 1953 and was called "Laizi" disease (Zheng *et al.*, 2005). Seventy-five cases of arsenosis were found in Zhijin County of Guizhou Province from March 1964 to February 1965 and the highest measured arsenic content of coal was 7180 ppm (Sun *et al.*, 1984). Further study in 1976 revealed 877 arsenosis cases in Xingren County, Guizhou Province. Zhou *et al.* (1993) reported additional arsenosis patients in Xingyi City and measured the As concentration in coal as high as 8300 ppm. From these detailed studies, it became obvious that the arsenosis was related in some way to the occurrence of As-rich coal.

Environmental samples of foodstuffs—corn, chili peppers, beans, sweet potatoes, and vegetables—as well as groundwater, surface water, rock, soil, and coal were analyzed. In affected villages, outdoor, indoor air, and indoor dust were also sampled and analyzed (Zheng *et al.*, 1996). Collaboration between the Institute of Geochemistry, Guiyang and the US Geological Survey, Reston, began in 1996 mainly because of the latter's expertise in coal geochemistry and petrography.

Splits of coal samples collected near affected villages analyzed for trace elements revealed high concentrations of As and other deleterious elements (Belkin *et al.*, 1997). Subsequent field work and analysis not only confirms the high concentration of arsenic and other trace elements, but also provides evidence that bears on the paragenesis of these enrichments (Belkin *et al.*, 1998).

Mitigation of arsenosis has been effectively carried out by local health officials aided by appropriate anti-epidemic stations (Zhou *et al.*, 1993). A field test for As in

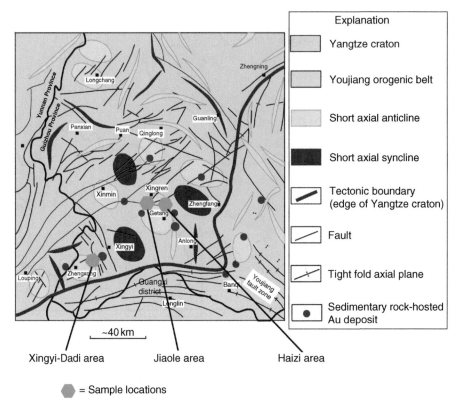

Figure 17.2   Geologic index map showing major structural and tectonic features, sedimentary rock-hosted gold deposits, and the three study areas. The area lies between $24°25'00''$ and $26°32'00''$ N latitude and $104°14'00''$ and $106°07'00''$ E longitude. Adapted from Peters (2002).

coal (Zhou *et al.*, 1993) was used to identify mines producing coal high in As so that these mines could be closed. An additional, more accurate As test kit was developed (Belkin *et al.*, 2003) but the burden of mitigation is still the effective cooperation of, and the communication between, the local health officials and the village leaders.

3. The Size of the Problem

Endemic arsenic contamination in drinking water is a major public health issue, especially in developing countries. Ng *et al.* (2003) conservatively estimated that ∼60 million people are at risk worldwide. Although most are in Bangladesh and India, significant populations at risk are located in Europe (e.g., Hungary, Romania), South America (e.g., Peru, Chile, Argentina), Central America (e.g., Mexico), and Asia (e.g., China, Vietnam). A recent compilation of arsenic contamination in different countries (Naidu *et al.*, 2006) discusses that As in potable groundwater is an issue of great concern for most of the USA (Ryker, 2006). Xie *et al.* (2006) reviewed the occurrences in China and described three manners of endemic arsenic poisoning: groundwater,

Figure 17.3 (A) General scenic view of southwestern Guizhou Province showing typical karst topography underlain by Mesozoic carbonates and in the foreground, an area of Permian clastics and coal. The Permian strata crop-out in eroded anticline windows through the Mesozoic carbonates. (B) Typical farm village house in southwestern Guizhou Province. Note the lack of a chimney on the structure. (C) Interior of a typical farm village house with a cook stove and chili peppers hung up to dry and for storage. (D) Small, nontimbered coal mine used for local domestic consumption. Prof. Baoshan Zheng standing to the left of the entrance.

usually occurring in restricted basins, mining and metallurgical activities, and domestic combustion of coal. Contaminated groundwater produces most of the cases with many millions at risk, especially in northern China (Guo *et al.*, 2001), whereas some of the more severe cases involve smelting and industrial releases.

With regard to arsenosis related to domestic coal combustion, at least 3000 patients have been diagnosed in southwest Guizhou Province, with Xingren

County having ∼2000 cases (Zheng, 1993). Liu *et al.* (2002) estimate that 70,000–200,000 people from six counties in Guizhou Province are considered at risk through the use of As-rich coal.

4. SYMPTOMS AND ETIOLOGY OF ARSENOSIS

Acute arsenic poisoning is often accidental in children, whereas intentional and covert poisonings predominate in adults (Fuortes, 1988). Chronic arsenic poisoning, arsenosis (also called arsenicosis), produces more gradual effects whose presence and severity are greatly influenced by life style, nutrition, and general individual health. Arsenic exists in inorganic and organic forms and in different oxidation states (−3, 0, +3, +5) and in the case of environmental exposure, toxicologists are primarily concerned with arsenic in the trivalent state (Hughes, 2002). Trivalent arsenic is the more toxic species in humans, whereas pentavalent arsenic, common in seafood, is less so. However, the methylation of inorganic arsenic is complex (Cullen *et al.*, 1989; Hughes, 2002) and the details of the various detoxification mechanisms and their effects on humans are beyond the scope of this chapter.

Many different systems within the body are affected by chronic exposure to inorganic arsenic. Among the characteristic features of exposure are skin lesions, dark colored crusts or patches that tend to be localized on the extremities. Blackfoot disease, a vasoocclusive disease described from rural Taiwan, leads to gangrene of the extremities and is observed in individuals chronically exposed to arsenic in their drinking water (Tseng, 1977). Table 17.1 shows the various effects noted in human populations. Inorganic arsenic is classified by the International Agency for Research on Cancer (IARC, 1980, 1987) and the US Environmental Protection Agency (EPA, 1988) as a known human carcinogen. However, there has been much discussion about the shape of the arsenic dose–response curve, the mechanisms of arsenic-cytogenic action are unclear, and no good animal model exists (Snow *et al.*, 2005, and references therein).

The symptoms observed in the affected areas of southwest Guizhou Province (Zhou *et al.*, 1993) have all the classic hallmarks of chronic arsenic exposure. The visible features start with an erythematous flush (Fig. 17.4A) which is quite noticeable in this cohort, as freckles are not a common feature in Asian populations. Various degrees of skin lesions, hyperpigmentation, keratosis, and hyperkeratosis (Fig. 17.4B), leading to some probable skin cancers (Fig. 17.4C) were observed. The most common feature observed is keratosis of the hands (Fig. 17.4D). Cancer and severe skin lesions have lessened dramatically in the last 20 years, due to the intervention by local public health officials and the recognition of the etiology of the arsenic poisoning disease.

The etiology of the chronic arsenic poisoning reported here is the ingestion of foodstuffs contaminated by arsenic-containing smoke from domestic combustion of coal using open, nonvented stoves (Fig. 17.3C). The use of coal for cooking and heating became predominant in the 1960s when population increases resulted in partial to severe deforestation. Southwestern Guizhou Province is a mountainous

Table 17.1 Effects observed in humans and laboratory animals after chronic arsenic exposure

System	Effect
Skin	Skin lesions—keratosis, hyperkeratosis
Cardiovascular	Blackfoot disease
Nervous	Peripheral neuropathy, encephalopathy
Hepatic	Cirrhosis, hepatomegaly
Hematological	Bone marrow depression
Endocrine	Diabetes
Renal	Proximal tubule degeneration, papillary and cortical necrosis

Adapted from Hughes (2002).

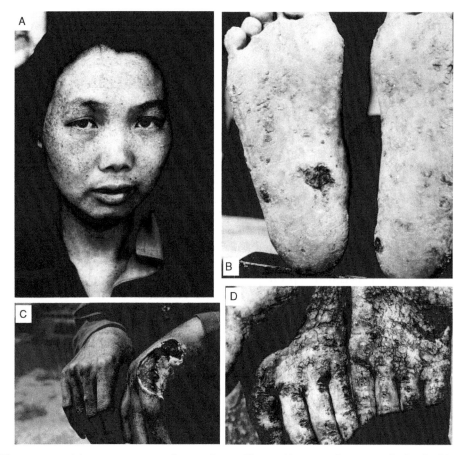

Figure 17.4 (A) Young woman from a farm village with an erythematous flush (freckles). (B) Dark spots on extremities such as feet are typical chronic arsenic poisoning symptoms. (C) Extreme illness from arsenic poisoning, skin cancer and subsequent infection. This photograph was taken in the early 1980s, before medical intervention. (D) Typical keratosis (perhaps hyperkeratosis in some areas) on hands.

plateau region that has a damp, generally cool, autumn climate. After harvest, the farmers bring in and hang up to dry various crops that are used throughout the season such as chili peppers (Fig. 17.3C) and corn. Burning As-rich coal in the open stoves volatilizes As and other trace elements and these elements are absorbed by the various crops drying in the rafters. Early reports of the arsenic content of chili peppers (Fig. 17.3C) found concentrations in some houses to average 500 ppm (Zheng, 1993; Zheng *et al.*, 1996; Zhou *et al.*, 1993) More recent studies of chili peppers dried in this way have arsenic concentrations 30–70 times that of the background levels of chili peppers in other China markets (<1 ppm); this apparent reduction in contamination attests to the effectiveness of the local health officials in arsenosis mitigation (Liu *et al.*, 2002). Although, the ingestion of arsenic-contaminated food is the primary vector, other arsenic sources are also important. Liu *et al.* (2002) suggests that the sources of arsenic in the affected villages are food (50–80%), air–dust (10–20%), water (1–5%), and direct skin contact (<1%). The arsenic poisoning issue is further compounded by the high concentrations of chromium, antimony, cadmium, mercury, and fluorine that also concentrate in these foods and that have coal as their source (Table 17.2).

5. METHODS

The major- and trace-element geochemistry of the coal samples was determined by instrumental neutron activation analysis (INAA; Baedecker and McKown, 1987), inductively coupled plasma atomic emission spectrometry (Bullock *et al.*, 2002), inductively coupled plasma mass spectrometry (Bullock *et al.*, 2002), and cold vapor atomic absorption spectrometry (CVAA; Aruscavage and Crock, 1987) from representative splits of the crushed and powdered coal. Either 16 mesh (1 mm) or selected blocks of coal were prepared as 2.5 cm polished pellets by routine techniques. Petrographic examination and preliminary mineral identification were done on the polished pellets with a JEOL JSM-840 scanning electron microscope (SEM) using a LaB_6 electron emitter equipped with a Princeton Gamma Tech energy-dispersive X-ray spectrometer (EDS). Quantitative electron microprobe analyses (EMPA) of major and minor elements were obtained in Reston, VA, with a JEOL JXA-8900 five spectrometer, fully automated electron microprobe using wavelength-dispersive X-ray spectrometry. The samples were examined with reflected light and by SEM backscattered electron imaging to define points for analysis. Analyses of the mineral phases were made at 20 keV accelerating voltage and 30 nA probe current, and counting times of 20–120 s, using a 1-μm diameter probe spot. Natural and synthetic sulfide standard reference materials were used. The analyses were corrected for electron beam/matrix effects, and instrumental drift and deadtime using a Phi-Rho-Z (CITZAF; Armstrong, 1995) scheme as supplied with the JEOL JXA-8900 electron microprobe. Relative accuracy of the analyses, based upon comparison between measured and published compositions of standard reference materials, is ~1–2% for concentrations >1 wt% and ~5–10% for

Table 17.2 Partial geochemistry of selected coals from southwestern Guizhou Province, PR China

Sample	As (ppm)	Sb (ppm)	Au (ppb)	Hg (ppm)
Jiaole				
RBF96-As-102	1695	73	nd	15
RBF96-As-103	2223	55	nd	14
As-d	1591	132	569	45
As-e	7931	165	347	8.5
J5	607	40	193	29
J10	405	29	28	2.0
JL-AS5-97A	239	38	324	17.6
Haizi				
RBF96-H2-105	35,040	209	nd	4.1
RBF96-H2-106	33,880	205	nd	3.2
H2	32,320	140	<130	5.8
H4	48	8	3.5	0.32
H7	318	13	6.1	0.48
H9	203	24	5.8	2.2
H10	124	9	7.1	0.64
HZ-AU7-97A	7820	364	nd	5.2
Xingyi-Dadi				
As-a	5.2	0.4	183	0.1
As-b	274	0.7	2.0	0.48
As-c	386	0.5	258	0.41
G4	1100	1.6	6.4	0.26
DD-2-97A	925	0.4	nd	0.24

Data on a whole-coal, as-determined basis.
As, Sb, and Au determined by instrumental neutron activation analysis (INAA).
Hg determined by cold vapor atomic absorption (CVAA).
nd = not determined.

concentrations <1 wt%. Element detection limits (wt%) for the mineral data discussed here at the three sigma level were As (0.02), Co (0.01), Cu (0.02), Fe (0.01), Mn (0.01), Ni (0.01), S (0.02), Sb (0.02), Se (0.01), and Zn (0.01). An empirical correction was made for the Fe Kb–Co Ka peak overlap.

6. GEOLOGICAL SETTING

Southwestern Guizhou Province is underlain by an extensive thick volume of Upper Paleozoic and Lower Mesozoic sedimentary rocks. The Permian strata, although areally much less extensive, contains coal-bearing argillaceous sedimentary rocks of the Longtan Formation (Fig. 17.2). Southwestern Guizhou Province and adjacent Yunnan Province and Guangxi autonomous region lie along the southwest margin of the Yangtze Precambrian craton and in the northern Nanpanjiang orogenic fold zone (Fig. 17.2). The area is as structurally complex as it is where the Tethyan-Himalayan and Pacific Ocean tectonic plates join (Yin and Nie, 1996; Zhang *et al.*, 1984). The sedimentation in this region was controlled by a basin that evolved from late Paleozoic to early Mesozoic time along a broad, restricted tidal flat in the Early Triassic (Yang, 2000). The Permian sedimentation comprised predominantly shallow clastics with abundant coal interbeds. Deeper water depositional conditions commenced at or near the P–T boundary and the coal sequences are overlain by a thick sequence of Mesozoic carbonate rocks that weather to form the classic karst topography of South China (Fig. 17.3A). Erosion of anticlinal domes produce windows where most of the Permian sediments crop out and Upper Permian coal beds are exposed.

6.1. Description of the coal

The Upper Permian coal fields in western Guizhou Province lie between the Chuan-Dian (Sichuan-Yunnan) old land and the Chuan-Qian (Sichuan-Guizhou) carbonate platform. Wang *et al.* (1993) classified the coal fields into three types: continental facies, mixed marine–continental facies, and marine facies. All the coal-bearing rocks occur stratigraphically between the Emeishan Basalt (or Maokou Limestone) and the Lower Triassic rocks. The coal fields of best quality and volume are paralic; their distribution was controlled by paleogeography and eustatic events during the late Permian. The coal studied in southwestern Guizhou Province occurs in the Longtan Formation or its correlatives and is anthracite and bituminous in rank (Belkin *et al.*, 2006). It crops out in structurally complex areas that appear as fault- and/or fold-controlled windows exposed by erosion through an extensive thickness of Triassic carbonates (Fig. 17.3A). The Longtan coal in southwestern and northern Guizhou Province is commonly anthracite whereas in other regions in south China, it is bituminous. The vitrinite reflectance of the studied coals ranges from 2.4 to 4.5 (R_{max}), and the anthracite tends to be somewhat friable.

6.2. Samples studied

Previous studies (Zheng, 1993) have identified a range of Permian coal environments and geochemistry. The samples studied represent these variations. We have collected ~50 samples from three areas, Haizi Township, Jiaole Township, and the Xingyi–Dadi area (Fig. 17.2). Some of the mines are small "dog holes" (Fig. 17.3D), and collection of kilogram samples in all localities was not possible. Selected samples were analyzed for proximate and ultimate analysis and all were analyzed for trace elements. In this report, we discuss only the trace-element results as they bear directly on chronic arsenic poisoning.

7. Geochemistry of the Coal

The coals have low, moderate, to high ash contents, 9–56 wt% (550 °C determination), but are extremely enriched in arsenic and other trace elements. For this study related to local adverse human health effects, we collected coals that may have high arsenic contents and therefore these should not be considered typical of Longtan coal geochemistry (Belkin *et al.*, 2006). Table 17.2 shows representative, selected trace-element data from the three collection areas. The coals have As concentrations typically >100 ppm, commonly >1000 ppm, and in a few selected cases >30,000 ppm (3 wt%) on a whole coal, as-determined basis. The coals that have ~3 wt% As are exceptional, as As contents that high are usually restricted to metalliferous samples from ore deposits. The coals are also highly enriched in Sb (up to 370 ppm), Hg (up to 45 ppm), and Au (up to 570 ppb); U is also high in some samples (up to 65 ppm). An examination of the regional distribution of trace-element concentrations and general geochemical relationships documents the highly variable nature of trace-element enrichment. Figure 17.5A and B shows the covariation of As and Sb, and U and Hg, respectively. The major differences in trace-element content are due to differences related to ore-forming fluid–coal–rock interactions and local structural, redox, and lithologic conditions. Local variability in the distribution of ore and associated elements is also observed in the gold mineralization in proximity to the sample collection sites (Fig. 17.2).

8. Mineralogy and Mode of Occurrence of Arsenic in Guizhou Coal

A wide variety of minerals was observed; however, this discussion will emphasize the mode of occurrence of the As-bearing phases; both trivalent and pentavalent arsenic minerals were observed.

Pyrite is the most common sulfide, occurring as framboids, euhedral crystals, and irregular shapes. The range of As in pyrite determined by EMPA is from the detection limit (~200 ppm) in unaltered framboids to about 4.5 wt% in grains adjacent to arsenopyrite (Table 17.3). Copper occurs in amounts up to 0.4 wt% in the most As-rich pyrites. Arsenopyrite occurs in a variety of habits, including large 150–250-μm crystals to narrow 1–5-μm veins, and small crystals (Fig. 17.6A). Se contents in arsenopyrite are fairly uniform (0.15–0.2 wt%), whereas Cu is more variable (from nil to 0.13 wt%). A third As-bearing sulfide, composed of As, Pb, and S (probable sartorite or a related species), is present rarely. Alteration of As-bearing phases was commonly observed. For example, Fig. 17.6B shows a wavelength-dispersive X-ray map of As, wherein surrounding the arsenopyrite is a rind of As-rich Fe-oxyhydroxide alteration.

Another group of As-bearing minerals contains arsenic in the 5^+ valence state as arsenate commonly substituting for the phosphate group. An unidentified As-bearing iron phosphate, usually associated with banded iron oxyhydroxide as veins or masses, has a P/As atomic ratio on the order of 4 (Belkin *et al.*, 1997). Jarosite

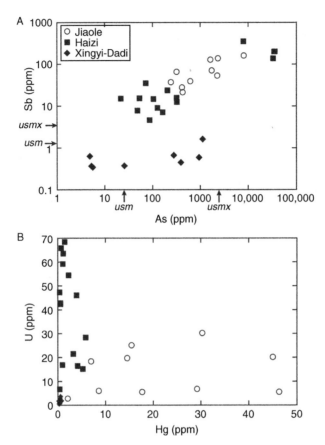

Figure 17.5 (A) log–log plot of the covariation between the As and Sb content of coals from the three study areas. Analysis by instrumental neutron activation analysis (INAA) and data reported on a whole-coal, as-determined basis. The arithmetic mean (*usm*) and maximum value (*usmx*) reported for US coal (Bragg *et al.*, 1997) are shown by arrows. (B) Covariation between U and Hg for the same samples as shown in (A). Analysis of U by INAA and Hg by cold vapor atomic absorption (CVAA).

$[K_2Fe_6^{3+}(SO_4)_4(OH)_{12}]$ was commonly observed as an alteration product of sulfides or as mixtures with iron oxyhydroxide and both frequently contained a few weight percent As. An additional As-rich phase was only observed as scattered micrometer-sized grains, containing only Fe and As ($\pm$O), as identified by SEM-EDS. The atomic ratio of Fe/As is about 1 and the phase is possibly scorodite, $FeAsO_4 \cdot 2H_2O$.

Some coals display evidence of movement of hydrothermal fluids, showing primary phases and their alteration products. Sample As-d shows a complete range of diagenetic, low-As, framboids progressing through various stages to framboid pseudomorphs composed totally of As-bearing iron oxyhydroxide. Veins of jarosite, arsenopyrite, and As-bearing iron oxyhydroxide are common is some samples. Some samples show evidence of fracturing, and subsequent filling by As-bearing iron oxyhydroxide.

The three samples from Haizi Township that have As concentrations in excess of 3 wt% (Table 17.2) are mineralogically peculiar. They contain small grains and veins

Table 17.3 Representative electron microprobe analysis of arsenopyrite and pyrite from southwestern Guizhou Province coal

Sample	(1) As-e	(2) As-e	(3) As-e	(4) As-e	(5) As-e	(6) As-e
n	2	3	3	3	1	2
S (wt%)	19.68	20.06	51.37	53.12	51.27	50.44
As	45.28	44.45	2.87	0.05	0.02	4.44
Se	0.19	0.15	0.02	0.01	bdl	0.03
Cu	0.04	0.09	0.38	bdl	0.02	0.29
Ni	bdl	bdl	bdl	bdl	0.01	bdl
Zn	0.02	0.02	0.01	0.01	0.02	bdl
Co	bdl	bdl	bdl	bdl	bdl	bdl
Fe	33.93	34.92	45.69	47.41	46.83	45.52
Mn	bdl	bdl	bdl	bdl	bdl	bdl
Sb	0.33	0.03	0.02	bdl	bdl	bdl
Sum	99.48	99.73	100.36	100.59	98.24	100.72
S (at%)	33.51	33.85	64.99	66.10	65.55	64.14
As	32.99	32.10	1.55	0.03	0.05	2.42
Se	0.13	0.10	0.01	—	—	0.01
Cu	0.04	0.08	0.24	—	0.01	0.19
Ni	—	—	—	—	0.01	—
Zn	0.02	0.01	0.01	—	0.01	—
Co	—	—	—	—	—	—
Fe	33.17	33.84	33.19	33.87	34.37	33.24
Mn	—	—	—	—	—	—
Sb	0.15	0.01	0.01	—	—	—

Analyses 1 and 2 = arsenopyrite; analyses 3–6 = pyrite.
Analysis 5 is an 8-μm framboid.
Analysis 6 is a pyrite sharing an edge with arsenopyrite.
n = number of analytical points averaged.
bdl = below detection limit.

of iron-oxyhydroxide and pyrite framboids, and the concentration of As in these phases is completely inadequate to account for the As abundance on a whole coal basis. However, in SEM backscattered electron image, a distinct banding character-ized by differing image brightness is easily observed (Fig. 17.6C). Some of this banding forms box-like arrangements, but in all cases, the bands appear to have sharp edges (Fig. 17.6C). The areas range from a few μm to tens and a few hundreds of μm in width. SEM-EDS results show that these bright bands are highly enriched in As. They also contain S, sometimes Fe, and traces of Al and Si (μm-sized clay particles). Wavelength dispersive X-ray mapping reveals that indeed these bright electron-backscattered image areas are highly enriched in As (Fig. 17.6D). Using routine SEM resolution, we could not resolve any discrete As-bearing phase in these areas at 50,000 times magnification. Thin fragments of sample H2 were prepared for exami-nation by an advanced field-emission TEM at the National Institute of Science and Technology, Gaithersburg, MD. No discrete As-bearing phase could be observed using this instrument at magnifications of 1 million times. Thus, we can rule out

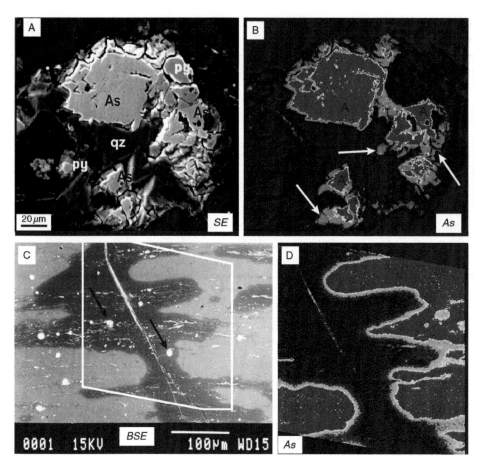

Figure 17.6 (A) Secondary electron SEM image of a group of sulfides and quartz in coal sample As-E. Arsenopyrite (As) and pyrite (py) are intergrown with quartz (qz). (B) Wavelength-dispersive X-ray map of As of the sample area shown in (A). Red (false color) areas (A) denote high As values and correspond to arsenopyrite. The arrows indicate blue zones of lower As concentration and represent alteration of the arsenopyrite. The pyrite has an As concentration below 0.1 wt%. (C) Backscattered electron SEM image of a polished block of coal (sample RBF96-H2–106) containing ~3 wt% As (whole-coal, as-determined basis). The arrows indicate pyrite framboids. (D) Wavelength-dispersive X-ray map of As from the area denoted by the white frame in (C). The red color shows those areas of high As concentration and they correspond to the high-Z areas in the backscattered image (C). Note the pyrite framboids indicated in (C) have a low abundance of As.

finely dispersed arsenopyrite, As-bearing pyrite, or any other As-phase as the localization of the As.

A reconnaissance study of two high-arsenic samples was conducted using high-energy X-rays from a synchrotron source in order to define the nature of bonding in the arsenic-bearing phases. Collection of diffraction spectrum intensity across the XANES (X-ray absorption near-edge structure) and EXAFS (extended X-ray absorption fine structure) regions of an absorption spectrum can provide three-dimensional information on the electronic state and chemical coordination for each crystallographic site of the chosen element. Results from this work demonstrate that the preponderance of the As is in the 5^+ valence state, thus we conclude that the As in the high ($\sim$3 wt%) samples is organically bound in the organic coal structure. The three samples also show abundant evidence for a multistage paragenesis and arsenic remobilization related to the passage of fluids. Figure 17.6C shows that the initial concentration and distribution of arsenic has been affected by fluids moving along fractures and mineral-rich zones. Although, we cannot know the exact distribution of arsenic prior to these fluids, we assume that the arsenic was evenly distributed on the scale of our SEM observations. Also, note that in Fig. 17.6C and D, primary pyrite framboids have low As concentrations suggesting As introduction after framboid formation and perhaps later than peat accumulation.

9. CHINESE SEDIMENTARY ROCK-HOSTED, CARLIN-TYPE, GOLD DEPOSITS

Carlin-type sedimentary rock-hosted gold deposits are widespread in the north-central part of the Great Basin of USA (Vikre *et al.*, 1997). The deposits are located along structurally controlled linear trends. The gold is submicrometer-sized often present as inclusions in an isotopically distinctive epigenetic As-rich pyrite. Typically, the pyrite contains 1–10 wt% As. The deposits are enriched in As, Sb, Ba, $\pm$Hg, and $\pm$Tl, and tend not to be enriched in Pb, Zn, and Cu. The host rocks are generally organic matter–bearing pyritic marls or massive limestones, but siliceous sedimentary rocks and mafic volcanics host ore as well. Alteration assemblages typically involve significant decalcification and silicification. The conditions of formation have been defined from fluid inclusion and isotope studies (Ilchik and Barton, 1997; Rose, 1996). Ore-stage temperatures are mostly between 190 and 300 °C, pressure from 300 to 800 bar, and the ore fluid composition 0–6 wt% NaCl equivalent with significant CO_2 and measurable H_2S. Associated igneous and metamorphic rocks are generally absent in the US localities.

Carlin-type gold deposits in the Guizhou and Yunnan Provinces and the Guangxi autonomous region are located along the western margins of the Precambrian Yangtze craton and were discovered in about 1980 (Cunningham *et al.*, 1988). A recent discussion provides an excellent review of the major deposits and their paragenesis (Peters *et al.*, 2007). The gold deposit distribution is controlled generally by regional rifts, whereas locally, anticlines, high-angle faults, stratabound-breccia bodies, and unconformity surfaces are favorable sites (Fig. 17.2). The host rocks range in age from Paleozoic to Mesozoic and tend to be impure limestone, siltstone, and argillite. Alteration assemblage types are silicification, decalcification, argillation, carbonation, and albitization. The gold is disseminated (submicrometer-sized) in all deposits and enrichments in Sb, As, Hg, and Tl are characteristic. The ore minerals are gold, electrum, arsenopyrite, stibnite, orpiment, realgar, and cinnabar. Ore fluids from fluid inclusion studies show trapping temperature from 165 to 290 °C, low salinity fluids of 3–9 wt% NaCl equivalent, and formation pressures from 50 to 560 bar (Li and Peters, 1996, 1998).

10. Metamorphism of Coal and Trace-Element Enrichment

To date, a gold ore-deposit has not been found in the coal, but it is very reasonable to assume that the Carlin-type ore-forming fluids, were the source responsible for the gold and enrichments in As, Sb, U, and Hg in the coal (Huang, 1993). The ore-fluid temperatures are appropriate for producing the vitrinite reflectance values (anthracite rank) that we observe. These fluids have produced anthracite coal from the bituminous Longtan coals found in areas devoid of mineralization. Furthermore, the element assemblage characteristic in the Carlin-type gold deposits in this area is essentially identical to that found enriched in the coal (Ashley *et al.*, 1991; Belkin *et al.*, 1998; Peters *et al.*, 2007; Qian *et al.*, 1995).

Redox processes strongly affect the transportation and precipitation of trace elements in natural fluids. The role of the coal and associated organic matter in transporting or precipitating ore and associated trace elements is uncertain. The more important redox reactions may not be of the trace elements themselves but the redox-depending reactions of the potential scavenger phases (e.g., iron oxides). The coal in southwestern Guizhou is a high-sulfur coal (Chou, 1996) and the presence of excess sulfur may have contributed to the sulfidization of the deposits. The fluid source for the gold deposits is speculative. Zhu *et al.* (1997) proposed that juvenile fluids associated with deep faults and mantle-derived rock bodies were responsible for the ore. Han and Sheng (1996) suggested that thermo–dehydration and connate waters were responsible. The age of the deposits is also not well determined. Although the gold deposits in southwestern Guizhou Province are controlled by Yanshan structures and hence may be of Yanshanian age (190–65 Ma), some workers believe that they formed in the Eocene after the Yanshan folding.

A tentative paragenesis for the arsenic and other trace-element enrichments in these coals could be (1) coal formation from a peat precursor, (2) introduction of trace-element components after or during coal bituminous rank attainment, (3) movement of later hydrothermal solutions through cracks and mineral-rich zones leaching and/or redepositing the arsenic and other trace elements, and (4) partial oxidation near surface. The origin of the arsenic-bearing solutions continues to be contentious, with various authors proposing either a direct magmatic influence or fluids (Nie and Xie, 2006) without any magmatic component. Peters *et al.* (2007) pointed out that although there are some igneous rocks in the Carlin-type gold deposit area in Guizhou Province, there is no compelling evidence to link the ore-forming event to igneous activity.

11. Mitigation of Chronic Arsenic Poisoning in Guizhou Province

During the investigation of chronic arsenic poisoning in southwestern Guizhou Province, Dr. D. Zhou developed an effective field test to analyze the As content of coal samples from small mines used by the villages for domestic coal combustion

(Belkin *et al.*, 2003). A more accurate test kit was developed from modification of a test kit used for As-water analysis (Belkin *et al.*, 2003). Although these test kits are important to identify the mines producing high-As coal, the main reason and cause of improved health is the education of the residents using the coal. This area is poor, and alternative methods for heating and cooking does not exist. However, recognition of the cause of poor health motivates the residents to select coal sources with minimum arsenic content.

12. CONCLUSIONS

The association of earth materials—coal—with adverse human health requires the expertise and collaboration of the geological, medical, and public health communities. Detailed examination of the coal involved in the incidence of arsenosis has led to a better understanding of As- and other trace-element enrichment processes and provides a predictive basis to examine the coal deposits of southwestern Guizhou Province and their suitability for domestic coal combustion. This case study has described the relationship between mineralogical, geochemical, and geological data and the etiology of chronic arsenic poisoning as studied in southwestern Guizhou Province.

ACKNOWLEDGMENTS

The authors gratefully acknowledge the constructive reviews of John E. Repetski (USGS) and Curtis A. Palmer (USGS). We dedicate this report to the memory of Ann P. Leibrick, a dedicated environmentalist.

REFERENCES

Armstrong, J. T. (1995). CITZAF: A package of correction programs for the quantitative electron microbeam X-ray analysis of thick polished materials, thin films, and particles. *Microbeam Analysis* **4,** 177–200.

Aruscavage, P. J., and Crock, J. G. (1987). Atomic absorption methods. *In* "Methods for Geochemical Analysis" (P. A. Baedecker, ed.), pp. C1–C6. U.S. Geological Survey Bulletin, 1770.

Ashley, R. P., Cunningham, C. G., Bostick, N. H., Dean, W. E., and Chou, I.-M. (1991). Geology and geochemistry of three sedimentary-rock-hosted disseminated gold deposits in Guizhou Province, People's Republic of China. *Ore Geol. Rev.* **6,** 133–151.

Baedecker, P. A., and McKown, D. M. (1987). Instrumental neutron activation analysis of geochemical samples. *In* "Methods for Geochemical Analysis" (P. A. Baedecker, ed.), U.S. Geological Survey Bulletin 1770, H1–H14.

Belkin, H. E., Zheng, B., Zhou, D., and Finkelman, R. B. (1997). Preliminary results on the geochemistry and mineralogy of arsenic in mineralized coals from endemic arsenosis areas in Guizhou Province, P.R. China. *In* Proceedings, Fourteenth Annual International Pittsburgh Coal Conference & Workshop, 23–27 September, 1997, Taiyuan, Shanxi, People's Republic of China, CD-ROM, p. 1–20.

Belkin, H. E., Warwick, P. D., Zheng, B., Zhou, D., and Finkelman, R. B. (1998). High arsenic coals related to sedimentary rock-hosted gold deposition in southwestern Guizhou Province, People's

Republic of China. *In* Proceedings, Fifteen Annual International Pittsburgh Coal Conference & Workshop p. 5. Pittsburgh14–17 September, 1998, CD-ROM.

Belkin, H. E., Kroll, D., Zhou, D.-X., Finkelman, R. B., and Zheng, B. (2003). A field test kit to identify arsenic-rich coals hazardous to human health. Natural Science and Public Health—Prescription for a better environment. Conference Abstracts, USGS Open-File Report 03–097 unpaginated.

Belkin, H. E., Tewalt, S. J., Hopkins, M. S., Finkelman, R. B., Zheng, B., Wu, D., Li, S., Zhu, J., and Wang, B. (2006). The world coal quality Inventory: Coal chemistry for the People's Republic of China. *In*, Proceedings of the 23rd Annual Meeting of the Society for Organic Petrology, Beijing, P.R. China, 15–22 Sept. 2006, Abstracts and Program, vol. 23, pp. 27–28.

Bragg, L. J., Oman, J. K., Tewalt, S. J., Oman, C. J., Rega, N. H., Washington, N. H., and Finkelman, R. B. (1997). The COALQUAL CD-ROM: Analytical data, sample locations, and descriptive information, analytical methods and sampling techniques, database perspective, and bibliographic references for selected U.S. coal samples U.S. Geological Survey Open-File Report 97–134, CD-ROM.

Bullock, J. H., Cathcart, J. D., and Betterton, W. J. (2002). Analytical methods utilized by the United States Geological Survey for the analysis of coal and coal combustion by-products. U.S. Geological Survey Open-File Report 02–389, version 1.0, p. 15.

Chou, C.-L. (1996). Origin of superhigh-organic-sulfur coal and evolution of organic sulfur species during coal maturation; a review. *Abst. Programs Geol. Soc. Am.* **28**(7), 209.

Cullen, W. R., McBride, B. C., Manji, H., Pickett, A. W., and Reglinski, J. (1989). The metabolism of methylarsine oxide and sulfide. *Appl. Organomet. Chem.* **3**, 71–78.

Cunningham, C. G., Ashley, R. P., Chou, I.-M., Huang, Z., Wan, C., and Li, W. (1988). Newly discovered sedimentary rock-hosted disseminated gold deposits in the People's Republic of China. *Econ. Geol.* **83**, 1462–1467.

EPA (1988). Special report on ingested inorganic arsenic. Skin cancer; nutritional essentiality, US Environmental Protection Agency, EPA/625/3–87/-13 .

Finkelman, R. B., Belkin, H. E., and Zheng, B. (1999). Health impacts of domestic coal use in China. *Proc. Natl. Acad. USA* **96**, 3427–3431.

Finkelman, R. B., Belkin, H. E., Centeno, J. A., and Zheng, B. (2003). Geological Epidemiology: Coal combustion in China. Chapter 6. *In* "Geology and Health: Closing the Gap." (H. C. W. Skinner and A. R. Berger, eds.), pp. 45–50. Oxford University Press, New York.

Fuortes, L. (1988). Arsenic poisoning. Ongoing diagnostic and social problem. *Postgrad. Med.* **83**(1), 233–234.

Guo, X., Fujino, Y., Kaneko, S., Wu, K., Xia, Y., and Yoshimura, T. (2001). Arsenic contamination of groundwater and prevalence of arsenical dermatosis in the Hetao plain area, Inner Mongolia, China. *Mol. Cell. Biochem.* **222**(1–2), 137–140.

Han, Z., and Sheng, X. (1996). Gold deposits in Southwest Guizhou and their metallogenetic model. *Guizhou Geol.* **13,** 46–52.

Huang, Y. (1993). A possible relation between the disseminated gold mineralization and the Late Permian zonation of coal metamorphism in Western Guizhou. *Geol. Guizhou (Quarterly)* **10**(4), 300–307 (in Chinese with English abstract).

Hughes, M. F. (2002). Arsenic toxicity and potential mechanisms of action. *Toxicol. Lett.* **135**, 1–16.

IARC (1980). Monographs on the evaluation of the carcinogenic risk of chemicals to man: Some metals and metallic compounds, Vol. 23, pp. 39–141. International Agency for Research on Cancer, Lyon.

IARC (1987). Arsenic and arsenic compounds (Group 1). *In* "IARC monographs on the evaluation of the carcinogenic risks to humans. Supplement 7" pp. 100–103. International Agency for Research on Cancer, Lyon.

Ilchik, R. P., and Barton, M. D. (1997). An amagmatic origin of Carlin-type gold deposits. *Econ. Geol.* **92**, 269–288.

Li, Z., and Peters, S. G. (1996). Geology and geochemistry of Chinese sediment-hosted (Carlin-type) gold deposits. Geological Society of America, Abstract with Program, 1996 Annual meeting, p. A153.

Li, Z., and Peters, S. G. (1998). Comparative geology and geochemistry of sediment-hosted (carlin-type) gold deposits in the People's Republic of China and in Nevada, USA. U.S. Geological Survey Open-file Report 98–0466, p. 160.

Liu, J., Zheng, B., Aposhian, H. V., Zhou, Y., Chen, M.-L., Zhang, A., and Waalkes, M. P. (2002). Chronic arsenic poisoning from burning high-arsenic-containing coal in Guizhou, China. *Environ. Health Perspect.* **110**(2), 119–122.

Naidu, R., Smith, E., Owens, G., Bhattacharya, P., and Nadebaum, P. (eds.) (2006)."Managing Arsenic in the Environment: From soil to human health." CSIRO Publishing, Collingwood, AU, p. 656.

Ng, J. C., Wang, J., and Shraim, A. (2003). A global health problem caused by arsenic from natural sources. *Chemosphere* **52**, 1353–1359.

Nie, A., and Xie, H. (2006). A study on Emei mantle plume activity and the origin of high-As coal in southwestern Guizhou Province. *Chinese J. Geochem.* **25**(3), 238–244.

Peters, S. G. (ed.) (2002). Geology, geochemistry, and geophysics of sedimentary rock-hosted gold deposits in P.R. China. U.S. Geological Survey Open-File Repost 02–131, version 1.0, CD-ROM.

Peters, S. G., Huang, J., Li, Z., and Jing, C. (2007). Sedimentary rock-hosted Au deposits of the Dian-Qian-Gui area, Guizhou, and Yunnan Provinces, and Guangxi District, China. *Ore Geol. Rev.* **31**, 170–204.

Qian, H., Chen, W., and Hu, Y. (1995). Features of As, Sb, Hg, Tl and their mineral assemblages of some disseminated gold deposits in Guizhou and Guangxi Provinces. *Geol. J. Univ.* **1**, 45–52 (in Chinese with English abstract).

Ryker, S. J. (2006). Extent and severity of arsenic contamination of groundwater used for drinking – water in the US. *In* "Managing Arsenic in the Environment: From soil to human health." (E. Smith, G. Owens, P. Bhattacharya, and P. Nadebaum, eds.), pp. 511–524. CSIRO Publishing, Collingwood, AU.

Rose, A. W. (1996). Hydrothermal alteration at Carlin-type gold deposits. Geological Society of America, Abstracts with program, Annual meeting, p. A22.

Snow, E. T., Sykora, P., Durham, T. R., and Klein, C. B. (2005). Arsenic, mode of action at biologically plausible low doses: What are the implications for low dose cancer risk? *Toxicol. Appl. Pharmacol.* **205**, S557–S564.

Sun, B., Nie, G., Zhang, Y., and Mai, Z. (1984). Investigation report of chronic arsenosis. *In* "Collection of data of the Sanitation and Anti-epidemic Station of Guizhou Province" Report No. **4**, pp. 284–289 (in Chinese).

Tseng, W.-P. (1977). Effects and dose-response relationships of skin cancer and Blackfoot disease with arsenic. *Environ. Health Perspect.* **19**, 109–119.

Vikre, P., Thompson, T. B., Bettles, K., Christensen, O., and Parratt, R. (eds.) (1997).Carlin-type gold deposits field conference, Society of Economic Geologists, Guidebook Series volume 28, p. 287.

Wang, L., Lo, J., Wang, C., and Wang, M. (1993). Geology and coal-accumulating regularity of the paralic coal fields of Late Permian in western Guizhou. *Geol. Guizhou (Quarterly)* **10**(4), 291–299 (in Chinese with English abstract).

Xie, Z. M., Zhang, Y. M., and Naidu, R. (2006). Extent and severity of arsenic poisoning in China. *In* "Managing Arsenic in the Environment: From soil to human health." (E. Smith, G. Owens, P. Bhattacharya, and P. Nadebaum, eds.), pp. 541–552. CSIRO Publishing, Collingwood, AU.

Yang, M. G. (2000). Regional geology of the South China Domain. *In* "Concise Regional Geology of China." (Y. Q. Cheng, ed.), pp. 172–218. Geological Publishing House, Beijing.

Yin, A., and Nie, S. (1996). A Phanerozoic palinspastic reconstruction of China and its neighboring regions. *In* "The Tectonic Evolution of Asia." (A. Yin and T. M. Harrison, eds.), pp. 442–485. Cambridge University Press, Cambridge.

Zhang, Z. M., Liou, J. G., and Coleman, R. G. (1984). An outline of plate tectonics in China. *Geol. Soc. Am. Bull.* **95**, 295–312.

Zheng, B. (1993). Enrichment regularity and origin of arsenic in coals in southwestern Guizhou Province. *In* "New Explorations in Mineralogy, Petrology, and Geochemistry" pp. 271–272. Seismology Press, Beijing (in Chinese).

Zheng, B., Yu, X., Zhang, J., and Zhou, D. (1996). Environmental geochemistry of coal and endemic arsenism in southwest Guizhou, P.R. China. 30th International Geologic Congress, Abstracts, 3:410.

Zheng, B., Wang, B., Ding, Z., Zhou, D., Zhou, Y., Zhou, C., Chen, C., and Finkelman, R. B. (2005). Endemic arsenosis caused by indoor combustion of high-As coal in Guizhou Province, P.R. China. *Environ. Geochem. Health* **27,** 521–528.

Zhou, D., Liu, D., Zhu, S., Li, B., Jin, D., Zhou, Y., Lu, X., Hu, X., and Zhou, C. (1993). Investigation of chronic arsenism caused by pollution of high arsenic coal. *J. Chin. Prev. Med.* **27,** 147–150 (in Chinese with English abstract).

Zhu, L., Jin, J., He, M., Liu, X., and Hu, R. (1997). On the possibility of the participation of juvenile fluids in the gold deposit formation in Southwestern Guizhou. *Geol. Rev. (Dizhi Lunping)* **43**(6), 586–592.

Index

Printed and bound by CPI Group (UK) Ltd, Croydon, CR0 4YY

03/10/2024

01040332-0008